Recent Advances in Deep Learning Applications

This book presents a collection of rigorously revised papers selected from the 22nd IEEE International Conference on Machine Learning and Applications (IEEE ICMLA 2023). It focuses on deep learning architectures and their applications in domains such as health care, security and threat detection, education, fault diagnosis, and robotic control in industrial environments. Novel ways of using convolutional neural networks, transformers, autoencoders, graph-based neural networks, and large language models for the above applications are covered in this book. Readers will find insights to help them realize novel ways of using deep learning architectures and models in real-world applications and contexts, making this book an essential reference guide for academic researchers, professionals, software engineers in the industry, and innovative product developers.

Key Features:

- Presents state-of-the-art research on deep learning
- Covers modern real-world applications of deep learning
- Provides value to students, academic researchers, professionals, software engineers in the industry, and innovative product developers.

Recent Advances in Deep Learning Applications

New Techniques and Practical Examples

Edited by
Uche Onyekpe, Vasile Palade,
and M. Arif Wani

CRC Press
Taylor & Francis Group
Boca Raton London New York

CRC Press is an imprint of the
Taylor & Francis Group, an **informa** business
A CHAPMAN & HALL BOOK

First edition published 2025
by CRC Press
2385 NW Executive Center Drive, Suite 320, Boca Raton FL 33431

and by CRC Press
4 Park Square, Milton Park, Abingdon, Oxon, OX14 4RN

CRC Press is an imprint of Taylor & Francis Group, LLC

ISBN: 9781032944623 (hbk)
ISBN: 9781032944678 (pbk)
ISBN: 9781003570882 (ebk)

DOI: 10.1201/9781003570882

Typeset in Times LT Std
by KnowledgeWorks Global Ltd.

Contents

Preface

In the continuously evolving field of artificial intelligence, deep learning stands as a beacon of modern innovation, bringing to reality what was once thought to be science fiction. Deep learning architectures are made up of networks with multiple layers, which possess the ability to independently identify important hierarchical patterns in various structured and unstructured data. Progress in deep learning has facilitated the generation of new representations and the extraction of insights, as well as the capacity to infer from raw data sources like images, videos, text, speech, time series, and other complex data types. These deep learning methods are currently being applied to address a variety of fascinating real-world problems, covering areas such as condition monitoring, fault diagnostics, medical diagnostics, self-driving cars, and numerous other fields.

This book puts forward 16 chapters, exploring recent advances, novel techniques, and architectures in the application of deep learning to diverse real-life domains. This book is broken into four parts, exploring the profound and real-world implications of deep learning in:

- computer vision
- natural language processing (NLP)
- predicting modeling, as well as
- methodological approaches for reinforcement learning, data reduction, etc.

The 16 chapters of the book are extensions of selected papers from the 23rd IEEE International Conference on Machine Learning Applications (IEEE ICMLA), Jacksonville – USA, 14–17th December 2023.

The chapters are summarized below:

Chapter 1 presents YOLOv8 automated image segmentation (YOLOv8AIS), an algorithm for automated image segmentation which uses YOLOv8 and active learning to reduce manual annotation and improve object detection. The effectiveness of the proposed algorithm was evaluated across various conditions and datasets.

Chapter 2 discusses energy-efficient deep learning approaches in detecting glaucoma in order to address the shortage of qualified opthamologists in rural regions. The study highlights the success of the deep learning approaches for accurately detecting glaucoma.

Chapter 3 explores the use of cost-effective and -efficient deep learning algorithms to solve the JPEG compression artifact removal (CAR) problem in pixel domain by mapping a low-quality compressed image to the corresponding high-quality image.

Chapter 4 discusses the modeling of face emotion perception using deep learning techniques. The research findings improve our understanding of the relationship between movements of the eye and gaze patterns, and emotion perception.

Chapter 5 demonstrates the potential of large language models (LLMs) in educational assessment by exploiting the semantic reasoning, contextual comprehension, and transfer learning capabilities of LLMs for automated short-answer grading and misconception detection. The study lays the foundation for future research on the use of LLMs to understand nuanced responses from students and to precisely detect misconceptions.

Chapter 6 explores how languages are learned by humans, specifically whether word classes are innate or learned. A deep recurrent neural network is trained to predict the next word(s) within a sequence. The findings from this research show that internal representations of the input sequences are clustered with respect to the word class of the predictions. This suggests that syntax rules and other abstract representations may naturally emerge as a consequence of predictive coding and processing during language acquisition. The outcome of this study provides a starting point for further investigations into the neural mechanisms underlying language acquisition and processing, and may be useful in NLP applications where an understanding of the organization of neural representations of language input can help improve language modeling, machine translation, etc.

Chapter 7 discusses the use of deep learning to advance threat detection capabilities within the realm of cybersecurity. The chapter also explores strategic data selection techniques, such as the K-means diversity-based query strategy, in optimizing model performance.

Chapter 8 presents the development of novel techniques and a comprehensive pipeline for automated contradiction detection, seamlessly integrating an information retrieval system, machine learning classifiers, and explainable AI. The information retrieval system is geared toward biomedical data and encompasses a robust datastore alongside syntactic and semantic similarity components.

Chapter 9 discusses the impact of natural language processing in the transformation of education through the development of digital platforms that enhance student's learning experiences. The chapter proposes the integration of human-like generation, evaluation, and feedback into the digital platform. The proposition has the capability to enhance personalized learning and demonstrates that automated metrics can align closely with human evaluations, thus enhancing the effectiveness of ed-tech tools.

Chapter 10 explores the use of transformer graph neural network (T-GNN) in real estate pricing to address challenges around sparse data availability, temporal, and geographical changes. The T-GNN substantially outperforms the nongraph neural network approaches, providing a robust solution due to its ability to handle missing input features, and its remarkable understanding of space and time.

Chapter 11 discusses the challenges with utilities as a result of the evolving energy mix and electricity market. The chapter proposes an approach for the detection and clustering of errors from model predictive controllers for fleets of residential devices.

Chapter 12 presents a hybrid technique that combines recurrent neural network and physics informed neural network to model and forecast healthcare demand and the dynamics of COVID-19. The proposed approach effectively estimates key parameters, unobserved states, and forecasts ICU demand, providing insights into the dynamics of the pandemic.

Chapter 13 addresses the issue of fraud within the Medicare program, which results in losses amounting to billions of dollars annually. The study proposes a novel data reduction technique, one class Gaussian mixture models, and demonstrates that fraud detection models trained on 20% of that original data can retain a high performance. The research offers a scalable and efficient technique for the detection of fraud in the presence of large-scale health datasets.

Chapter 14 proposes a unique deep learning–based hybrid approach made of convolutional and recurrent layers to locate harmful, flammable, or polluting gas leaks using information on molecular movement, obstacles, and wind direction. The proposed technique outperforms existing approaches significantly in locating the source of the gas leak.

Chapter 15 presents "oracle" conditioning and iterative neural networks, an approach inspired by dynamical systems and recurrent neural networks, to enhance the accuracy of predictive models in sparse data domains. The study addresses the challenge associated with the training of deep neural networks with a small dataset, which is a very common issue with machine learning problems in the physical science field. The effectiveness of the proposition is demonstrated on a lander pattern analysis and electromagnetic source localization problem.

Chapter 16 discusses the challenges associated with the use of a hierarchical multiagent reinforcement learning framework to analyze simulated air defense scenarios. A two-level decision-making process is proposed to address the challenge (i.e., low-level policies control individual units, while a high-level commander policy issues macro commands in view of the overall mission targets). The approach, which is validated empirically, demonstrates a safe-to-fail and cost-effective method for exploring real-world defense scenarios and strategy planning.

The editors of the book would like to that the authors who submitted extended versions of their IEEE ICMLA 2024 publications for inclusion in this book. It is our hope that the knowledge contained in this book can inspire and enlighten both experts and novices in the field of artificial intelligence.

Uche Onyekpe, Edinburgh, UK
Vasile Palade, Coventry, UK
M. Arif Wani, Srinagar, India
August, 2024

Editors

Dr. Uche Onyekpe is a machine learning expert at Ofcom (Office of Communications, UK), where he focuses on developing assessment/audit strategies for AI algorithms used by online platforms such as Instagram, TikTok, and X. He also serves as the director of the African Institute for Artificial Intelligence, a nonprofit organization dedicated to advancing AI across the African continent.

Dr. Onyekpe previously held academic positions at York St. John University and Coventry University on machine learning. His professional experience spans various sectors, including health, construction, and transport, where he has led projects at the intersection of AI and these fields. He has published numerous research papers in these areas and has several years of experience working as a consultant within robotics and social care. He has delivered keynote talks at reputable seminars and events on machine learning and applications.

Vasile Palade is a professor of artificial intelligence and data science in the Centre for Computational Science and Mathematical Modelling at Coventry University, UK. He previously held several academic and research positions at the University of Oxford–UK, University of Hull–UK, and the University of Galati–Romania. His research interests are in machine learning, with a focus on neural networks and deep learning, and with main applications to computer vision, natural language processing, autonomous driving, smart cities, and health, among others. Professor Palade is author and co-author of more than 300 papers in journals and conference proceedings as well as several books on machine learning and applications. He is an associate editor for several reputed journals, such as *IEEE Transactions on Neural Networks and Learning Systems* and *Neural Networks*. He has delivered keynote talks to reputed international conferences on machine learning and applications.

Prof. M. Arif Wani completed his M.Tech. in computer technology at the Indian Institute of Technology, Delhi, and his Ph.D. in computer vision at Cardiff University, UK. He is a professor at the University of Kashmir, having previously served as a professor at California State University, Bakersfield.

His research interests are in the area of machine learning, with a focus on neural networks, deep learning, computer vision, pattern recognition, and classification tasks. He has published many papers in reputed journals and conferences in these areas. Dr. Wani has co-authored the book *Advances in Deep Learning* and co-edited many books on machine learning and deep learning applications.

Contributors

Michael Ajao-Olarinoye
Coventry University
Coventry, United Kingdom

Alessandro Antonucci
IDSIA, USI-SUPSI
Lugano, Switzerland

Andrea Arikawa
University of North Florida
Jacksonville, United States

Maria Barger
Worcester Polytechnic Institute
Worcester, United States

James Becker
Montana State University
Bozeman, United States

Ladislau Bölöni
University of Central Florida
Orlando, United States

Joel Brynielsson
KTH Royal Institute of Technology and FOI Swedish Defence Research Agency
Stockholm, Sweden

Amanda Carp
AI Vision Sweden AB
Sweden

Tianjie Chen
Pennsylvania State University Harrisburg
United States

Giacomo Del Rio
IDSIA, USI-SUPSI
Lugano, Switzerland

Ayan Dutta
University of North Florida
Jacksonville, United States

Ender Mete Eksioglu
Istanbul Technical University
Istanbul, Turkey

John Hancock
Florida Atlantic University
Boca Raton, United States

Fei He
Coventry University
Coventry, United Kingdom

Justin Hill
Southern Company
Greater Birmingham, Alabama Area
United States

Meisam J. Seikavandi
IT University of Copenhagen
Copenhagen, Denmark

Reid A. Johnson
Zillow Group
Seattle, United States

Maria Jung Barrett
IT University of Copenhagen
Copenhagen, Denmark

Md Faisal Kabir
Penn State University
Harrisburg, United States

Indika Kahanda
University of North Florida
Jacksonville, United States

Hasan Huseyin Karaoglu
Istanbul Technical University
Istanbul, Turkey

Nazmul Kazi
University of North Florida
Jacksonville, United States

Taghi Khoshgoftaar
Florida Atlantic University
Boca Raton, United States

Jinyoon Kim
Pennsylvania State University Harrisburg
United States

Kishore Surendra
Surendra University Hospital
Hamburg, Germany

Andy Krause
Zillow Group
Seattle, United States

Patrick Krauss
University Erlangen-Nuremberg
Erlangen, Germany

O. Patrick Kreidl
UNF
Jacksonville, United States

Iliya Kulbaka
University of North Florida
Jacksonville, United States

Prajwol Lamichhane
University of North Florida
Jacksonville, United States

Viswadeep Lebakula
Oak Ridge National Laboratory
Oak Ridge, Tennessee, United States

Xudong Liu
University of North Florida
Jacksonville, United States

Andreas Maier
University Erlangen-Nuremberg
Erlangen, Germany

Claus Metzner
University Erlangen-Nuremberg
Erlangen, Germany

Faraz Moghimi
Zillow Group
Seattle, United States

Seyed Mousavi
Coventry University
Coventry, United Kingdom

Zindoga Mukandavire
Institute of Applied Research and Technology, Emirates Aviation University
Dubai, United Arab Emirates United Arab Emirates

Jeff Munk
Oak Ridge National Laboratory
Oak Ridge, Tennesse, United States

Krish Nachnani
The Harker School
United States

Chukwuka Victor Obionwu
Fakultät für Informatik, Otto-von-Guericke-Universität
Magdeburg, Germany

Randy Paffenroth
Worcester Polytechnic Institute
Worchester, United States

Vasile Palade
Coventry University
Coventry, United Kingdom

Harsh Pathak
Worcester Polytechnic Institute
Worchester, United States

Sandeep Reddivari
University of North Florida
Jacksonville, United States

Swapnoneel Roy
University of North Florida
Jacksonville, United States

Michael Rüegsegger
Armasuisse S+T
Switzerland

Achim Schilling
University Erlangen-Nuremberg
Erlangen, Germany

Adrian Schneider
Armasuisse S+T
Switzerland

Ardian Selmonaj
IDSIA, USI-SUPSI
Lugano, Switzerland

Paul Stoewer
University Erlangen-Nuremberg
Erlangen, Germany

Oleg Szehr
IDSIA, USI-SUPSI
Lugano, Switzerland

Agnes Tegen
FOI Swedish Defence Research Agency
Stockholm, Sweden

Eve Tsybina
Oak Ridge National Laboratory
Oak Ridge, Tennessee, United States

Karthikeyan Umapathy
University of North Florida
Jacksonville, United States

Petra A. Wark
Coventry University
Coventry, United Kingdom

PART ONE

Deep Learning for Computer Vision

Automated Image Segmentation Using Self-Iterative Training and Self-Supervised Learning with Uncertainty Scores

1

Jinyoon Kim, Tianjie Chen, and Md Faisal Kabir

1.1 INTRODUCTION

Machine learning algorithms can be categorized as either supervised or unsupervised learning, depending on the task of the object [1]. In the case of supervised learning, the availability of extensive, high-quality labeled datasets play a pivotal role in determining the performance of models. However, obtaining such extensive labeled data is challenging because manual annotation is an expensive and labor-intensive procedure. To mitigate these issues, various approaches have been developed, including active learning [2–4], semi-supervised learning [5–7], and automated data labeling [8–10]. Active learning [11] refers to the use of learning algorithms that proactively choose which data

DOI: 10.1201/9781003570882-2

to learn from to maximize the usefulness gained from each labeled instance. However, this method encounters obstacles such as inconsistent performance, sensitivity to hyperparameters, and potential difficulties with certain datasets and tasks.

On the other hand, semi-supervised learning [12] presents a notable benefit in scenarios where labeled data is limited by utilizing both labeled and unlabeled data. It takes advantage of the abundance of unlabeled data, thus reducing the reliance on manual image segmentation. Automated data labeling [13] aims to either fully or partially automate the image segmentation process to significantly reduce the time needed for manual annotation. An example of such a system is the Amazon automated data labeling [10], which simplifies the creation and modification of training datasets. It underscores the drive to streamline the segmentation process in complex tasks like instance segmentation [14], significantly increasing the overall effectiveness of computer vision models.

The objective of this study is to incorporate the state-of-the-art YOLOv8 model [15] with an automated image segmentation algorithm and apply it to various datasets such as the plant leaves disease dataset of PlantVillage and the skin lesion dataset from HAM10000 patients. Additionally, we applied a vision transformer model, OneFormer [16], to compare the performance of our algorithm on a different computer vision architecture. This tailored adaptation of the YOLOv8 model magnifies its capacity to discern and identify intricate patterns, greatly enhancing the efficacy of auto-segmentation algorithms.

To further strengthen our approach, we have synthesized various aspects of computer vision techniques into a single framework, resulting in the embedding of the automated image segmentation (AIS) algorithm. Our iterative training algorithm yields a significant increase in the performance of the auto-segmentation process. This not only underscores the synergy of current technologies but also charts an interesting trajectory for future research in semi-supervised learning and computer vision.

The notable contributions of this study are as follows:

- Trained the algorithm on a dataset of over 20,000 plant images
- Used YOLOv8 for advanced object detection, improving segmentation and speeding up training
- Implemented a classification correction mechanism and uncertainty data selection to optimize model accuracy
- Incorporated semi-supervised and self-supervised learning, combining labeled, unlabeled data, and iterative segmentation improvement for better performance
- Integrated active learning to refine segmentation and adjust model training process, ensuring adaptability to new data
- The auto-segmentation system enhances both data annotation and model training, improving object detection accuracy and resilience
- Tested the system's performance with the HAM10000 skin lesion dataset, which presents more challenging features and skewed class distribution
- Evaluated the system's ability to work with minimal labeled data and proposed a method to address its limitations
- In addition to YOLOv8, tested the system with a vision transformer model to assess its performance

The chapter is divided into five sections. Following the introduction section, Section 1.2 discusses related works on semi-supervised learning, active learning, and the auto-segmentation system, offering an understanding of their impact on this research. Section 1.3 outlines the specific materials and methods employed, including the dataset used for the experiment and a detailed description of the algorithm's step-by-step process. The experimental outcomes and the environmental configuration are presented and analyzed in Section 1.4. Finally, Section 1.5 concludes the chapter, summarizing the study and suggesting areas for future research improvement.

1.2 RELATED WORKS

This section embarks on a journey through various research studies and approaches that serve as the foundation for the proposed YOLOv8-AIS model. By understanding the valuable insights offered by these works, one can better appreciate the context and rationale behind the design choices in YOLOv8 AIS. The surveyed literature ranges from deep active learning techniques and semi-supervised learning to advanced object detection and weakly supervised data creation methods.

The comprehensive research, "A Comparative Survey of Deep Active Learning" [3], shapes the foundation of this research with pivotal deep active learning (DAL) methods. Pseudosegmentation and uncertainty sampling, central techniques in our model, are further enriched by "Deep Active Learning for Named Entity Recognition" [2], despite not adopting its Convolutional Neural Network (CNN) - CNN LSTM (Convolutional Neural Network Long Short-Term Memory) structure. In YOLOv8-AIS, these pseudo-labels, produced from predictions, become interim ground truths, while uncertainty sampling zeroes in on complex cases. Seamlessly integrating with YOLOv8's instance segmentation, our model stands at the confluence of deep and active learning, enhancing performance and label accuracy. This unique blend elevates the efficiency of automated segmentation, marking a significant leap in machine learning.

The pivotal studies "Semi-supervised Active Learning for Instance Segmentation via Scoring Predictions" [5] and "Learning from Noisy Large-Scale Datasets with Minimal Supervision" [6] have steered advancements in semi-supervised learning. [5] introduced a groundbreaking active learning framework that synergizes initial labeled data with self-labeled data for refined segmentation. Meanwhile, [6] emphasized useful insights for the semi-supervised learning, even though its core focus was on noisy annotations that were not concerned in this research. Building upon these foundations, the YOLOv8-AIS algorithm assimilates the robust object detection of YOLOv8 and the spirit of [5]. Unlike [6]'s dual network approach, YOLOv8-AIS champions a singular advanced model. Its active learning loop perpetually hones the model and improves labeled data quality, setting a novel paradigm in processing expansive datasets.

Snorkel, introduced in [9], facilitates rapid training data creation via weak supervision. By allowing users to design segmentation functions that generate noisy labels, Snorkel consolidates them into probabilistic labels through a generative model, reducing manual annotation efforts for large datasets. Conversely, our method taps into semi-supervised learning, capitalizing on the YOLOv8 model and the abundance of unlabeled data for training. This direct approach, unlike Snorkel's reliance on human expertise for multiple segmentation functions, is more straightforward and demands less human intervention.

OneFormer [16], experimented alongside the YOLOv8 object detection model in this research, introduces a novel approach to universal image segmentation. It employs a single transformer model that achieves state-of-the-art performance across semantic, instance, and panoptic segmentation tasks through a single training process. The framework's task conditioned joint training strategy incorporates a task token, dynamically adjusting to the specific segmentation task at hand. Additionally, it utilizes a query-text contrastive loss to improve task and class distinction. This innovative design significantly reduces resource requirements, making high-quality segmentation more accessible and efficient.

In overview, this research reflects the integration and adaptation of several existing methodologies in the fields of active learning, semi-supervised learning, and advanced object detection models with instance segmentation. It draws from the strengths of these methods while addressing their limitations to optimize the use of plentiful unlabeled data. The model thus aims to create a balanced solution, reducing the human effort required in image segmentation while acquiring satisfactory performance. Ultimately, it contributes a new perspective to the ongoing discussions around automated image segmentation and machine learning.

1.3 MATERIALS AND METHODS

1.3.1 YOLOv8

The you only look once (YOLO) models [17, 18] revolutionized object detection by unifying location identification and classification, both of which were traditionally separated from each other. This approach allows YOLO models to analyze an entire image during training and prediction in one go. This global view facilitates the recognition of overall contextual patterns within the image, enhancing both the accuracy of object detection and classification.

YOLOv8 [19] incorporates an advanced loss function blending mean squared error (MSE) for bounding box regression and binary cross-entropy (BCE) for object probability. It also adopts a novel neural network architecture that leverages both feature pyramid network (FPN) and path aggregation network (PAN) to significantly boost predictive accuracy. These improvements in YOLOv8 enhance its scalability and adaptability, making it an integral part of the YOLOv8-AIS algorithm for efficient automated image segmentation.

1.3.2 OneFormer

Vision transformers have recently gained prominence for their effectiveness across various vision tasks. Following this trend, we also tested our proposed algorithm using a vision transformer model built on a framework called the OneFormer. Originated from the success of transformers in natural language processing, vision transformers adapt their architecture for visual data to process images as sequences of patches. OneFormer builds on this foundation by introducing significant advancements such as a unified framework for addressing multiple segmentation tasks (semantic, instance, panoptic) simultaneously with a single model. Compared to the Mask2Former framework, it offers improved performance in task adaptability and segmentation precision through innovations include task conditioned training and query-text contrastive loss [20].

1.3.3 Dataset Description

The PlantVillage dataset is a publicly accessible collection of leaf images representing various plant species, each labeled with specific disease conditions or as healthy. As detailed in Table 1.1, the PV-Tomato and PV-Apple datasets, derived from the PlantVillage [21] dataset, were used in the experiments.

The PV-Apple dataset, fully manually annotated, serves as a performance benchmark. In contrast, only the initial training and test subsets of the PV-Tomato dataset are manually annotated, with the rest left for algorithmic labeling. Both datasets were instance-segmented using polygons via the "Roboflow" tool, as shown in Figure 1.1. This annotation format, widely used in studies, was chosen to enhance the model's final performance and to enable the creation of a comprehensive auto-labeled instance segmentation dataset, leveraging YOLOv8's inherent capabilities.

The HAM10000 dataset consists of over 10,000 dermoscopy images, curated to include a broad spectrum of pigmented skin lesions for machine learning–based automated diagnosis. As shown in Table 1.2, it has an imbalanced class distribution. Unlike the PlantVillage dataset, the HAM10000 dataset includes preexisting segmentation masks, facilitating its use in experiments. Its varied nature, compared to baseline datasets like PV-Tomato and PV-Apple, provides a comprehensive view of the system's effectiveness across different dataset types and distributions, highlighting its versatility in research.

TABLE 1.1 PlantVillage dataset structure used for experiment

DATASET	*PV-TOMATO DATASET*	*PV-APPLE DATASET*
Manual annotation	Only initial train and test	Entire dataset
Number of images	18,160 images	3164 images
Proportion of initial training	25/100 images per class (250/1000 images)	50 images per class (200 images)
Proportion of active learning	15,348 images	2649 images
Proportion of test data	10% of total (1812 images)	10% of total (315 images)

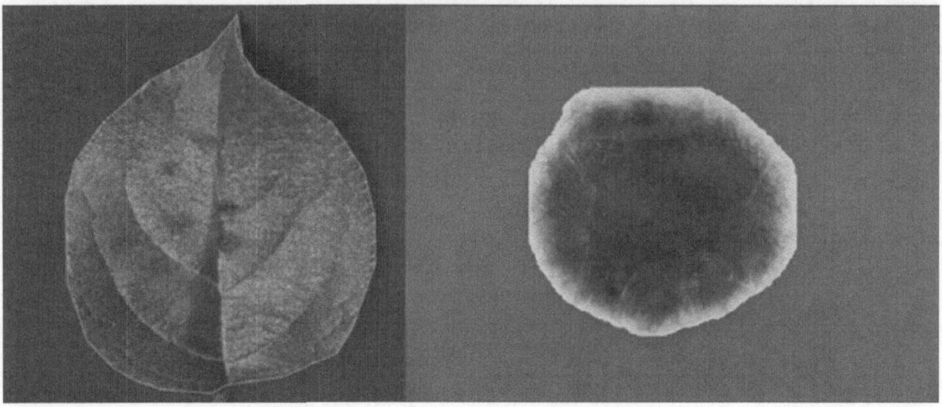

FIGURE 1.1 Annotations of Instance segmentation in precise polygon shapes,, highlighting the distinct texture of the object.

TABLE 1.2 Utilized image data distribution for each class from HAM10000 dataset

CLASSES	*NUMBER OF IMAGES*
Actinic keratosis	327
Basal cell carcinoma	514
Benign keratosis-like lesions	1099
Dermatofibroma	115
Melanoma	1113
Melanocytic nevi	6705
TOTAL	9758

1.3.4 Methodology

The YOLOv8-AIS methodology utilizes an iterative active learning process. In each cycle of training, prediction, and segmentation, the model refines its object identification and segmentation capabilities. This process progressively enriches the labeled dataset and thereby boosts the model's performance.

Algorithm 1 Confidence score-based YOLOv8-AIS Algorithm Pseudo-Code

Input: Labeled data D_L, unlabeled data D_U, test data D_T, YOLOv8 model M, AIS parameters, λ

Output: Trained model M, metrics, auto-labels for dataset *LOOP Process*:

1: **for** each cycle **do**

2: Train M on D_L

```
Predict labels for D_U with confidence scores
Define heap size for classes as λ percent of unlabeled images
for each class do
   Maintain a min-heap for predictions
   Replace heap top for predictions with higher confidence
end for
Auto-label images from heap using YOLOv8 segmentation
Correct misclassified labels then update D_L and D_U
Assess M on D_T
end for
return M, metrics, auto-labels
```

1. Active Learning Process and Prediction: As illustrated in Figure 1.2 and detailed in Algorithm 1, the proposed model for automated image segmentation employs an iterative active learning approach. The initial model is trained on manually segmented data, then performs inference on all unlabeled images in the dataset to calculate their confidence scores. During each inference, images are processed in batches, with the YOLOv8 model, trained on the available labeled data, predicting labels for these images. Each image is assigned a confidence score, which serves as an indicator of the model's certainty regarding the assigned label. These confidence scores play a crucial role in the auto-segmentation process as they help determine which images are selected for segmentation in each iteration. Based on the confidence scores, the algorithm then selects a (λ) percentage of the inference results

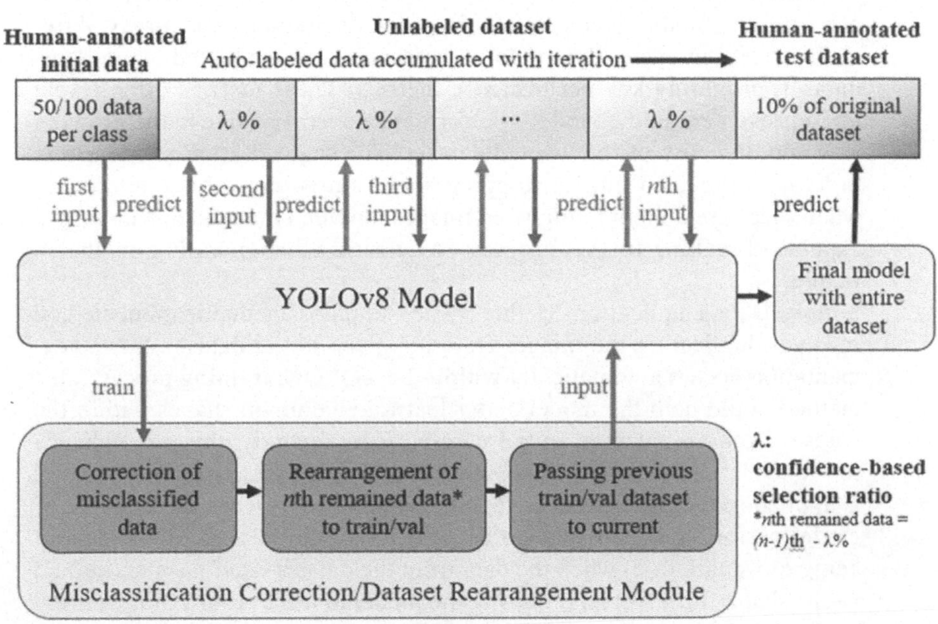

FIGURE 1.2 An overview of the overall architecture.

with the highest uncertainty scores. This (λ) variable is used to establish the size of the heap for each class in the dataset. The results with the highest uncertainty scores are then added to the training and validation splits of the dataset, completing one iteration of the training process.

2. Heap Structure, Auto-Segmentation, and Label Correction: The algorithm processes images and maintains a heap, a structure that stores inference results such as confidence scores with predicted classes and the actual class information for each class. The heap size is determined by the λ percent calculated earlier. As predictions are being made, predicted labels with the lowest confidence scores, which signify a high level of uncertainty, are added to the heap. These labels can replace existing entries with higher confidence scores if the heap reaches its maximum capacity. Once the prediction phase is completed, all labels within the heap are used for auto-segmentation. For each misclassification of the class information of the selected segmentation labels, the output class is corrected by reverting it to the original class according to its saved actual class information, ensuring the accuracy of the labels. Next, the freshly corrected and auto-labeled images are then merged into the current labeled dataset for future training iterations, directing the active learning mechanism of the AIS algorithm towards the most complex instances.
3. Model Retraining, Performance Evaluation, and Stopping Conditions: Following the auto-segmentation and correction process, the labeled images are removed from the unlabeled data pool. The model is then retrained on the updated labeled dataset. With each iteration, the quantity and diversity of the labeled data used for training increase, thereby allowing the model to continually improve its predictive accuracy. After each retraining cycle, the model's performance is evaluated using a test dataset, providing key performance metrics. These metrics often reveal an improvement in the model's performance over time due to the growing size and diversity of the labeled dataset. The segmentation and learning process is repeated until a stopping condition is met. This could occur when there are no more unlabeled images, or when the model is no longer capable of making further progress in detecting the remaining unlabeled images.
4. Simple Data Duplication Method: The simple data duplication method involves duplicating the images from the given initial dataset. Since augmentation occurs automatically within the YOLOv8 training process, this method could help the model better learn the common shapes within the dataset from an enlarged initial dataset. Consequently, this can enhance the model's performance in subsequent training iterations. This method is applied to the dataset under specific conditions: 1) when the collapse of model performance is expected because of the initial size of the dataset being too small, or 2) when the data from the dataset itself has unclear and vague shapes or features, requiring the model to learn from a more numerous initial dataset.

1.3.5 OneFormer Model Application

After conducting experiments with YOLOv8-AIS, we applied our algorithm to the OneFormer vision transformer model to assess its performance on this task. The main objective was to determine if the OneFormer model could adapt to the initial dataset sufficiently to capture common shapes and features, allowing it to utilize the learned weights for subsequent training iterations. This would enable successful inferences to be integrated into the existing training dataset, similar to the YOLOv8 approach.

The OneFormer model differs fundamentally from YOLOv8 and comes from a different framework, necessitating a distinct iterative training process to achieve a similar function. The dataset, originally in YOLOv8's instance segmentation annotation format (text files), was converted into a semantic segmentation mask format for use with OneFormer. Once trained on these image and mask pairs, the model inferred mask annotations for a λ proportion of unannotated data, which were then added to the existing training dataset.

1.4 EXPERIMENTS

1.4.1 Environmental Setup

1. Experimental Environment: The study used an NVIDIA GeForce RTX 3070 Laptop GPU for the YOLOv8 model. For the OneFormer model, online environments from VAST.AI were employed due to the significant computational resources required. The research utilized refined apple and tomato leaf images from the Plant Village dataset and the HAM10000 skin lesion dataset. Data preprocessing and YOLOv8-AIS implementation were done in Python using PyCharm. The OneFormer model was implemented via a Jupyter notebook on VAST.AI, with a server equipped with 48 GB of disk space and 2 NVIDIA RTX 4090 GPUs.
2. Data Set Construction: Each dataset, PV-Tomato, PV-Apple, and HAM10000, had 10% reserved as a test split from the start. An initial dataset size was selected for training, while the remaining data was used for active learning without segmentation information, only classification data.
3. Class-Wise Initial Data Set Sizes and Their Purposes: Different training set sizes (25 and 100 images per class) were used for the Tomato dataset to assess how the initial size influences the algorithm's performance. The PV-Apple dataset, initialized with 50 images per class, served as a performance benchmark against fully manually annotated datasets. A balance between the classes in the training dataset was maintained to avoid model bias and enhance generalization capabilities. For the HAM10000 dataset, 60 images per class were used to investigate how the model performs on datasets

with more vague features and shapes, as well as significantly imbalanced data distribution compared to the Tomato and Apple datasets.

4. λ Variable Analysis: The impact of different λ values (10% and 20%) on model performance was tested using PV-Tomato and PV-Apple datasets. λ represents the fraction of the unlabeled dataset auto-labeled and added during each iteration. For minimal initial datasets in the HAM10000 dataset, λ was fixed at 30% due to observations that varying λ did not significantly impact training outcomes in baseline cases.
5. Evaluating Performance with Minimal Initial Data Set Sizes Under Varied Conditions: In the second stage of experiments, the initial data size for each class was reduced to 20 for both the PV-Tomato and PV-Apple datasets. The PV-Apple dataset was tested under various conditions, including the minimal initial dataset size (ALL), missing half of the classes with the lowest accuracy (TOP2), and missing all but the class with the highest accuracy (TOP1). These tests evaluated whether the model could overcome missing data using the misclassification correction algorithm. The PV-Tomato dataset focused on using the simple data duplication method to mitigate performance drops when the initial dataset size was critically low, with two cases: the minimal initial dataset size of 20 images per class (ALL) and the same dataset with simple data duplication applied (ALL DUPLICATED).
6. Evaluating Model Performance on a New and Ambiguously Shaped Data Set: The HAM10000 dataset, with its imbalanced distribution and vague shapes, was used to assess whether the simple data duplication method could improve predictive accuracy under challenging conditions.
7. OneFormer Settings: Due to high computational requirements, the OneFormer model was tested solely on the PV-Apple dataset. The λ value was fixed at 30%. The training dataset was the same as the minimal size PV-Apple dataset used for YOLOv8-AIS but converted into a semantic segmentation mask format. Two experiments were conducted: one with 10 epochs using the original minimal dataset and another with a tripled dataset size via simple data duplication.
8. Performance Evaluation: The mAP@0.5 score was used as the performance metric, widely recognized in object detection and instance segmentation for its comprehensive evaluation capabilities. Scores were measured at every iteration of the training process based on the initial test split.

1.4.2 Experiment on Semi-Supervising Capability

The adaptability and performance of the YOLOv8 AIS model under different initial conditions and λ values were examined using the PV-Tomato dataset. Initial training conditions included sets of 25 and 100 images per class with λ values of 0.1 (10%) and 0.2 (20%). The mAP@0.5 scores, reflecting the model's performance, showed that initial results were dependent on the training set size, with the model trained on 100 images per class achieving a higher initial mAP@0.5 score due to more labeled data.

TABLE 1.3 mAP@0.5 scores for tomato (λ = 0.1), tomato (λ = 0.2) on each initial data type (25/100)

INITIAL DATA	*25 PER CLASS*		*100 PER CLASS*	
DATASET	*λ = 0.1*	*λ = 0.2*	*λ = 0.1*	*λ = 0.2*
Initial	0.6966	0.6966	0.9165	0.9165
Iteration 1	0.9176	0.9410	0.9359	0.9486
Iteration 2	0.9590	0.9557	0.9609	0.9574
Iteration 3	0.9395	0.9631	0.9494	0.9574
Iteration 4	0.9546	0.9613	0.9574	0.9615
Iteration 5	0.9457	0.9628	0.9619	0.9639
Iteration 6	0.9604		0.9598	
Iteration 7	0.9622		0.9598	
Iteration 8	0.9637		0.9598	
Iteration 9	0.9625		0.9621	
Iteration 10	0.9552		0.9654	

As active learning iterations progressed, all models showed steady improvement in mAP@0.5 scores. Notably, the model trained with 25 images per class caught up with the performance of the model trained with 100 images per class, demonstrating the YOLOv8 AIS algorithm's effectiveness in leveraging unlabeled data. Specifically, the best mAP@0.5 scores for the model trained with 25 images per class were 0.9637 and 0.9631 for λ values of 0.1 and 0.2, respectively. The model trained with 100 images per class achieved mAP@0.5 scores of 0.9654 for λ = 0.1 and 0.9639 for λ = 0.2. Although variations in the λ parameter were expected to impact training, the results indicated that changes in λ had no significant effect, highlighting the model's resilience.

These results, summarized in Table 1.3 and illustrated in Figure 1.3, emphasize the robustness and adaptability of the proposed algorithm in semi-supervised environments, as well as the impact of λ on the rate of improvement and the number of iterations required for the mAP@0.5 score to converge.

1.4.3 Comparison to Human Annotation

The analysis extended to the PV-Apple dataset provided a benchmark for comparing the YOLOv8-AIS model's performance with manually annotated labels. Starting with 50 images per class and using λ values of 0.1 and 0.2, the mAP@0.5 score was used to measure performance. The model trained on the fully labeled dataset achieved an mAP@0.5 score of 0.9649, reflecting the accuracy typically obtained with meticulous human annotation. However, the YOLOv8-AIS model, starting from a lower score, significantly improved through active learning, reaching an mAP@0.5 score of 0.9751 with λ = 0.1 and 0.9756 with λ = 0.2. These results highlight the model's ability to not only automate the annotation process but also enhance the overall performance of object detection.

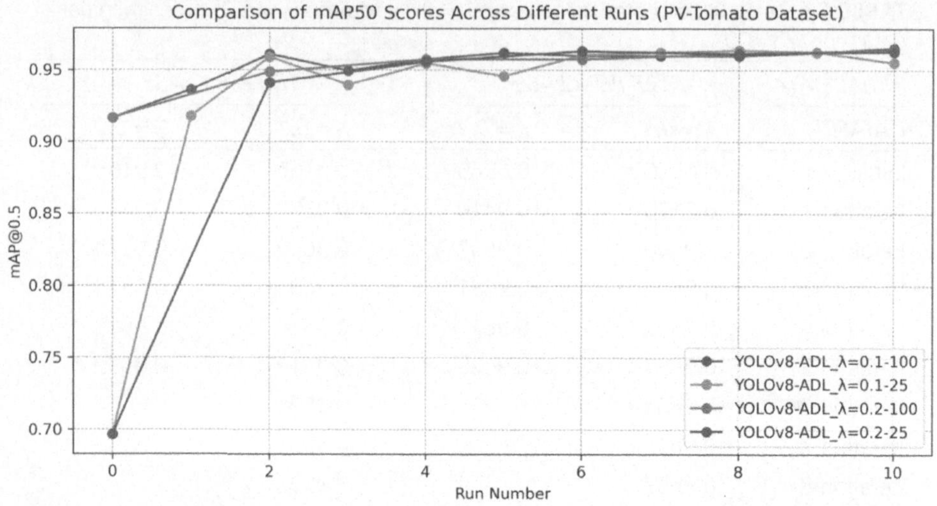

FIGURE 1.3 YOLOv8-AIS performance on PV-Tomato.

The initial dataset size once again influenced early training outcomes, while changes in the λ value had minimal impact on overall performance. These observations are detailed in Table 1.4 and illustrated in Figure 1.4. The findings underscore the model's ability to effectively learn from semi-supervised datasets by incorporating auto-labeled data, consistently enhancing performance. This insight suggests that future research should explore the model's efficacy with smaller, imbalanced datasets and those with more complex features and textures.

TABLE 1.4 mAP@0.5 scores for apple (λ = 0.1), apple (λ = 0.2) and apple (manual) with fixed initial data (50)

INITIAL DATA	*25 PER CLASS*		*100 PER CLASS*
Initial	0.9083	0.9083	0.9649
Iteration 1	0.9690	0.9709	
Iteration 2	0.9743	**0.9756**	
Iteration 3	**0.9751**	0.9625	
Iteration 4	0.9728	0.9750	
Iteration 5	0.9673	0.9731	
Iteration 6	0.9682		
Iteration 7	0.9722		
Iteration 8	0.9624		
Iteration 9	0.9729		
Iteration 10	0.9688		

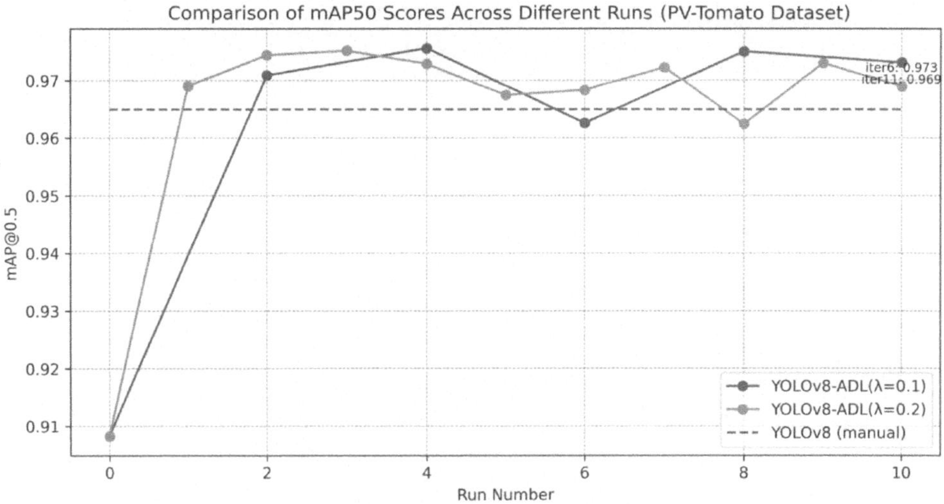

FIGURE 1.4 YOLOv8-AIS and YOLOv8(manual) performance on PV-Apple.

1.4.4 Experiments on Minimal Dataset Sizes and Dataset Imbalance

Building on our previous research [22], this study tested the system's adaptability to datasets with minimal initial data or imbalances due to missing class instances. Initially, the PV-Apple dataset had 50 images per class, reduced to 20 images per class for the initial training phase. This reduced dataset led to a significant initial performance decrease, but models trained on iteratively labeled datasets showed remarkable improvements (Table 1.5, Figure 1.5 as 'apple-all' (ALL)). This indicates the YOLOv8 model's effectiveness in leveraging minimal datasets when the data is clear and distinct.

We also tested the impact of intentionally imbalanced datasets by excluding classes with the lowest mAP@50 scores. Two scenarios were tested: one with half the classes removed (TOP2) and another with only the top-performing class retained (TOP1). Both scenarios initially showed low performance but rapidly improved with training iterations. This suggests the model can learn features from missing class data if the dataset's classes share similar features, aided by our correction algorithm. The outcomes are detailed in Table 1.5 and Figure 1.5.

TABLE 1.5 mAP@0.5 scores for ALL, TOP1, and TOP2 test cases

LOOP	*ALL*	*TOP1*	*TOP2*
Loop1	0.1917	0.1930	0.1886
Loop2	0.8847	0.7779	0.8896
Loop3	0.9777	0.9830	0.9749
Loop4	0.9936	0.9925	0.9910

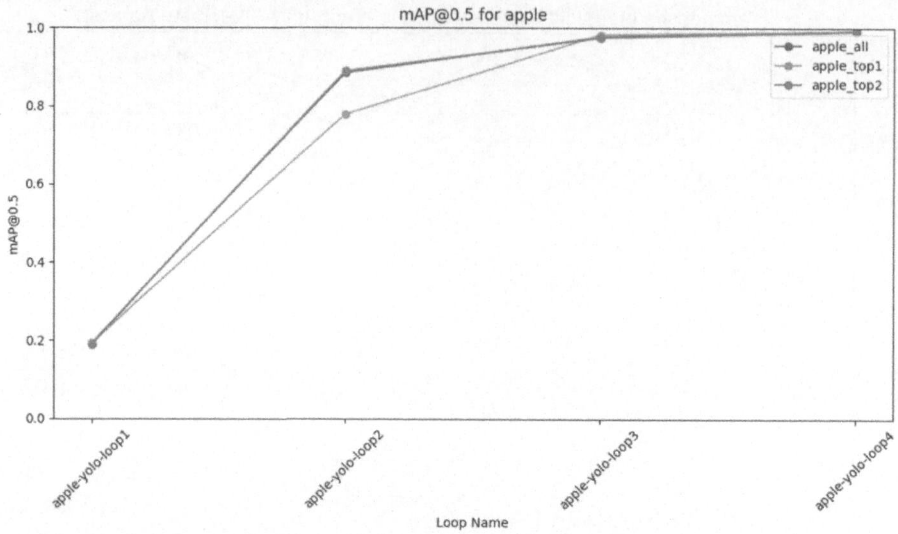

FIGURE 1.5 YOLOv8-AIS on PV-Apple minimal and imbalanced version.

Remarkably, the model trained on the minimal dataset outperformed the model trained on the larger dataset, which had mAP@50 scores of 0.9688 and 0.9731 for λ values 0.1 and 0.2, respectively. With the lambda value fixed at 30% and using the minimal dataset, the model achieved scores of 0.9936 (ALL), 0.9925 (TOP1), and 0.9910 (TOP2), showing nearly a 2% overall performance improvement.

1.4.5 Mitigating Performance Collapse Beyond Critical Thresholds through Data Duplication

Following the successful adaptation to minimal and imbalanced datasets, we also explored scenarios where the system failed to accurately segment common shapes from the initial data, resulting in limited improvements in subsequent training iterations. This issue arose during experiments with the PV-Tomato dataset, where reducing the number of images per class from 25/100 to 20 led to performance improvements across iterations, but the final results were unsatisfactory due to the model's inability to accurately segment shapes from the initial dataset.

To address these challenges, we employed the simple data duplication method, duplicating the initial dataset without adding new segmented data. This duplication provided sufficient information for the model to learn common shapes and features across classes. While models trained on the duplicated datasets did not match the performance levels seen in previous PV-Tomato experiments, their performance significantly improved compared to the baseline established with the original minimal dataset (ALL DUPLICATED). These improvements are detailed in Table 1.6 and illustrated in Figure 1.6.

In the ALL case using the raw minimal dataset, an mAP@50 score of 0.9145 was achieved, nearly a 5% decrease from the original best performance of 0.9637 for $\lambda = 0.1$

TABLE 1.6 mAP@0.5 scores for ALL and ALL DUPLICATED test cases in the tomato version

LOOP	*ALL*	*ALL_DUPLICATED*
Loop1	0.1471	0.2871
Loop2	0.8004	0.6114
Loop3	0.8239	0.8977
Loop4	0.9080	0.9466
Loop5	0.8970	0.9651
Loop6	0.9132	0.9654
Loop7	0.9340	0.9667
Loop8	0.9145	N/A

and 0.9631 for $\lambda = 0.2$ in the original PV-Tomato experiment with 25 images per class, and 0.9654 for $\lambda = 0.1$ and 0.9639 for $\lambda = 0.2$ with 100 images per class. However, the ALL DUPLICATED case reached a performance level similar to the original experiments, achieving an mAP@50 score of 0.9667.

This method encounters limitations when the shapes of objects vary significantly among classes or are ambiguous, compared to other datasets with specific and clear shapes. Therefore, the inherent characteristics of the dataset itself are not well suited for iterative training based on a small initial dataset proportion. These limitations were revealed when YOLOv8-AIS was tested on the HAM10000 dataset. This dataset contains classes with varying and ambiguous segmentation processes, alongside a natural data distribution imbalance among the classes. This led to varied performance in the

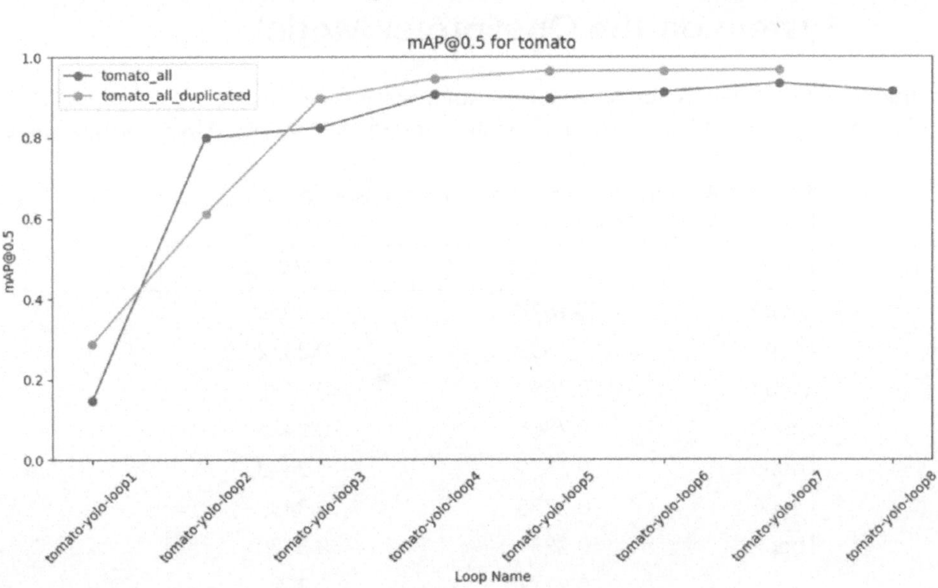

FIGURE 1.6 YOLOv8-AIS on PV-Apple minimal and imbalanced version.

TABLE 1.7 mAP@0.5 scores by class

CLASS	*MAP50*
All	0.317
Actinic keratoses	0.138
Basal cell carcinoma	0.157
Benign keratosis-like lesions	0.272
Dermatofibroma	0.098
Melanoma	0.310
Melanocytic nevi	0.925

YOLOv8-AIS model. While the model excelled in classes with high data distribution and clear shapes and features, it performed poorly in classes with low data distribution and ambiguous shapes and features, which are difficult to capture from the small initial dataset. These disparities resulted in catastrophic performance across all classes, yielding poor average results. The performance gap between the classes can be reviewed in Table 1.7. Unfortunately, this poor performance could not be significantly improved even through the data duplication method. The iterative training processes' results for both test cases—one for the baseline (ALL) and another for the version with the duplicated method (ALL DUPLICATED)—are detailed in Table 1.8, and their visualizations as graphs are illustrated in Figure 1.7.

1.4.6 Experiments on the Application of the System on the OneFormer Model

While the YOLOv8-AIS has demonstrated impressive performance across various conditions and datasets, it had not been previously tested whether our AIS algorithm could

TABLE 1.8 Iteration performance comparison for two sets of the HAM10000 dataset version

LOOP	*FIRST SET*	*SECOND SET*
Loop1	0.1878	0.2083
Loop2	0.1859	0.2912
Loop3	0.2697	0.3235
Loop4	0.2866	0.3424
Loop5	0.2651	0.3473
Loop6	0.2730	0.3432
Loop7	0.2851	0.3570
Loop8	0.2777	0.3695
Loop9	0.2829	0.3167

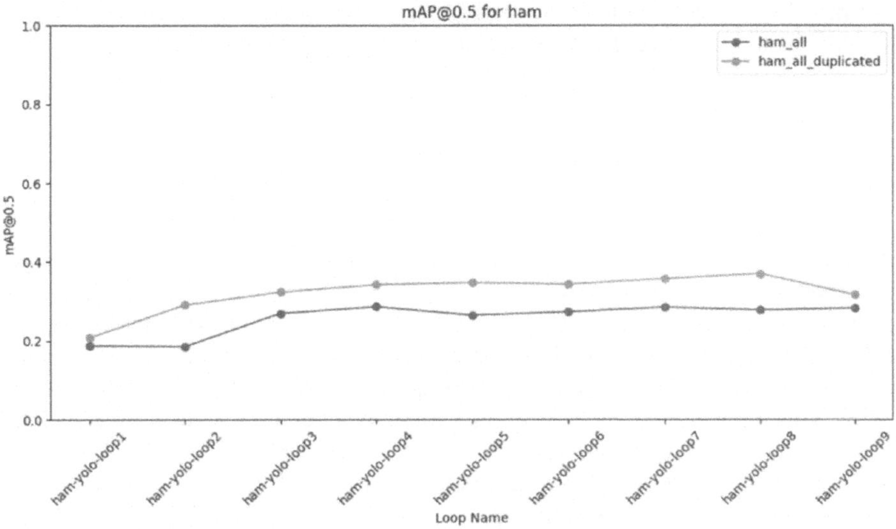

FIGURE 1.7 YOLOv8-AIS on PV-Apple minimal and imbalanced version.

extend beyond YOLOv8 instance segmentation models. The OneFormer application of the model was experimented within two test cases, as introduced in the OneFormer settings section. However, the results presented in Figure 1.8 have shown that the transformer-based model is not as versatile as YOLOv8 in environments where it must make accurate predictions based on a very limited proportion of the entire dataset.

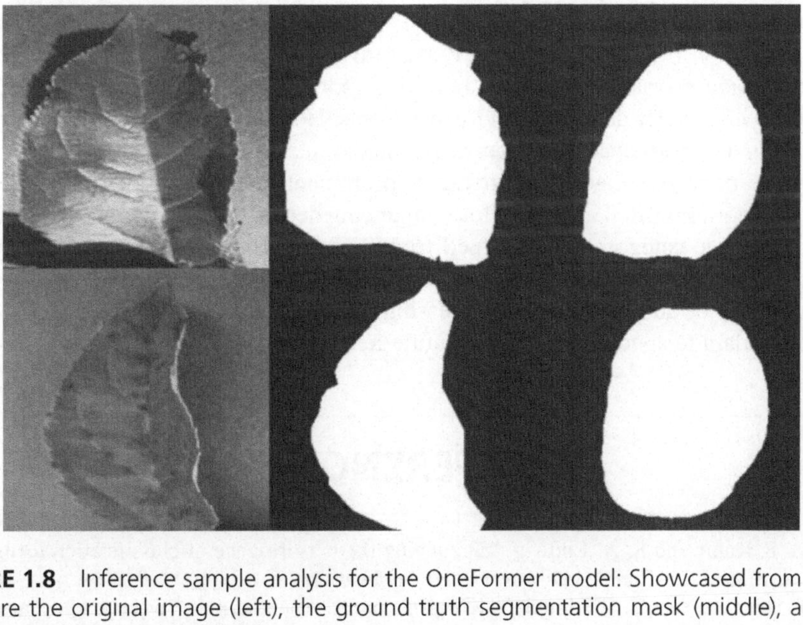

FIGURE 1.8 Inference sample analysis for the OneFormer model: Showcased from left to right are the original image (left), the ground truth segmentation mask (middle), and the model's predictive output (right).

This limitation resulted in the segmentation performed in later iterations failing to capture the basic features and shapes of objects in the images in both test cases.

1.5 CONCLUSION

This study demonstrates the effectiveness of the proposed automated image segmentation method in enhancing object detection models like YOLOv8. Incorporating auto-labeled data into the training set at each iteration significantly improved model performance. The improvement rate was found to be more influenced by the initial dataset size, which determines the model's starting accuracy, than by the λ value dictating the proportion of auto-labeled data. This highlights the importance of the initial dataset in enabling the model to capture essential features and shapes at the start of training.

The proposed method, when implemented with YOLOv8, successfully improved model performance and created accurately segmented datasets, even when the initial dataset size was minimal or missing data for certain classes. In cases where the model struggled due to a reduced dataset, the simple data duplication method restored performance by effectively increasing the dataset size.

However, limitations arose when applying the algorithm to a broader range of datasets, particularly the HAM10000 dataset. When faced with skewed data or ambiguous features, the model struggled to generate accurate segmentation for subsequent iterations. This suggests that the algorithm's success is heavily dependent on the model's inherent performance on specific datasets. These limitations could be mitigated by using cleaner datasets or by improving the model's structure to handle diverse datasets better.

Applying our algorithm to the OneFormer model also highlighted that the algorithm's success relies on models that efficiently capture features and shapes from small initial datasets. Vision transformer models, requiring more data and computational resources, struggled with smaller datasets in our experiments. However, as these models improve in efficiency and performance, they may better handle smaller datasets in the future.

In summary, our algorithm provides an efficient method for creating segmented image data with minimal manual effort, enhancing detection performance even on datasets with unique categories not learned from pretrained weights. Nonetheless, its success largely depends on the model's performance and the cleanliness of the dataset. This limitation may be addressed by preprocessing image data or by further improving vision models to adapt to specific datasets in future research.

REFERENCES

1. M. F. Kabir and S. A. Ludwig, "Enhancing the performance of classification using super learning," Data-Enabled Discovery and Applications, vol. 3, pp. 1–13, 2019.
2. Y. Shen, H. Yun, Z. C. Lipton, Y. Kronrod, and A. Anandkumar, "Deep active learning for named entity recognition," arXiv preprint arXiv:1707.05928, 2017.

3. X. Zhan, Q. Wang, K. Huang, H. Xiong, D. Dou, and A. B. Chan, "A comparative survey of deep active learning," 2022.
4. P. F. Jacobs, G. M. de Buy Wenniger, M. Wiering, and L. Schomaker, "Active learning for reducing labeling effort in text classification tasks," 2021.
5. J. Wang, S. Wen, K. Chen, J. Yu, X. Zhou, P. Gao, C. Li, and G. Xie, "Semi-supervised active learning for instance segmentation via scoring predictions," 2020.
6. A. Veit, N. Alldrin, G. Chechik, I. Krasin, A. Gupta, and S. Belongie, "Learning from noisy large-scale datasets with minimal supervision," in Proceedings of the IEEE Conference on Computer Vision and Pattern Recognition, pp. 839–847, 2017.
7. X. Yang, Z. Song, I. King, and Z. Xu, "A survey on deep semi-supervised learning," 2021.
8. X. Chen, B. Mersch, L. Nunes, R. Marcuzzi, I. Vizzo, J. Behley, and C. Stachniss, "Automatic labeling to generate training data for online lidar-based moving object segmentation," 2022.
9. A. Ratner, S. Bach, H. Ehrenberg, J. Fries, S. Wu, and C. Ré, "Snorkel: Rapid training data creation with weak supervision," in Proceedings of the VLDB Endowment. International Conference on Very Large Data Bases, vol. 11, no. 3. NIH Public Access, 2017.
10. "Amazon automate data labeling," 2023, retrieved May 5, 2023. [On line]. Available: https://docs.aws.amazon.com/sagemaker/latest/dg/sms automated-labeling.html
11. H. Hino, "Active learning: Problem settings and recent developments," 2020.
12. Y. Ouali, C. Hudelot, and M. Tami, "An overview of deep semi supervised learning," 2020.
13. H. Vishwakarma, H. Lin, F. Sala, and R. K. Vinayak, "Good data from bad models: Foundations of threshold-based auto-labeling," 2022.
14. A. M. Hafiz and G. M. Bhat, "A survey on instance segmentation: State of the art," International Journal of Multimedia Information Retrieval, vol. 9, pp. 171–189, 2020.
15. Roboflow, "Yolov8," retrieved Jan 10, 2023. [Online]. Available: https://roboflow.com/model/yolov8
16. J. Jain, J. Li, M. Chiu, A. Hassani, N. Orlov, and H. Shi, "Oneformer: One transformer to rule universal image segmentation," 2022.
17. J. Redmon, S. Divvala, R. Girshick, and A. Farhadi, "You only look once: Unified, real-time object detection," 2016.
18. J. Terven and D. Cordova-Esparza, "A comprehensive review of yolo: From yolovl and beyond," 2023.
19. D. Reis, J. Kupec, J. Hong, and A. Daoudi, "Real-time flying object detection with yolov8," 2023.
20. B. Cheng, A. G. Schwing, and A. Kirillov, "Per-pixel classification is not all you need for semantic segmentation," 2021.
21. Kaggle, "Plant village dataset," retrieved September, 2016.
22. J. Kim and M. F. Kabir, "Automated data labeling for object detection via iterative instance segmentation," in 2023 International Conference on Machine Learning and Applications (ICMLA), pp. 845–850, 2023.

2 Energy-Efficient Glaucoma Detection

Leveraging GAN-Based Data Augmentation for Advanced Diagnostics

Krish Nachnani

2.1 INTRODUCTION

Glaucoma is a common eye disease that causes vision loss and blindness. It is the second-leading cause of blindness behind cataracts. [1, 2] outline the causes of glaucoma, observing that when an abnormality in the aqueous humor causes fluid to build up, the increased pressure damages the optic nerve. By the year 2040, 111.8 million people worldwide are expected to have the disease [3]. Age, race, and family history of glaucoma put an individual at a higher risk of developing it [4]. Glaucoma is especially dangerous because there are no symptoms during the preliminary stages. In developed countries, 50% of people with the disease are unaware that they have it, and this figure is as high as 90% in underdeveloped parts of the world [1, 3].

Effective treatment of glaucoma is blocked on a number of fronts [5] explains that general ophthalmologists in India lack the skills required for glaucoma screening. As a result, they are unable to refer cases to specialists. In rural areas of the United States, there are just 0.58 ophthalmologists per 100,000 individuals [6]. With the assistance of automated

 DOI: 10.1201/9781003570882-3

screening, where high-risk patients are identified via algorithms, the limited physician resources can be more effectively utilized and more patients overall can receive proper treatment. Also, there is potential for improved diagnostic accuracy. In [7], researchers reveal that when breast cancer doctors combined their observations with artificial intelligence (AI), the results were 2.6% more accurate than doctors alone. For these reasons, we explore efficient techniques for automated glaucoma screening. We aim to shorten the process of diagnosing a patient and allow doctors to address more patients.

The impact of the problem generates additional requirements. For example, glaucoma screening is essential not only in urban areas, but also in rural locations. Many communities most in need of diagnostic assistance have very limited resources in other ways, such as internet access or computer power. Solutions that can effectively automate glaucoma diagnosis while operating in a small resource footprint are even more practical for solving the broad problem.

Our study makes the following contributions. We evaluate the effectiveness of energy-efficient convolutional neural networks (CNNs) in glaucoma classification. We also evaluate a combination of featurization techniques and machine learning algorithms for the same problem, providing an alternative method for efficient diagnostics. Both are compared with the state-of-the-art ResNet architecture so that the relative merit of the energy-efficient techniques can be better understood. Also, deep neural networks require large amounts of training data to better understand the underlying features and improve classification capability [8]. Therefore, we utilize common data augmentation techniques to produce geometrically altered images and use decoder-encoder generative adversarial networks (DE-GANs) to create synthetic training images to augment the dataset [9]. We chose MobileNetV2 for the initial featurization approach because of the inherent efficiency of the architecture and our focus on energy-efficient approaches. MobileNetV2 has been shown to deliver a drastic reduction in the number of FLOPs (floating point operations) without sacrificing accuracy [10], and can be used within resource-constrained devices.

Our results reveal that a variety of CNNs are able to effectively diagnose glaucoma. Furthermore, machine learning algorithms over featurized data perform better than ResNet, an energy-expensive algorithm. MobileNetV2, which is an alternative approach to creating efficiency, does not perform quite as well as ResNet [10–12]. Overall, we find featurized machine learning algorithms to be an optimal method to detect and diagnose glaucoma.

The remainder of the chapter is structured as follows. We first compare related work, then outline the study's methodology, before showing and discussing the results, and finally, concluding the discussion.

2.2 RELATED WORK

Over the years, a plethora of methodologies have emerged for the classification of glaucoma [13–15] offer an exhaustive examination of these techniques, emphasizing the pivotal role of feature extraction and the use of CNN architectures in diagnosing glaucoma.

Techniques surrounding data augmentation and generative adversarial networks have also been prevalent. [16] explore existing methods of data augmentation, including generative models.

Several types of traditional machine learning model–based approaches that rely on hand-crafted features extracted from images are popular. In [17], researchers utilize superpixel classification. The method is tested on the RIM-One database and applies the support vector machine (SVM) classifier to achieve an average accuracy of 98.6% [18] employs a multiparametric optic disk detection and localization method through the use of region-based statistical and textural features. Using retinal fundus images from the DRIONS, MESSIDOR, and ONHSD databases, the Random Forest classifier produces accuracies of 99.3%, 98.8%, and 99.3%, respectively. [19] evaluates a multibranch neural network (MB-NN) model using data from Tongren hospital in China, achieving an accuracy of 91.5%. Prior studies employ SVMs to enhance glaucoma detection from imaging data, leveraging features such as retinal nerve fiber layer (RNFL) thickness and measurements of the rim and cup area for classification [20–22]. Despite the decent performance of SVMs and other conventional machine learning techniques in glaucoma diagnosis [23, 24], their adoption in clinical practice remains limited.

With the advent of deep learning models in computer vision that use CNNs combined with the technique of transfer learning, huge improvements are seen in model performances. Several researchers have come up with approaches to use deep learning for glaucoma classification [25]. In [26], a two-stage approach is implemented. In the first stage, the optic disk is extracted from the image. The second stage pertains to classifying the image through a CNN. The study uses several datasets, including the ORIGA dataset, to produce a maximum accuracy of 100%. [27] modifies the Frangi method and uses feature extraction. On the STARE and DRIVE databases, this method reaches 94% accuracy. An innovative machine-to-machine (M2M) approach for evaluating fundus photographs in glaucoma is introduced in [28]. This method trains a deep learning model on color fundus photographs, labeled with the corresponding RNFL thickness from spectral-domain optical coherence tomography (SDOCT) – a recognized standard for structural damage assessment in glaucoma.

Studies have also focused on efficient architectures for algorithm-powered healthcare systems in other healthcare domains [29] describes that deep neural networks have potential scalability when used through edge devices [30] outlines the issues with resource-heavy machine learning algorithms in healthcare and suggests solutions. In [31], researchers explain that medically oriented AI can benefit rural communities. [32] illustrates that many patients in rural areas are forced to travel long distances for care. However, AI-driven platforms make remote consultations possible. [33] develops and evaluates a MobileNetV2 model to detect COVID-19, stressing the efficient nature of the architecture.

Many papers target the problem of accuracy improvement at its root, which is augmenting the dataset with additional images. [34, 35] propose a method named neural augmentation, where a neural network learns the ideal augmentation to improve a classifier. [36] evaluate geometric and photometric augmentations on coarse-grained data using a simple CNN, discovering cropping as providing the best increase in performance. [37] analyze the capability of generative adversarial networks in the domain of medical imaging for image reconstruction, segmentation, and other traditional and novel applications.

Our study differs from the above work in several dimensions. We focus heavily on the use of computationally efficient techniques and compare them with ResNet. We evaluate two energy-efficient approaches to glaucoma detection: first, utilizing MobileNetV2, and second, utilizing MobileNetV2-based featurization followed by machine learning training. We further assess the validity of the energy-efficient methods by testing against augmented data, which includes both traditional geometric transformations and generative networks. DE-GANs have not been applied to the problem of augmenting a glaucoma dataset in the past.

2.3 METHODOLOGY

Figure 2.1 displays the study's methodology, which is explained in the following paragraphs.

The experiments in this study are run on the ORIGA database, which contains 650 images taken with a fundus camera. Researchers from the Singapore Eye Research Institute gathered and annotated the images [38]. Figure 2.2 shows two examples of the scans. In our first step, the dataset is balanced by equating the number of images in both categories, which is achieved by undersampling the healthy image category. As a result, the dataset is reduced to 276 images. The images are then split into train, validation, and test folders. We first reserve ~20% of the dataset for testing. The remaining ~80% are further split into 80/20 for training and validation. This results in 162 training images, 41 validation images, and 71 test images.

To avoid overfitting and improve the generalization of classification models, we augment the original dataset in two ways: augmentation using geometric transformations

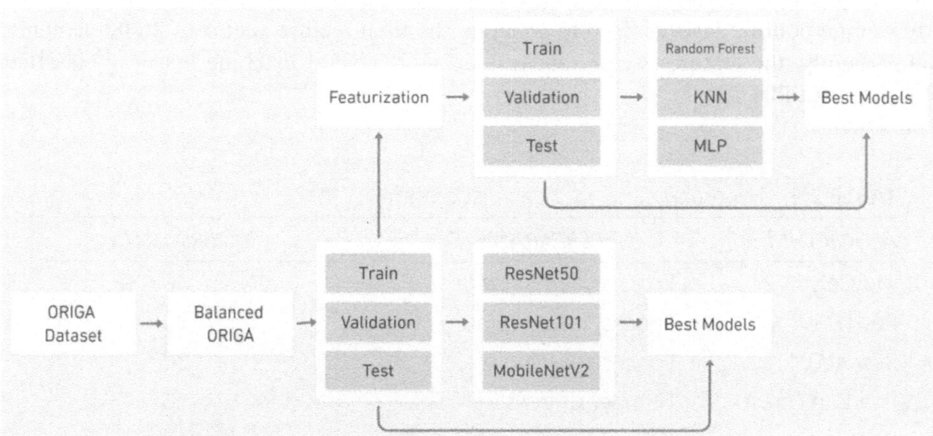

FIGURE 2.1 Experimental methodology.

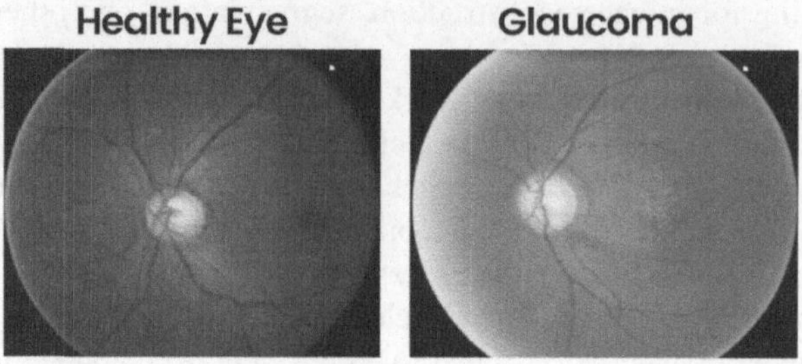

FIGURE 2.2 Sample scans from the ORIGA database.

and augmentation using DE-GANs. In the first approach, we use transformations such as rotation, zoom, and flipping to expand the original training dataset. We add 165 transformed images to the training set, resulting in a total of 439 images in the geometrically augmented dataset. In the second separate approach, we use a DE-GAN to generate an additional 200 synthetic images, which are added to the training set. In total, the GAN-augmented dataset has 474 images. Details about producing the generated images are described in the following subsection.

We conduct experiments with the original dataset and the two augmented variations. Figure 2.1 displays the two methods through which the experiments were conducted on the original dataset. The first method utilizes MobileNetV2, ResNet50, and ResNet101, all examples of established state-of-the-art CNNs. All three networks are pretrained with ImageNet. The second method applies featurization followed by the Random Forest, K-Nearest Neighbors (KNN), and MLP classifiers, which are frequently used machine learning algorithms. Table 2.1 shows the hyperparameters and number of experiments conducted for each algorithm. In the case of MLP, three architectures are investigated, utilizing one layer of 100 neurons, two layers of 100 neurons each, and three layers of 100 neurons each, respectively.

Figure 2.3 displays the process of featurization. Feature vectors are derived for each image by generating output from a MobileNetV2 model pretrained with ImageNet. An average pooling layer is used to generate the final feature vector of 2048 elements. Subsequently, the feature vectors are trained with several machine learning classifier algorithms (Table 2.1).

TABLE 2.1 Hyperparameters for each algorithm

ALGORITHM	*HYPERPARAMETERS*	*EXPERIMENTS*
MobileNetV2	Epochs, learning rate	30
ResNet50	Epochs, learning rate	30
ResNet101	Epochs, learning rate	30
Random Forest	Number of trees	6
KNN	K	17
MLP	Epochs, learning rate, layers	108

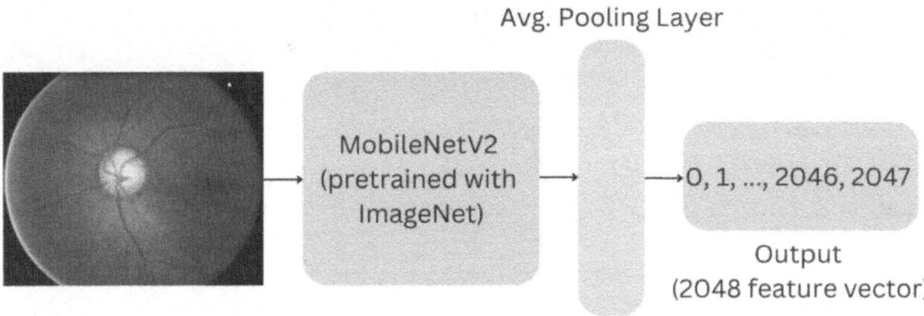

FIGURE 2.3 Featurization.

After the best-performing models in each architecture (MobileNetV2, ResNet50, ResNet101, Random Forest, KNN, and MLP) are established, we perform follow-on experiments with the augmented datasets. However, we first explain the process of augmentation before describing the procedure for running experiments on the augmented datasets.

The first technique to augment images is geometric augmentation, as described in the dataset subsection. The second technique uses a DE-GAN – an enhanced naive GAN – that utilizes a pretrained decoder-encoder architecture to map the input random Gaussian noise vector to an informative (posterior) noise vector (Zhong et al.). In creating the decoder-encoder architecture, we first train a variational auto-encoder (VAE) with the original dataset. Then, the decoder and encoder from the trained VAE are swapped. The Gaussian noise vectors pass through the pretrained decoder to obtain a latent space image, and then through the pretrained encoder to obtain the new posterior noise. Since the decoder's input and the encoder's output are of the same dimensions, this structure does not alter the dimensions of the noise vectors.

It has been empirically proven that when these noise vectors are input into the generator, the quality of the generated images improves and the GAN's training process accelerates. In addition to employing the Decoder-Encoder structure, DE-GANs use hidden-space loss and adversarial loss during training, since using hidden-space loss improves the robustness of the training algorithm. The adversarial loss function is shown in Equation 1. P(x) represents the probability distribution of the dataset which consists of the real images, and D(x) is the score given to a real image "x" by the discriminator. P(z) is the probability distribution of the noise, and D(G(z)) is the score given to a generated image G(z). The hidden space loss function, shown in Equation 2, calculates the divergence in extracted features using the activation maps of latent variables. The term $h(x_{real})$ represents the activation map when a real image is sent to the discriminator while $h(x_{gen})$ represents the activation map for a generated image. The combined loss function is shown in Equation 3. $\lambda 1$ and $\lambda 2$ are the respective weights for the adversarial and hidden space loss functions.

$$L_{adv} = E_{x\sim p(x)} log\ log\ \left(D(x)\right) + E_{z\sim p(z)} log\left(1 - D\left(G(z)\right)\right)$$

Equation 1: Adversarial Loss Function

$$L_{hid} = \frac{1}{N}\sum_{i=1}^{N}\left\| h^{(i)}x_{real} - h^{(i)}x_{gen} \right\|$$

Equation 2: Hidden-Space Loss Function

$$L = \lambda_1 L_{adv} + \lambda_2 L_{hid}$$

Equation 3: Combined Loss Function

The complete workflow is as follows: an input Gaussian noise vector is mapped to a posterior noise vector and then provided to the generator, which passes on the generated image to the discriminator to be judged. Both the generator and discriminator are CNNs. The Adam optimizer is used for training the DE-GAN, which has a beta_1 value of 0.5 and a default beta_2 value of 0.999. The learning rate used for producing the healthy eye images is 0.07. For the glaucoma images, the learning rate is 0.005. For both categories, it takes 5,000 epochs for the model to generate realistic images.

In all cases, accuracy and confusion matrices are used to assess results. The experiments are run in Google Colaboratory. Every image is sampled into 224 by 224 pixels. The CNN architectures are used from Tensorflow and the machine learning algorithms are used from SciKit Learn [39, 40]. The DE-GAN code is written in Tensorflow.

The model hyperparameters that result in the best-performing models on the original dataset are retained when conducting experiments on the augmented dataset. All the model architectures used previously (MobileNetV2, ResNet50, ResNet101, Random Forest, KNN, MLP) are trained using the retained hyperparameter values and evaluated on the test dataset.

2.4 RESULTS

2.4.1 Experiments and Results – ORIGA Dataset

Each convolutional neural network is evaluated in turn, followed by featurization experiments, which are listed by the machine learning algorithm used. All MLP variants are described together.

2.4.1.1 Experiments and results – MobileNetV2

Table 2.2 displays the MobileNetV2 experiments exploring different settings for learning rate and epochs. On average, the epochs values paired with a learning rate of 0.005 result in the greatest accuracy. Also, for a learning rate value of 0.00001, lower epochs values (20, 40, and 60) produce higher accuracies in comparison to higher epochs values (80 and 100).

TABLE 2.2 MobileNetV2 – Accuracy for different learning rate and epoch values

	EPOCHS				
LEARNING RATE	*20*	*40*	*60*	*80*	*100*
0.00001	62.50	69.99	62.50	55.00	55.00
0.00005	64.99	60.00	64.99	62.50	69.99
0.0001	67.50	64.99	64.99	67.50	55.00
0.0005	67.50	69.99	67.50	64.99	64.99
0.001	64.99	64.99	64.99	69.99	64.99
0.005	69.99	67.50	67.50	69.99	69.99

2.4.1.2 Experiments and results – ResNet50

Table 2.3 shows that, in ResNet50, the algorithm achieves its highest accuracy of 82% twice with 80 epochs, yet once with 100 epochs. Two of out three of these cases occur with a learning rate value of 0.005.

2.4.1.3 Experiments and results – ResNet101

Table 2.4 displays the ResNet101 experiments. When the learning rate is 0.00005, the accuracies of the experiments for four out of five epochs values are the same: 73%. The highest accuracy of 77% occurs once at the highest learning rate, 0.005, and once at the lowest learning rate, 0.00001.

2.4.1.4 Experiments and results – Featurization plus Random Forest

Table 2.5 displays the Random Forest experiments. The accuracies are almost completely independent of the number of trees: every experiment demonstrates accuracy values between 65% and 73%.

TABLE 2.3 ResNet50 – Accuracy for different learning rate and epoch values

	EPOCHS				
LEARNING RATE	*20*	*40*	*60*	*80*	*100*
0.00001	67.50	67.50	67.50	69.99	75.00
0.00005	72.50	69.99	69.99	77.49	77.49
0.0001	77.49	72.50	75.00	75.00	80.00
0.0005	72.50	77.49	80.00	82.49	80.00
0.001	80.00	77.49	80.00	80.00	80.00
0.005	80.00	80.00	80.00	82.49	82.49

TABLE 2.4 ResNet101 – Accuracy for different learning rate and epoch values

	EPOCHS				
LEARNING RATE	*20*	*40*	*60*	*80*	*100*
0.00001	62.50	60.00	60.00	77.49	69.99
0.00005	64.99	72.50	72.50	72.50	72.50
0.0001	69.99	69.99	69.99	75.00	67.50
0.0005	75.00	72.50	72.50	69.99	75.0017
0.001	75.00	75.00	75.00	72.50	75.00
0.005	72.50	60.00	75.00	75.00	77.49

2.4.1.5 Experiments and results – Featurization plus KNN

Table 2.6 displays the KNN experiments. Again, the accuracies vary minimally, with a high of 71% and a low of 58%. This highest accuracy of 71% occurs when K is 11, 13, and 15.

2.4.1.6 Experiments and results – Featurization plus MLP

The three MLP architectures are evaluated in turn with hyperparameters tuned for learning rate and epochs.

Table 2.7 displays the experiments of an MLP algorithm with one hidden layer, which consists of 100 neurons. For a learning rate of 0.005, the experiment with 20 epochs performs considerably better than 40, 60, 80, and 100 epochs, each of which has an accuracy of 44%. The best accuracy is 83%.

Table 2.8 displays MLP results with two hidden layers, each consisting of 100 neurons. The highest accuracy is 85%, which is achieved with a learning rate of 0.001 and 80 epochs.

Table 2.9 shows MLP results for three hidden layers, each of which consists of 100 neurons. Nine experiments achieve an accuracy of 83%, while the highest accuracy is 85%.

TABLE 2.5 Random Forest – Accuracy versus number of trees

NUMBER OF TREES	*ACCURACY*
10	68.29
15	68.29
25	65.85
40	70.73
100	70.73
200	73.17

TABLE 2.6 KNN – Accuracy versus K

K	*ACCURACY*
1	58.53
3	58.53
5	63.41
7	60.97
9	68.29
11	70.73
13	70.73
15	70.73
17	65.85
19	65.85
21	63.41
23	65.85
25	68.29
27	68.29
29	68.29
31	65.85
33	65.85

Figure 2.4 shows confusion matrices for the MLP algorithms. In each case, the false negative rate is low: both MLP (100) and MLP (100, 100) obtain rates of 4.9% and MLP (100, 100, 100) obtains a rate of 2.4%. Given that false negatives are highly detrimental, MLP (100, 100, 100) is preferred because it is not only at the highest accuracy but also has lower false negatives.

Table 2.10 displays detailed metrics for the best-performing model, MLP (100, 100, 100).

TABLE 2.7 MLP (100) – Accuracy for different learning rate and epoch values

	EPOCHS				
LEARNING RATE	*20*	*40*	*60*	*80*	*100*
0.00001	53.65	56.09	63.41	68.29	70.73
0.00005	68.29	75.60	78.04	78.04	78.04
0.0001	73.17	78.04	80.48	80.48	80.48
0.0005	68.29	73.17	82.92	75.60	78.04
0.001	75.60	75.60	78.04	78.04	78.04
0.005	75.60	43.90	43.90	43.90	43.90

TABLE 2.8 MLP (100, 100) – Accuracy for different learning rate and epoch values

	EPOCHS				
LEARNING RATE	*20*	*40*	*60*	*80*	*100*
0.00001	56.09	56.09	56.09	53.65	58.53
0.00005	58.53	68.29	75.60	75.60	73.17
0.0001	58.53	75.60	73.17	75.60	73.17
0.0005	75.60	75.60	82.92	82.92	82.92
0.001	75.60	82.92	82.92	85.36	82.92
0.005	56.09	73.17	78.04	75.60	80.48

TABLE 2.9 MLP (100, 100, 100) – Accuracy for different learning rate and epoch values

	EPOCHS				
LEARNING RATE	*20*	*40*	*60*	*80*	*100*
0.00001	56.09	56.09	56.09	56.09	53.65
0.00005	56.09	73.17	75.60	75.60	78.04
0.0001	73.17	73.17	75.60	78.04	75.60
0.0005	75.60	82.92	80.48	82.92	82.92
0.001	80.48	80.48	82.92	80.48	80.48
0.005	68.29	80.48	85.36	82.92	82.92

TABLE 2.10 MLP (100, 100, 100) – Classification report

	PRECISION	*RECALL*	*F1-SCORE*	*SUPPORT*
Glaucoma	0.81	0.96	0.88	23
No Glaucoma	0.93	0.72	0.81	18

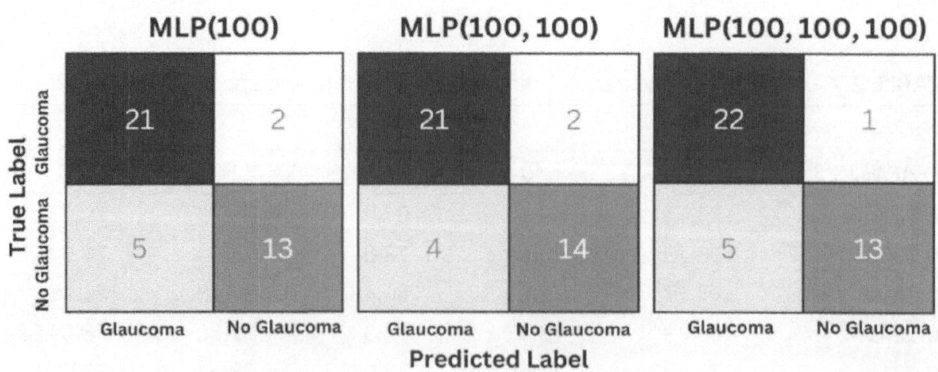

FIGURE 2.4 Confusion matrix for three MLP algorithms.

TABLE 2.11 Test accuracies on three distinct datasets

ALGORITHM	*LEARNING RATE, EPOCHS*	*ORIGA*	*GEOMETRIC*	*GAN*
MobileNetV2	0.005, 100; 0.001, 80	56.3%	50.7%	59.2%
ResNet50	0.005, 100; 0.0005, 80	56.3%	65.5%	57%
ResNet101	0.005, 100; 0.00001, 80	64.1%	63.4%	56.3%
MLP(100, 100, 100)	0.005, 60	64.8%	66.2%	66.2%

2.4.2 Experiments and Results – Augmented Dataset

Table 2.11 displays the test accuracies for MobileNetV2, ResNet50, ResNet101, and MLP (100, 100, 100) on the ORIGA dataset, the geometrically augmented dataset, and the GAN-augmented dataset. Since there were many best-performing models for each CNN architecture on the validation set, we selected two of the highest accuracy-producing models per algorithm in order to provide a more complete and robust understanding of the algorithm's performance. The test accuracies shown in Table 2.11 are the averages of the test accuracies of these two models. ResNet50 increases by 9.2% between the ORIGA dataset and the geometrically augmented dataset. Another prominent change in the algorithms' accuracies occurs between the GAN-augmented dataset and the ORIGA dataset for ResNet101, with a decrease in accuracy of 7.8%.

Figure 2.5 shows the confusion matrices for the test accuracies of MLP (100, 100, 100) on the geometrically augmented dataset and the GAN-augmented dataset. Both

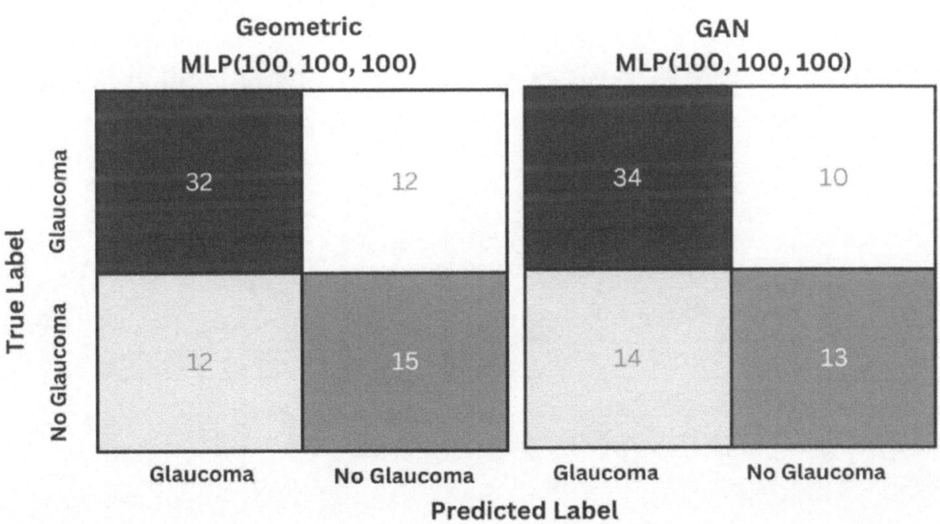

FIGURE 2.5 Confusion matrix for MLP (100, 100, 100)'s test accuracy on augmented datasets.

models have a test accuracy of 66.2%. However, the model on the geometrically augmented dataset has a false negative rate of 0.27, and the model on the GAN-augmented dataset has a false negative rate of 0.23. As a result, MLP (100, 100, 100) on the GAN-augmented dataset is preferred.

Across all the experiments conducted in this study, the best-performing model is MobileNetV2-featurization followed by MLP (100, 100, 100) on the GAN-augmented dataset, with a test accuracy of 66.2%. ResNet, for the most part, is close behind, but MobileNetV2 frequently produces test accuracies in the low 50% range.

2.5 DISCUSSION

2.5.1 Findings on ORIGA Dataset

Figure 2.6 indicates that machine learning algorithms with featurized data outperforms CNNs. MobileNetV2, ResNet50, and ResNet101 generated 70%, 82%, and 77% accuracy. The Random Forest and KNN classifiers achieve accuracies of 73% and 71%, while MLP (100), MLP (100, 100), and MLP (100, 100, 100) achieve 83%, 85%, and 85% accuracy, respectively.

Comparing Table 2.3 (ResNet50 – the best-performing CNN) with Tables 2.7, 2.8, and 2.9 (MobileNetV2 featurization followed by MLP), we can see that the featurized

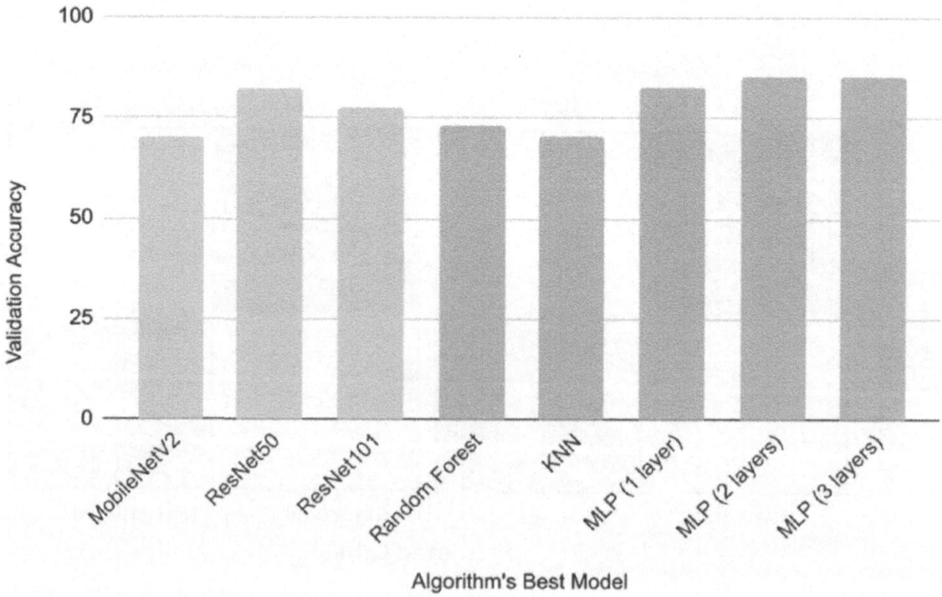

FIGURE 2.6 Accuracy versus algorithm's best model.

data approach does better, even though its components are notably more training speed efficient than the powerful ResNet [41]. However, it is possible that the cause is the small dataset. Prior work [26] has demonstrated that getting CNNs to deliver good results on the ORIGA dataset is challenging. While the results are not directly comparable with [26], since it uses different data processing techniques, we do note that the accuracy of 85% is within the general range of that reported in [26] for ORIGA.

Table 2.6 displays a pattern in the relationship between the value of K and the accuracy of the experiment: the highest accuracies come in clusters of three. For example, when K is 11, 13, and 15, the algorithm achieves the highest accuracy of 71%. Similarly, the second-highest accuracy of 68% is achieved jointly by the K values of 25, 27, and 29.

Tables 2.7, 2.8, and 2.9 demonstrate the positive effect of an increased number of hidden layers in the MLP classifier. The highest accuracy with one hidden layer is 83%, and the highest accuracy with two and three hidden layers is 85%. Each result is higher than the highest accuracies achieved by MobileNetV2, ResNet50, or ResNet101. However, increasing the number of hidden layers even more may result in further accuracy improvements. This is a target for future work.

The MLP classifier performs the best out of the machine learning algorithms. Since MLP is a feedforward artificial neural network, it is the most capable of learning from the large feature vectors generated by MobileNetV2 [42]. This relationship may account for MLP's high accuracy and demonstrates the exceptional performance of neural networks in comparison to other machine learning approaches.

2.5.2 Findings on the Augmented Datasets

In Table 2.11, we observe that both forms of augmentation benefit MLP's test accuracy, with MLP under the GAN-augmented dataset being the best-performing model in the study. On the other hand, the CNN-based models give mixed results. MobileNetV2, although temperamental, shows an improvement with the GAN-augmented dataset. The more energy-consuming models – ResNet50 and ResNet101 – have higher accuracies under the geometrically augmented dataset than the GAN-augmented dataset, while the energy-efficient models perform better under the GAN-augmented dataset. ResNet101, the largest out of the algorithms tested in this study, is the only one that does not benefit from either form of augmentation. As a result of these findings, we can conclude that MLP outperforms state-of-the-art CNN architectures even when faced with a larger dataset. In addition, a more thorough hyperparameter search with the augmented datasets may result in an even higher classification accuracy.

Figure 2.7 shows example images generated by the DE-GAN. The images appear pixelated because they are 64 by 64 pixels – much smaller than the original images, which are around 2467 by 2048 pixels. The generated images capture key components of the original image, such as the optic nerve, but are not able to capture the finer details, such as the veins. Training the DE-GAN to produce viable images required an extensive hyperparameter search. We also notice mode collapse during the training of the healthy

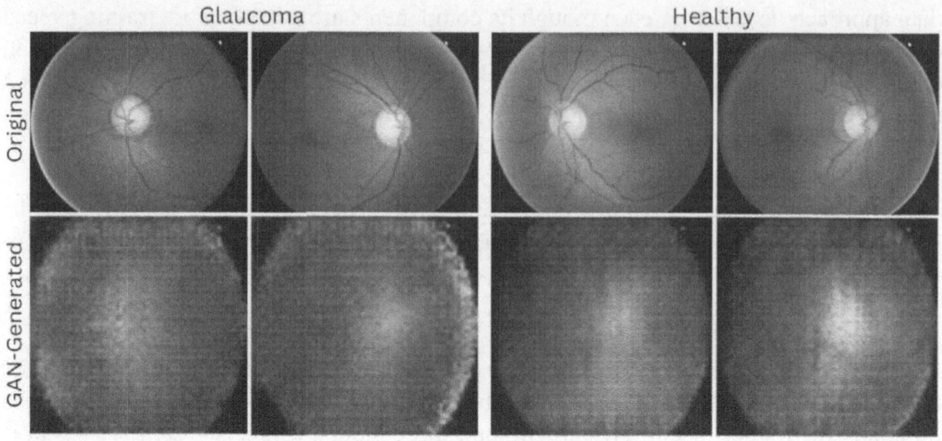

FIGURE 2.7 Original images versus GAN-generated images.

eye category, as the position of the optic nerve in relation to the rest of the eye remains the same, unlike the images produced for the glaucoma category. In the future, we aim to fine-tune the generated images to be of higher resolution and to capture finer details in the images.

Overall, the featurization approach yields good results, but we also only evaluate one featurization approach (MobileNetV2). Variants of the featurization approach, such as using ResNet, may improve the results, but at the cost of less energy efficiency. We consider these options in our future work.

2.6 CONCLUSION

This study finds that machine learning algorithms provided with featurized data outperform CNNs on the ORIGA dataset. It appears to be able to deliver better results on the relatively small dataset available, and also at better energy efficiency. When using a geometrically augmented dataset and a GAN-augmented dataset, we continue to observe that the energy-efficient solution outperforms the state-of-the-art ResNet. We achieve the best test accuracy (66.2%) across all experiments in this study by using the featurization approach with GAN augmentation. These findings provide valuable insights into the effectiveness of various model architectures and augmentation strategies in the context of the ORIGA dataset. In the future, we plan to utilize a Wasserstein GAN with a gradient penalty as another approach to augmenting the dataset, as it is more stable during training and less prone to mode collapse (Arjovsky et al., 2017). Further variants of the featurization approach, such as other MLP topologies with different numbers of layers and neurons per layer, and featurizing with ResNet may also be worth testing, as are other featurization approaches.

REFERENCES

1. 2020, www.cdc.gov/visionhealth/resources/features/glaucoma-awareness.html. Accessed 13 July 2023.
2. Kaleem, Mona. "Glaucoma." JHM, n.d. [Online], www.hopkinsmedicine.org/health/conditions-and-diseases/glaucoma. Aug. 21, 2020. Accessed 13 July 2023.
3. 2018, www.glaucomapatients.org/basic/statistics/. Accessed 13 July 2023.
4. Allison, Karen, et al. "Epidemiology of glaucoma: The past, present, and predictions for the future." Cureus, vol. 12, 2020, https://doi.org/10.7759/cureus.11686
5. Gawas, Lisika, et al. "Glaucoma screening skills among general ophthalmologists – How general should it be?" Indian Journal of Ophthalmology, vol. 70, no. 10, pp. 3534–3539, 2022., http://doi.org/10.4103/ijo.IJO_672_22
6. LeBrun, Nancy. "7 Things to Know about the Ophthalmologist Shortage." Healthgrades, 31 May 2023, www.healthgrades.com/pro/7-things-to-know-about-the-ophthalmologist-shortage. Accessed 13 July 2023.
7. Kiros, Hana. "Doctors Using AI Catch Breast Cancer More Often than Either Does Alone." MIT Technology Review, 22 July 2022, www.technologyreview.com/2022/07/11/1055677/ai-diagnose-breast-cancer-mammograms/. Accessed 13 July 2023.
8. Sun, Chen, et al. "Revisiting unreasonable effectiveness of data in deep learning era." 2017 IEEE International Conference on Computer Vision (ICCV), Oct. 2017, https://doi.org/10.1109/iccv.2017.97
9. Zhong, Guoqiang, et al. "Generative adversarial networks with decoder–encoder output noises." Neural Networks, vol. 127, July 2020, pp. 19–28, https://doi.org/10.1016/j.neunet.2020.04.005
10. Sandler, Mark, et al. "MobileNetV2: Inverted residuals and linear bottlenecks." 2018 IEEE/CVF Conference on Computer Vision and Pattern Recognition, 2018, https://doi.org/10.1109/cvpr.2018.00474
11. He, Kaiming, et al. "Deep residual learning for image recognition." 2016 IEEE Conference on Computer Vision and Pattern Recognition (CVPR), 10 Dec. 2015, https://doi.org/10.1109/cvpr.2016.90
12. Feng, Zijian. "An Overview of ResNet Architecture and Its Variants." Built In, 16 Sept. 2022, builtin.com/artificial-intelligence/resnet-architecture. Accessed 13 July 2023.
13. Barros, Daniele M., et al. "Machine learning applied to retinal image processing for glaucoma detection: Review and perspective." BioMedical Engineering OnLine, vol. 19, no. 1, 15 Apr. 2020, https://doi.org/10.1186/s12938-020-00767-2
14. Ran, An Ran, et al. "Deep learning in Glaucoma With optical coherence tomography: A review." Nature News, Nature Publishing Group, 7 Oct. 2020, www.nature.com/articles/s41433-020-01191-5
15. Thompson, Atalie C., et al. "A review of deep learning for screening, diagnosis, and detection of glaucoma progression." Translational Vision Science & Technology, vol. 9, no. 2, 2020, p. 42, https://doi.org/10.1167/tvst.9.2.42
16. Shorten, Connor, and Khoshgoftaar, Taghi M. "A survey on image data augmentation for dep learning - Journal of big data." SpringerOpen, Springer International Publishing, 6 July 2019, journalofbigdata.springeropen.com/articles/10.1186/s40537-019-0197-0
17. Mohamed, Nur Ayuni, et al. "An automated glaucoma screening system using cup-to-disc ratio via simple linear iterative clustering superpixel approach." Biomedical Signal Processing and Control, vol. 53, Aug. 2019, https://doi.org/10.1016/j.bspc.2019.01.003
18. Rehman, Zaka Ur, et al. "Multi-parametric optic disc segmentation using superpixel based feature classification." Expert Systems With Applications, vol. 120, 15 Apr. 2019, pp. 461–473, https://doi.org/10.1016/j.eswa.2018.12.008

19. Chai, Yidong, et al. "Glaucoma diagnosis based on both hidden features and domain knowledge through deep learning models." Knowledge-Based Systems, vol. 161, 1 Dec. 2018, pp. 147–156, https://doi.org/10.1016/j.knosys.2018.07.043
20. Burgansky-Eliash, Zvia, et al. "Optical coherence tomography machine learning classifiers for glaucoma detection: A preliminary study." Investigative Ophthalmology & Visual Science, vol. 46, 2005, pp. 4147–4152, https://iovs.arvojournals.org/article.aspx?articleid=2182263
21. Belghith, Akram, Bowd, Christopher, Medeiros, Felipe A, Balasubramanian, Madhusudhanan, Weinreb, Robert N, and Zangwill, Linda M. "Learning from healthy and stable eyes: A new approach for detection of glaucomatous progression." Artificial Intelligence in Medicine, vol. 64, 2015, pp. 105–115, https://pubmed.ncbi.nlm.nih.gov/25940856/
22. Shigueoka, Leonardo Seidi, et al. "Automated algorithms combining structure and function outperform general ophthalmologists in diagnosing glaucoma." PLoS One, vol. 13, 2018, p. e0207784, https://journals.plos.org/plosone/article?id=10.1371/journal.pone.0207784
23. Yousefi, Siamak, et al. "Detection of longitudinal visual field progression in glaucoma using machine learning." American Journal of Ophthalmology, vol. 193, 2018, pp. 71–79, https://pubmed.ncbi.nlm.nih.gov/29920226/
24. Goldbaum, Michael H, et al. "Comparing machine learning classifiers for diagnosing glaucoma from standard automated perimetry." Investigative Ophthalmology & Visual Science, vol. 43, 2002, pp. 162–169, https://pubmed.ncbi.nlm.nih.gov/11773027/
25. Chen, Xiangyu, et al. "Glaucoma detection based on deep convolutional neural network." 2015 37th Annual International Conference of the IEEE Engineering in Medicine and Biology Society (EMBC), 2015, https://doi.org/10.1109/embc.2015.7318462
26. Bajwa, Muhammad Naseer, et al. "Two-stage framework for optic disc localization and glaucoma classification in retinal fundus images using deep learning." BMC Medical Informatics and Decision Making, vol. 19, no. 1, 17 July 2019, https://doi.org/10.1186/s12911-019-0842-8
27. Budai, A., et al. "Robust vessel segmentation in fundus images." International Journal of Biomedical Imaging, 12 Dec. 2013, pp. 1–11, https://doi.org/10.1155/2013/154860
28. Medeiros, Felipe A, Jammal, Alessandro A, and Thompson, Atalie C. "From machine to machine: An OCT-trained deep learning algorithm for objective quantification of glaucomatous damage in fundus photographs." Ophthalmology, vol. 126, 2019, pp. 513–521, https://pubmed.ncbi.nlm.nih.gov/30578810/
29. Al-Marridi, Abeer, et al. "AI-based techniques on edge devices to optimize energy efficiency in m-health applications." Energy Efficiency of Medical Devices and Healthcare Applications, Academic Press, 2020, pp. 1–23, https://doi.org/10.1016/b978-0-12-819045-6.00001-7
30. Jia, Zhenge, et al. "The importance of resource awareness in artificial intelligence for healthcare." Nature Machine Intelligence, vol. 5, 12 June 2023, https://doi.org/10.1038/s42256-023-00670-0
31. Guo, Jonathan, and Li, Bin. "The application of medical artificial intelligence technology in rural areas of developing countries." Health Equity, vol. 2, no. 1, 1 Aug. 2018, pp. 174–181, https://doi.org/10.1089/heq.2018.0037
32. Frąckiewicz, Marcin. "Unlocking the Potential of AI for Rural Healthcare Access." TS2 SPACE, 3 May 2023, ts2.space/en/unlocking-the-potential-of-ai-for-rural-healthcare-access/. Accessed 13 July 2023.
33. Ukwandu, Ogechukwu, et al. "An evaluation of lightweight deep learning techniques in medical imaging for high precision COVID-19 diagnostics." Healthcare Analytics, vol. 2, 23 Aug. 2022, https://doi.org/10.1016/j.health.2022.100096
34. Halevy, Alon, et al. "The unreasonable effectiveness of data." IEEE Intelligent Systems, vol. 24, no. 2, Mar. 2009, pp. 8–12, https://doi.org/10.1109/mis.2009.36

35. Perez, Luis, and Wang, Jason. "The Effectiveness of Data Augmentation in Image Classification Using Deep Learning." arXiv.Org, 13 Dec. 2017, arxiv.org/abs/1712.04621
36. Taylor, Luke, and Nitschke, Geoff. "Improving Deep Learning with Generic Data Augmentation." 2018 IEEE Symposium Series on Computational Intelligence (SSCI), Nov. 2018, https://doi.org/10.1109/ssci.2018.8628742
37. Yi, Xin, et al. "Generative adversarial network in medical imaging: A review." Medical Image Analysis, vol. 58, Dec. 2019, p. 101552, https://doi.org/10.1016/j.media.2019.101552
38. Zhang, Zhou, et al. "Origa(-Light): An Online Retinal Fundus Image Database for Glaucoma Analysis and Research." 2010 Annual International Conference of the IEEE Engineering in Medicine and Biology, 2010, https://doi.org/10.1109/iembs.2010.5626137.
39. Abadi, Martín, et al. "TensorFlow: A System for Large Scale Machine Learning." Conference on Operating Systems Design and Implementation, 2016, https://www.usenix.org/system/files/conference/osdi16/osdi16-abadi.pdf
40. Pedrogosa, Fabian, et al. "Sci-kit learn: Machine learning in python." Journal of Machine Learning Research, vol. 12, 2011, pp. 2825–2830, https://jmlr.csail.mit.edu/papers/v12/pedregosa11a.html
41. DeepLabCut. "What Neural Network Should I Use? (Trade Offs, Speed Performance, and Considerations)." GitHub, 26 May 2021, github.com/DeepLabCut/DeepLabCut/wiki/What-neural-network-should-I-use%3F-(Trade-offs,-speed-performance,-and-considerations)
42. Banoula, Mayank. "An Overview on Multilayer Perceptron (MLP) [Updated]." Simplilearn.Com, 29 May 2023, www.simplilearn.com/tutorials/deep-learning-tutorial/multilayer-perceptron
43. M. Arjovsky, S. Chintala, and L. Bottou, "Wasserstein Generative Adversarial Networks," in *Proc. 34th Int. Conf. Mach. Learn. (ICML)*, vol. 70, pp. 214–223, 2017. [Online]. https://proceedings.mlr.press/v70/arjovsky17a.html

3 Deep JPEG Compression Artifact Removal with Harmonic Networks

Hasan H. Karaoglu and Ender M. Eksioglu

3.1 INTRODUCTION

Digital images are compressed for reasons such as less memory and faster transmission. JPEG [1] is one of the most popular modern image compression methods. Typical steps of JPEG compression scheme are to split an image into nonoverlapping 8×8 image blocks, to perform 2D forward discrete cosine transform (DCT) on them, to quantize the resulting block DCT spectrums with a suitable quantization table, to perform inverse block DCT on the quantized spectral coefficients, and to tile the compressed blocks. We note that better energy compaction property of the DCT compared to the other signal transforms is the reason for its use in JPEG algorithm. Since the compression process yields artifacts such as blockiness, ringing, banding, and blurring due to the quantization and block discontinuities [2], achieving the high-quality image from a compressed image is a crucial task for nice photographic images and computer vision applications such as image segmentation and depth estimation. However, compression artifact removal (CAR) is an ill-posed problem [2], meaning that obtaining a high-quality image is not a unique process. As noted earlier, although compression artifacts originate from transform domain, there are few approaches that solve the problem with deep networks in transform domain. Some of them, such as [3, 4], choose wavelet transform as the transform of interest. This is because the discrete wavelet transform (DWT)

 DOI: 10.1201/9781003570882-4

provides subband images as transform coefficients, which are suitable for convolution layers to seek a correlation between neighboring transform coefficients. The other transform-based deep techniques utilize DCT on small nonoverlapping image blocks [5, 6]. However, such treatment does not make use of the correlation of pixels on block boundaries, whereas in smooth regions, for example, it is reasonable and effective to use pixels at block boundaries. It is also known from the signal processing literature that processing with overlapping blocks significantly improves an algorithm's decompression performance. In the light of these facts, we propose DCT, discrete sine transform (DST), and discrete Hartley transform (DHT) domains; three networks to tackle JPEG artifacts. When designing our networks, we make use of local harmonic transform spectra. We show that harmonic transform spectra of overlapping image blocks can be arranged in such a way that the coefficients with the same spectral component form a channel. This can be quickly implemented with a fast convolution layer on graphics processing unit (GPU), which we refer to as harmonic filterbanks. We propose a deep convolutional neural network (CNN) as a nonlinearity function and train them in an end-to-end manner. Our experimental study indicates the effectiveness and efficiency of our networks by giving quantitative and qualitative results.

3.2 THE HISTORY OF JPEG COMPRESSION ARTIFACT REMOVAL

In this section, we will give a short history of CAR problem. Since its development is similar to that of image denoising, we will only give a brief history. For more details, the reader may refer to our denoising section and the comprehensive survey [2].

There is no settled observation model for JPEG CAR problem since the characteristics of compression noise depend on clean signals. However, some researchers assume the observation model as an additive signal independent (white) Gaussian noise; others consider it as an additive colored Gaussian noise with the assumption that there is no other noise source.

One can observe that scientifc efforts to reduce compression artifacts take their inspiration from image restoration, especially image denoising. Deep learning techniques make the relationship between the two noise removal problems (i.e., denoising and compression artifact removal), even closer. Existing CAR algorithms in the literature can be categorized into three parts: model-based, learning-based, and hybrid algorithms. The authors of [7] have proposed one of the earliest algorithms and have utilized the spatial-invariant Gaussian smoothing filter for the problem. Ramamurthi and Gersho [8] filter the compressed image adaptively after the classifcation of image blocks as an edge or monotone block. Among the advanced filtering techniques, [9] has taken into account the slope of neighboring blocks and minimized the total expected slope difference by describing it as an optimization problem. Some handicaps of these simple approaches ignore local statistics of image blocks and to result in poor images with blurred artifacts. With the purpose of smoothness of adjacent blocks,

Yang et al. [10, 11] have imposed set constraints and solve the derived formula by projecting onto the sets.

Regularization and prior modeling techniques and subsequent model-based algorithms try to solve the problem by imposing structural or statistical information (i.e., regularizer or prior), such as nonlocal self-similarity [12], low-rankness [13, 14], and transform sparsity [15–17], on uncompressed images. The downsides of regularization techniques are (1) the lack of one global regularizer for photographic images that contain many complex patterns, such as textures, flat areas, and edges, etc.; and (2) the need of high computation costs due to a hard optimization procedure.

Like most inverse problems, deep learning dramatically improves the decompression performance of the CAR algorithm. Artifact Removal CNN (ARCNN) [18], Denoising CNN (DnCNN) [19], and Memory Network (MemNet) [20] are representative successful (deep) learning–based CAR algorithms. ARCNN [18] is the seminal work whose architecture is a three-layer vanilla CNN. DnCNN [19] is another CNN algorithm using batch normalization and residual learning paradigms for the first time. MemNet [20] uses long- and short-term memory connections and gate mechanisms to attack the problem. However, they mostly tackle the problem by seeking a map from a compressed image to the ground truth image in a spatial domain. Besides, the decompression capability of deep CAR algorithms enhances designing more complex and modular deep network architectures at present. Despite the fact that the source of JPEG artifacts originates from the nonlinear mapping of transform coefficients (i.e., quantization), the established practice is to utilize them as auxiliary information [5, 6].

The hybrid approaches' target is to combine the virtues of model- and learning-based CAR methods. Deep plug-and-play (PnP) CNNs [21] utilize Gaussian denoiser networks to solve the JPEG CAR problem. Deep unrolled networks, which are yet another hybrid approach for the problem, solve it by unrolling classical iterative optimization algorithms. TNRD [22] is the first unrolled neural network with a sum of radial basis functions. DUN [23] is a hybrid method that unfolds the iterative algorithm by modeling JPEG residuals with a convolutional dictionary. However, their drawbacks are that while PnP CNN priors require strong pretrained Gaussian networks with all noise levels, unrolled networks demand high computational resources for more iterations.

3.3 HARMONIC TRANSFORMS

The DCT, DST, and DHT [24][1] are Fourier-related signal transforms and their aim is to separate a signal into harmonic cosine and/or sine basis vectors, respectively. Unlike the DFT [24], the three transforms generate real coefficients and have been widely used in many practical applications, especially signal denoising and compression due to energy compaction property, which collects most of a signal energy on a few of its harmonic transform coefficients [24].

TABLE 3.1 Harmonic transform definitions. In DHT definition, $cas(x) \triangleq cos(x) + sin(x)$

TRANSFORM	*1D TRANSFORM KERNEL*	*COEFFICIENT* α_K	*SYMMETRY*	*NOTES*
DCT	$\gamma_k \cos\left[\frac{(2k+1)m\pi}{2M}\right]$	$\begin{cases} \sqrt{\frac{1}{M}}, & k=0 \\ \sqrt{\frac{2}{M}}, & 1 \le k \le M-1 \end{cases}$	No	Even symmetric. DFT
DST	$\gamma_k \sin\left[\frac{(k+1)(m+1)\pi}{M+1}\right]$	$\sqrt{\frac{2}{M+1}}$	Yes	Odd symm. DFT
DHT	$\gamma_k \operatorname{cas}\left[\frac{2km\pi}{M}\right]$	$\sqrt{\frac{1}{M}}$	Yes	Real + imaginary DFT

2D forward transform of an image $x(k, l)$ for $0 \le k \le M-1$ and $0 \le l \le N-1$ can be written by

$$\tilde{X}(m, n) = \sum_{l=0}^{N-1} \sum_{k=0}^{M-1} x(k, l)\; w(k, l, m, n).$$

Here, the elements of $\tilde{X}(m, n)$ are called 2D forward transform coefficients for the frequency indices $0 \le m \le M-1$ and $0 \le n \le N-1$. The term $w(k, l, m, n)$ in (1) is called a transformation kernel and can be picked depending on the transform utilized. 2D harmonic transforms are implemented in a separate way [24]. Hence, the 1D transform kernels given in Table 3.1 can be easily extended to the 2D case. The 2D basis images of the three transforms of size 8×8 are shown in Figure 3.1.

There exist some similiarities and differences between the three harmonic transforms, as shown in Table 3.1. Unlike the DHT, the other two harmonic transforms belong to eight definitions for different symmetry and boundary conditions [24], but we will use the most popular definitions of the DCT and DST in this chapter. The three transforms

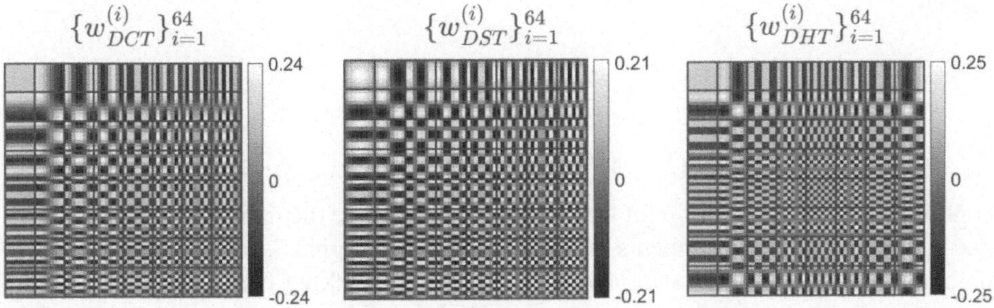

FIGURE 3.1 2D harmonic basis images. Blue lines are inserted for visualization.

yield real spectral coefficients. While the DST and DHT are symmetric harmonic transforms, the DCT is not symmetric. The assumption of the DCT on a 1D N-point signal is $2N$-point periodic and even symmetric; the DST assumes that the signal is $(2N + 1)$-point periodic and odd symmetric. On the other hand, the DHT assumes that the signal is N-point periodic. Besides, the only transform among the three transforms that do not have a DC basis is the DST. The importance of the transforms with DC components lies in keeping the average energy of the signal.

3.4 HARMONIC FILTERBANKS

In this section, among existing transform usages, we prefer the filterbank approach that utilizes harmonic basis images as a convolution kernel [25] to handle the JPEG CAR problem. Transform-based CAR algorithms on overlapping image blocks can be written by

$$\hat{\boldsymbol{x}}_k = \boldsymbol{W}^T \Upsilon(\boldsymbol{W}\boldsymbol{y}_k).$$

Here, $\boldsymbol{y}_k = \boldsymbol{R}_k\boldsymbol{y} \in \mathbb{R}^n and\ \hat{\boldsymbol{x}}_k = \boldsymbol{R}_k\hat{\boldsymbol{x}} \in \mathbb{R}^n$ are the kth image blocks extracted from the compressed image $\boldsymbol{y} \in \mathbb{R}^N$ and restored image $\hat{\boldsymbol{x}} \in \mathbb{R}^N$, respectively. We note that n and N represent block and image sizes. The matrix $\boldsymbol{W} \in \mathbb{R}^{n \times n}$ is a harmonic transform matrix; each row $\boldsymbol{w}_r^T$ coincides with a basis vector and $\boldsymbol{W}^T$ is transposed from $\boldsymbol{W}$. The function $\Upsilon(.)$ represents a fixed or learnable nonlinear function. We recall that the image block extraction matrix, $\boldsymbol{R}_k \in \mathbb{R}^{n \times N}$, whose entries contain only 0 and 1 terms, extracts the kth image block from the corresponding image. Assuming that the image $\boldsymbol{y}$ is padded circularly, the decompressed image, $\hat{\boldsymbol{x}}$, is obtained by aggregating and averaging all of the restored blocks.

$$\hat{\boldsymbol{x}} = \frac{1}{n}\sum_{i=1}^{N} \boldsymbol{R}_i^T \boldsymbol{W}^T \Upsilon(\boldsymbol{W}\boldsymbol{R}_i\boldsymbol{y}).$$

If we write $\boldsymbol{W}_k = \boldsymbol{W}\boldsymbol{R}_k \in \mathbb{R}^{n \times N}$, the image $\hat{\boldsymbol{x}}$ can be represented in a more compact form by

$$\hat{\boldsymbol{x}} = \boldsymbol{L}^T \Upsilon(\boldsymbol{L}\boldsymbol{y}).$$

Here, $\boldsymbol{L} = {}^{1}\!/\!_{\sqrt{n}}\left[\boldsymbol{W}_1^T\ \boldsymbol{W}_2^T \ldots \boldsymbol{W}_N^T\right] \in \mathbb{R}^{nN \times N}$. The harmonic transform spectrum, $\boldsymbol{L}\boldsymbol{y}$, contains magnitudes of different frequencies to synthesize the image of interest. Global and block transform coefficients of highly correlated signals such as images tend to be uncorrelated [26], as shown in Figure 3.2. Hence, seeking correlation between the spectrum coefficients and performing convolution on them is an ineffective attempt. Instead, as pointed out in [25, 27], packing the coefficients with the same frequency as a

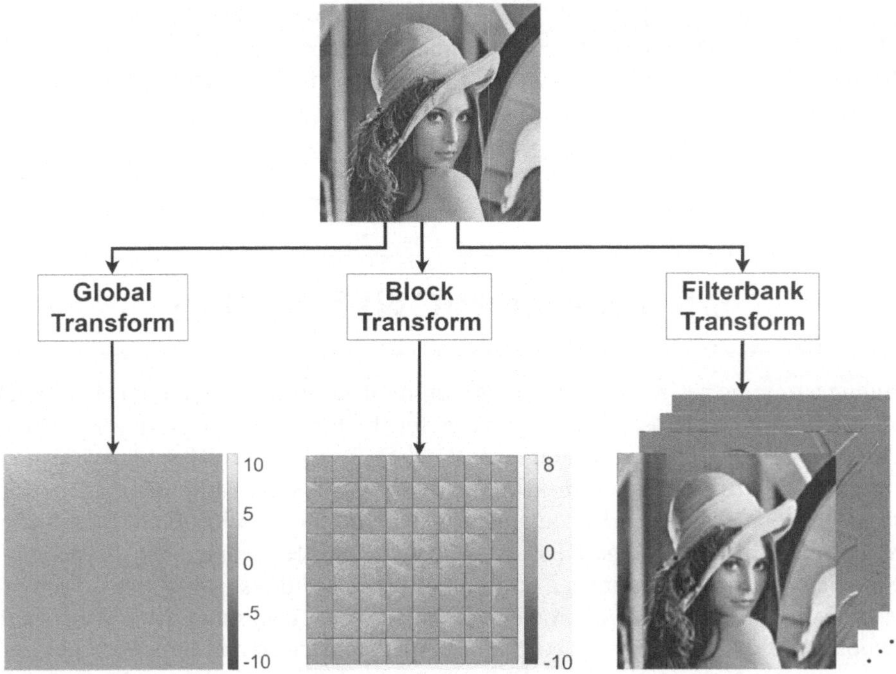

FIGURE 3.2 Three DCT magnitude spectrums of Lena. Black lines for block spectrum are inserted for visualization.

channel provides us feature maps that are suitable to be processed by convolution layers. This ordering scheme only permutes rather than unchanges the spectrum coefficients. The formal way to perform this spectral permutation step is to multiply $\boldsymbol{Ly}$ with a suitable permutation matrix, $\boldsymbol{P} \in \mathbb{R}^{nN \times nN}$.

$$\hat{\boldsymbol{x}} = \boldsymbol{L}^T \boldsymbol{P}^T \Upsilon(\boldsymbol{PLy}) = \boldsymbol{S}^T \Upsilon(\boldsymbol{Sy}).$$

In this equation, $\boldsymbol{S} \in \mathbb{R}^{nN \times N}$ and $\boldsymbol{S}^T \in \mathbb{R}^{N \times nN}$ are called forward and inverse harmonic filterbank (FB) transforms. The inner term, $\boldsymbol{Sy}$, in (5) can be explicitly written by $\tilde{\boldsymbol{y}} = \boldsymbol{Sy} = \left[\tilde{\boldsymbol{y}}_1^T \tilde{\boldsymbol{y}}_2^T \ldots \tilde{\boldsymbol{y}}_n^T\right]$. Each $\tilde{\boldsymbol{y}}_r$ for $1 \le r \le n$ is called a subband image. As justifed in [25, 27], the rth subband image, $\tilde{\boldsymbol{y}}_r$, can be obtained by convolving the rth harmonic basis, $\boldsymbol{w}_r^T$, with the compressed image, $\boldsymbol{y}$, for the circular padding condition.

$$\tilde{\boldsymbol{y}}_r = \frac{1}{\sqrt{n}}\left(\boldsymbol{w}_r^T \otimes \boldsymbol{y}\right).$$

Here, $\otimes$ denotes a convolution operation. Similarly, the inverse transform can be efficiently implemented with a deconvolution layer with the same basis images as

$$\hat{\boldsymbol{x}} = \frac{1}{\sqrt{n}} \sum_{r=1}^{n} \boldsymbol{w}_r \otimes \Upsilon(\tilde{\boldsymbol{y}}).$$

If we choose the DCT and DHT basis images for a harmonic FB transform, the first subband image, $\tilde{y}_1$, is called a DC subband image and is denoted by $\tilde{y}_{DC}$. The remaining subband images, $\tilde{y}_r$, for $2 \leq r \leq n$ are called AC subband images. When we pick the DST basis images for FB usage, we have no DC subband image due to the lack of a DC basis image of the DST. Hence, DST FB generates only AC subband images.

3.5 HARMONIC NETWORKS

Building harmonic network architectures requires determining two design choices. The first criterium is to choose 2D basis images for the harmonic transform. As depicted earlier, we have three transform options: DCT, DST, and DHT. The other design criterium is to determine the nonlinearity function $\Upsilon(.)$. The literature on image restoration problems, including CAR, contains many studies on the selection of the function. Fixed shape scalar functions, $\Upsilon(.): \mathbb{R} \rightarrow \mathbb{R}$, with tunable parameter(s) to operate on the whole or each subband image, such as soft and hard thresholding, have been proposed for the image denoising problem. However, since determining the shape of the function in advance directly affects the performance, Hel-Or and Ben-Artzi [25] have proposed learning the shape of each function by modeling with a sum of some special functions. Maharjan et al. [28] have proposed a CNN as the nonlinearity function for each band obtained by block DCT. In doing so, it has increased the computational burden of the algorithm and has not taken advantage of the correlation between pixels at adjacent block boundaries produced by block DCT. On the other hand, there is a clear relationship between neighboring spectral coefficients in each band. However, this case is not taken into account in deep networks. Keeping in mind that CNNs are powerful function approximators and inspired by [27], we propose a deep CNN as the nonlinearity function. During the network design stage, we try to keep the architecture of our networks simple and efficient. We make use of residual blocks for this purpose. This is because residual blocks facilitate the training of deep networks and preserve the training stability. Apart from this, our networks have simple CNN structures compared to the deblocking networks.

The architecture of the proposed networks is visualized in Figure 3.3. It is worth noting that the main distinction between the three harmonic networks is that it has no splitting and concatenation layers for the DSTNet. This is because the DST yields no average energy. Hence, DSTNet has more learnable parameters than DCTNet and DHTNet since all the subband images of DSTNet are given to a series of residual blocks. The forward FB transform block takes the compressed image, $\mathbf{y}$, and produces the subband images, $\tilde{\mathbf{y}}$. The DC subband is left untouched because of the importance of conserving average energy in signal and image processing applications (valid for DCTNet and DHTNet). The AC feature maps are given to one convolution (Conv) + rectified linear unit (ReLU) block followed by five residual blocks (RBs) and one Conv layer. Each RB contains four Conv + BNorm + ReLU layers and one sum connection, where BNorm is batch normalization for short. The feature maps yielded by the last Conv layer

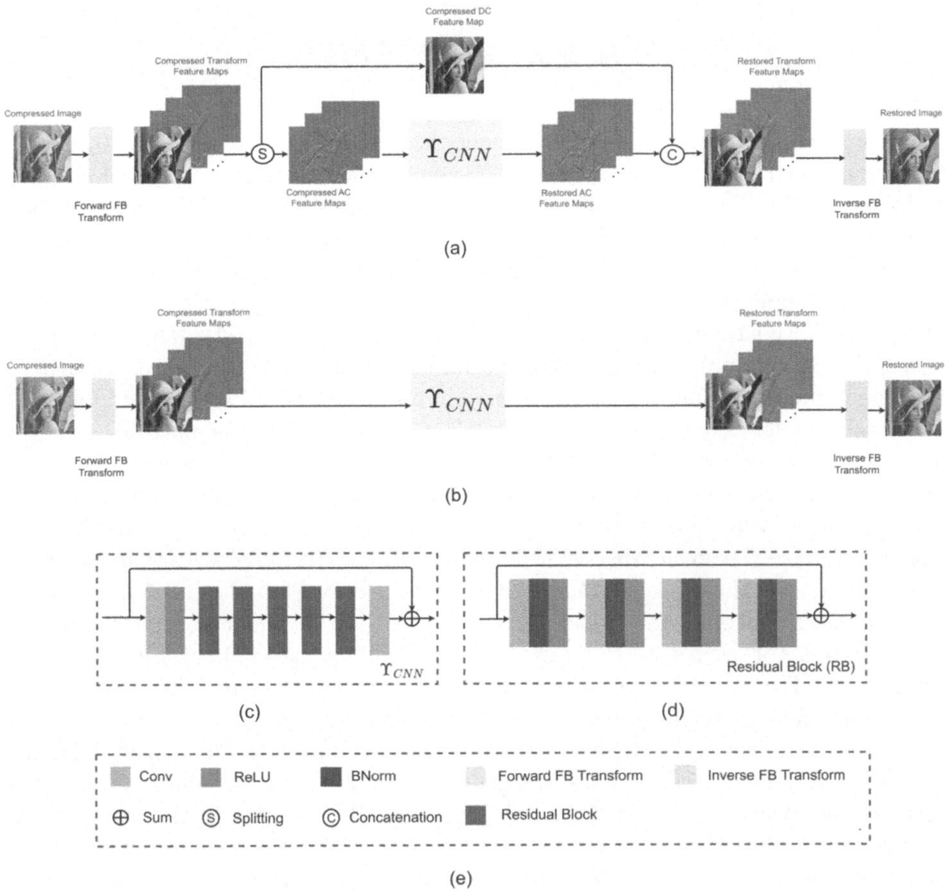

FIGURE 3.3 Harmonic network architectures. (a) DCTNet and DHTNet. (b) DSTNet. (c) CNN block. (d) Our residual block. (e) Building blocks of harmonic network architectures.

are concatenated with the DC feature map, $\tilde{\boldsymbol{y}}_{DC}$ (valid for DCTNet and DHTNet). In the last step, the resulting feature maps are mapped to an image domain by an inverse harmonic FB transform block.

Assuming that we have N compressed and ground truth training pairs, $\left\{\tilde{\boldsymbol{y}}^{(i)}, \boldsymbol{x}^{(i)}\right\}_{i=1}^{N}$, the loss function of the proposed networks to train in supervised learning is written by

$$\mathcal{L}(\Theta) = \frac{1}{2N} \sum_{i=1}^{N} \left\| \boldsymbol{x}^{(i)} - \mathcal{N}\left(\tilde{\boldsymbol{y}}^{(i)}; \Theta\right) \right\|_2^2 .$$

Here, $\mathcal{L}(.)$ is the loss function to be optimized. $\mathcal{N}(.)$ and Θ denote our harmonic networks and their learnable parameters.

3.6 EXPERIMENTS

We have trained four networks with four quality factors (QFs; i.e., $Q = 10, 20, 30,$ and 40) for each harmonic network for reducing JPEG compression artifacts on gray-scale images. BSDS500 [29] has been used to create compressed and uncompressed training patches. $133K$ uncompressed patches of size 60×60 have been extracted from 400 training images to make more use of the dataset. During the training stage, data augmentation steps such as rotation and flipping have also been adopted. JPEG low-quality compressed images for the QFs given above have been generated by the MATLAB JPEG encoder. The training and testing phases of the proposed networks have been conducted on the MatConvNet [30] deep learning framework. MatConvNet has been built upon Matlab (2019a) on a desktop computer with an Intel Core i7-8700k CPU 3.2 GHz, 64-bit operating system, 16 GB memory, and a Nvidia GeForce RTX2080 Ti GPU. The loss function to train our networks in (8) has been optimized via Adam [31] with a minibatch size of 64. All learnable parameters have been initialized with He initialization scheme [32]. The learning rate has been scheduled from $1e-2$ to $1e-5$ with an exponential decaying scheme. The weight decay parameter has been set to $1e-4$. The forward and inverse harmonic FB transform layers of the proposed networks use a total of 49 basis images of size 7×7. All of the remaining convolution layers use 49 filters of size $3 \times 3 \times 49$ for DCTNet and DHTNet and 48 filters of size $3 \times 3 \times 48$ for DSTNet. The training time of our networks has taken 34 hours with the hardware whose specifcations are given above.

We evaluate our harmonic networks on two benchmark datasets, namely, Classic5 (5 test images) and LIVE1 (24 test images), which are shown in Figure 3.4. The test sets have not been included in the training datasets. We have used the publicly available codes for all of the compared methods. As quantitative performance metrics, the peak signal-to-noise ratio (PSNR, measured in dB), the structural similarity index (SSIM), and the peak signal-to-noise ratio for blocking effects (PSNR-B, measured in dB) are selected. The algorithms selected for comparison are ARCNN [18], TNRD [22], DnCNN [19], MemNet [20], QGAC [33], IACNN [34], and DUN [23].

FIGURE 3.4 Test images for Classics5 (first row) and LIVE1 (second row) datasets.

3.7 RESULTS

The average PSNR, SSIM, and PSNR-B results on the two test sets are reported in Tables 3.2, 3.3, and 3.4 for the four QFs. Besides, Table 3.5 reports the total number of learnable parameters and the parameter gains in percentage terms compared to the number of parameters of our harmonic DCTNet. As seen from the PSNR table (Table 3.2), our three networks surpass ARCNN and TNRD dramatically at the cost of increased parameter number in terms of average PSNR. Our networks have an average performance gain of 0.25 dB over DnCNN, despite having 28.55% fewer parameters than DnCNN for all quality factors. While the proposed networks have 28.34% fewer parameters than MemNet, they show competitive performance results to MemNet, despite multisupervision scheme, short-term memory, long-term memory, and recursive gate units. Our harmonic networks beat IACNN by 0.2 dB on average, with about 4 million fewer parameters. Despite having 96% fewer parameters than QGAC, a more sophisticated network, the proposed networks are about 0.1 dB behind QGAC. As a final comparison, our networks are outperformed by the hybrid unfolded network DUN by about 0.25 dB on average for 10 million fewer learnable parameter gains. From the PSNR-B table (Table 3.4), our networks have an average performance gain of 0.25 and 0.35 dB over DnCNN and IACNN. While our networks show better performance than QGAC in terms of PSNR-B, they are outperformed by the DUN algorithm by a small margin.

TABLE 3.2 The average PSNR (dB) results of different methods. The best three results are highlighted in bold, italic, and underlined, respectively

DATASETS	*Q*	*JPEG [15]*	*SA-DCT [16]*	*ARCNN [18]*	*TNRD [22]*	*DnCNN [19]*	*MemNet [20]*
Classic5	10	27.82	28.88	29.03	29.28	29.40	<u>29.69</u>
	20	30.12	30.92	31.15	31.47	31.63	<u>31.90</u>
	30	31.48	32.14	32.51	32.78	32.91	-
	40	32.43	33.00	32.68	-	33.77	-
LIVE1	10	27.77	28.65	28.96	29.15	29.19	<u>29.45</u>
	20	30.07	30.81	31.29	31.46	31.59	<u>31.83</u>
	30	31.40	32.08	32.67	32.84	32.98	-
	40	32.35	32.99	32.74	-	33.96	-
DATASETS	*Q*	*QGAC [33]*	*IACNN [34]*	*DUN [23]*	*DCTNet (Ours)*	*DSTNet (Ours)*	*DHTNet (Ours)*
Classic5	10	*29.84*	29.53	**29.95**	29.67	29.64	29.67
	20	*31.98*	31.87	**32.11**	31.89	31.84	31.89
	30	*33.22*	33.08	**33.33**	33.15	33.16	<u>33.18</u>
	40	-	33.91	**34.11**	<u>34.02</u>	33.99	*34.04*
LIVE1	10	*29.53*	28.80	**29.61**	29.43	<u>29.45</u>	29.42
	20	*31.86*	31.76	**31.98**	<u>31.83</u>	31.81	<u>31.83</u>
	30	33.23	33.14	**33.38**	<u>33.26</u>	33.25	*33.27*
	40	-	34.06	**34.32**	<u>34.25</u>	34.24	*34.28*

TABLE 3.3 The average SSIM results of different methods. The best three results are highlighted in bold, italic, and underlined, respectively

DATASETS	*Q*	*JPEG [1]*	*SA-DCT [16]*	*ARCNN [18]*	*TNRD [22]*	*DnCNN [19]*	*MemNet [20]*
Classic5	10	0.7595	0.8071	0.7929	0.7992	0.8026	0.8107
	20	0.8344	0.8663	0.8517	0.8576	0.8610	0.8658
	30	0.8666	0.8914	0.8806	0.8837	0.8861	-
	40	0.8849	0.9055	0.9019	-	0.9141	-
LIVE1	10	0.7730	0.8093	0.8076	0.8111	0.8123	0.8193
	20	0.8512	0.8781	0.8733	0.8769	0.8802	0.8846
	30	0.8851	0.9078	0.9043	0.9059	0.9090	-
	40	0.9041	0.9240	0.9196	-	0.9346	-
DATASETS	*Q*	*QGAC [33]*	*IACNN [34]*	*DUN [23]*	*DCTNet (Ours)*	*DSTNet (Ours)*	*DHTNet (Ours)*
Classic5	10	**0.8370**	0.8124	*0.8343*	0.8109	0.8102	0.8113
	20	**0.8850**	0.8729	*0.8848*	0.8659	0.8649	0.8662
	30	**0.9070**	0.9007	*0.9061*	0.8899	0.8896	0.8903
	40	-	*0.9141*	**0.9179**	0.9037	0.9030	0.9041
LIVE1	10	**0.8400**	0.8207	*0.8370*	0.8214	0.8209	0.8215
	20	**0.9010**	0.8861	*0.8997*	0.8862	0.8850	0.8864
	30	*0.9250*	0.9210	**0.9251**	0.9139	0.9131	0.9142
	40	-	*0.9313*	**0.9384**	0.9290	0.9282	0.9294

TABLE 3.4 The average PSNR-B (dB) results of different methods. The best three results are highlighted in bold, italic, and underlined, respectively

DATASETS	*Q*	*JPEG [1]*	*SA-DCT [16]*	*ARCNN [18]*	*TNRD [22]*	*DnCNN [19]*	*MemNet [20]*
Classic5	10	25.21	28.16	28.76	29.04	29.13	29.31
	20	27.50	29.75	30.59	31.05	31.19	31.29
	30	28.94	30.83	31.98	32.24	32.38	-
	40	29.92	31.59	33.03	33.22	33.23	-
LIVE1	10	25.33	28.01	28.77	28.88	28.90	29.04
	20	27.57	29.82	30.79	31.04	31.07	31.14
	30	28.92	30.92	32.22	32.28	32.34	-
	40	29.96	31.79	33.25	33.28	33.28	-
DATASETS	*Q*	*QGAC [33]*	*IACNN [34]*	*DUN [23]*	*DCTNet (Ours)*	*DSTNet (Ours)*	*DHTNet (Ours)*
Classic5	10	*29.43*	29.27	**29.61**	29.33	29.30	29.28
	20	31.37	31.18	**31.61**	*31.49*	31.40	31.39
	30	32.42	32.18	**32.67**	32.59	*32.62*	32.60
	40	-	33.04	33.36	*33.40*	33.39	**33.41**
LIVE1	10	*29.15*	28.71	**29.25**	29.05	29.03	29.00
	20	31.27	31.05	**31.42**	*31.31*	31.29	31.26
	30	32.50	32.13	**32.65**	32.60	*32.63*	32.61
	40	-	33.15	33.51	33.55	*33.56*	**33.58**

TABLE 3.5 The number of parameters and gain (in percent) of different methods. Parameter gain is computed with respect to our harmonic networks. Negative gain denotes that the compared network has fewer paramaters than our baseline harmonic networks

	ARCNN [18]	*TNRD [22]*	*DnCNN [19]*	*MemNet [20]*
Parameters.	106K	26K	669K	667K
Gain	–350.94%	–1738.46%	28.55%	28.34%
	QGAC [33]	*IACNN [34]*	*DUN [23]*	*Harmonic Nets (Ours)*
Params.	12000K	4321K	10490K	478K
Gain	96.02%	88.94%	95.44%	0%

All these comparisons show that our networks are more suitable for parameter-efficient platforms such as mobile devices. Figures 3.5 and 3.6 give the visual results for test image Barbara and Bikes for Q = 20. From image Barbara, it can be said that our DCTNet is better than DnCNN for recovering the textural details. The other good result belongs to our DSTNet, which beats DnCNN with a 0.15 dB PSNR difference. When

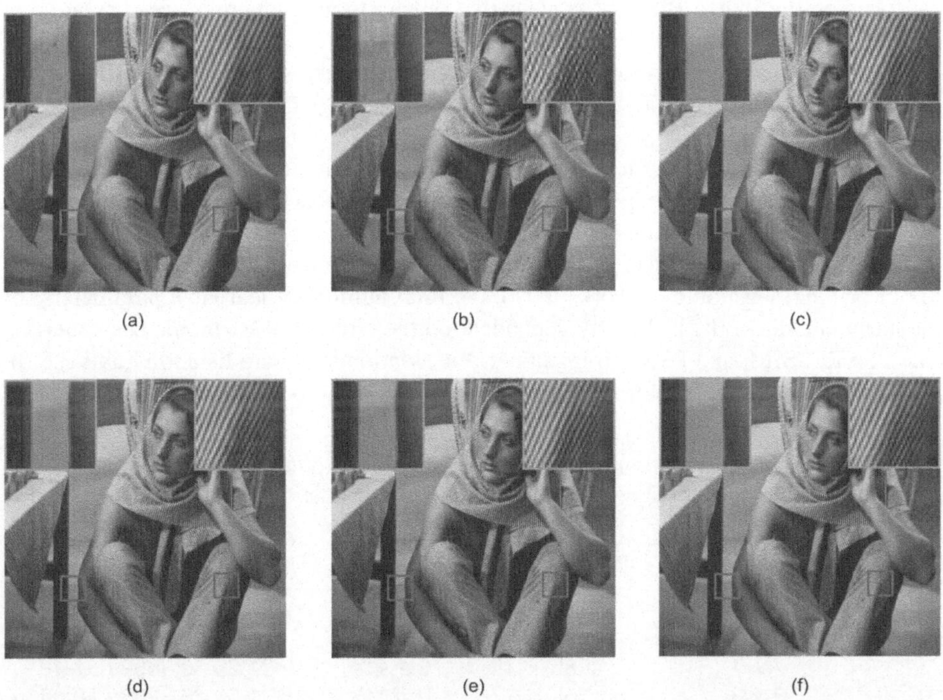

FIGURE 3.5 Visual results of image Barbara with Q = 20. (a) Clean, PSNR/SSIM/PSNR-B. (b) JPEG, 28.34/0.8535/25.66. (c) DnCNN [30], 30.57/0.8915/29.98. (d) DCTNet, 31.20/0.8999/30.50. (e) DSTNet, 31.07/0.8988/30.36. (f) DHTNet, 31.18/0.9000/30.32.

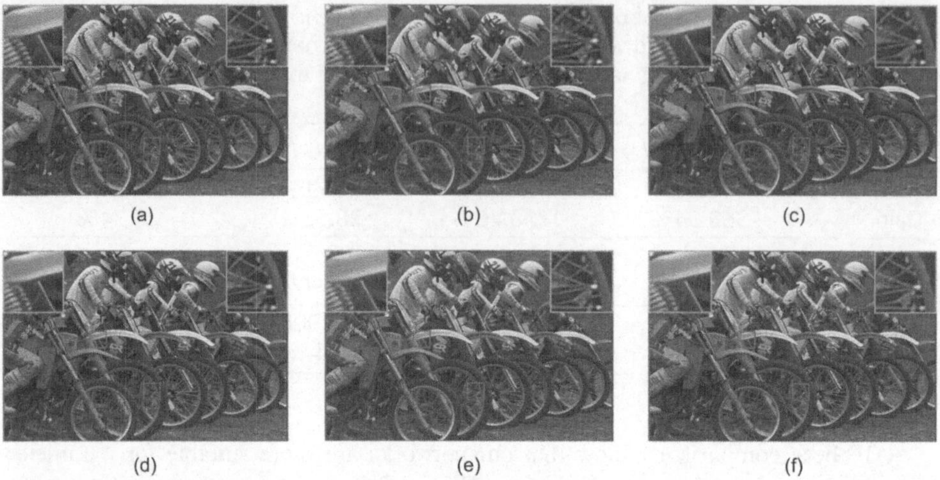

FIGURE 3.6 Visual results of image Bikes with Q = 20. (a) Clean, PSNR/SSIM/PSNR-B. (b) JPEG, 28.06/0.8377/25.21. (c) DnCNN [30], 30.29/0.8860/29.82. (d) DCTNet, 30.70/0.8932/30.04. (e) DSTNet, 30.65/0.8923/29.98. (f) DHTNet, 30.70/0.8933/29.98.

we focus on the image Bikes, it is clear that our harmonic networks are very good at recovering fine details.

It can be readily inferred that our harmonic networks show similar SSIM results in Table 3.3 for the two test sets. However, we also note that IACNN is better than the proposed networks.

In order to show the efficiency and robustness of the proposed methods, we have also conducted additional experiments examining the impact of varying the number of 2D harmonic basis images on deblocking performance. We train our networks for 5×5 and 3×3 2D harmonic basis images. We note that the networks with 7×7 corresponds to our original harmonic networks. Also, the total number of learnable parameters for the harmonic networks is directly dependent on the number of harmonic basis images. The PSNR, SSIM, and PSNR-B results of this ablation study are listed in Table 3.6. In all 5×5 networks, the performance loss compared to the original harmonic networks

TABLE 3.6 Comparison of our harmonic networks with a different number of 2D harmonic basis images for Q = 20. Number of parameters denotes total number of model parameters

NETWORK	*SIZE*	*# PARAMETERS*	*DATASET*	*PSNR/SSIM/PSNR-B*
DCTNet	3 × 3	16K	Classics5	31.50 / 0.8591 / 30.92
	5 × 5	125K	Classics5	31.79 / 0.8643 / 31.23
	7 × 7	478K	Classics5	31.89 / 0.8659 / 31.49
	3 × 3	16K	LIVE1	31.50 / 0.8803 / 30.94
	5 × 5	125K	LIVE1	31.74 / 0.8847 / 31.13
	7 × 7	478K	LIVE1	31.83 / 0.8862 / 31.31

(Continued)

TABLE 3.6 *(Continued)* Comparison of our harmonic networks with a different number of 2D harmonic basis images for Q = 20. Number of parameters denotes total number of model parameters

NETWORK	*SIZE*	*# PARAMETERS*	*DATASET*	*PSNR/SSIM/PSNR-B*
DSTNet	3 × 3	16K	Classics5	31.44 / 0.8578 / 30.90
	5 × 5	125K	Classics5	31.78 / 0.8642 / 31.29
	7 × 7	479K	Classics5	31.85 / 0.8654 / 31.40
	3 × 3	16K	LIVE1	31.48 / 0.8793 / 30.94
	5 × 5	125K	LIVE1	31.75 / 0.8848 / 31.20
	7 × 7	479K	LIVE1	31.82 / 0.8859 / 31.29
DHTNet	3 × 3	16K	Classics5	31.49 / 0.8590 / 30.92
	5 × 5	125K	Classics5	31.79 / 0.8643 / 31.27
	7 × 7	478K	Classics5	31.89 / 0.8662 / 31.39
	3 × 3	16K	LIVE1	31.51 / 0.8804 / 30.95
	5 × 5	125K	LIVE1	31.74 / 0.8847 / 31.14
	7 × 7	478K	LIVE1	31.83 / 0.8864 / 31.26

is 0.1 dB, while the performance loss is 0.4 dB in the 3×3 DCTNet, DSTNet, and DHTNet. However, it is noteworthy that the 3×3 DCTNet outperforms TNRD and ARCNN, despite having far fewer parameters.

3.8 CONCLUSIONS

There exist some limitations for our harmonic networks. The first one is the dimension and number of 2D harmonic basis images. This brings high computational complexity and slow inference time of our networks in platforms where memory and inference speed is of concern. It also prevents us from extending the proposed networks to multidimensional data such as color images and video signals. To handle this obstacle, we should use feature selection modules such as bottleneck layers in ResNet architecture. Our networks require new design choices for different coding standards apart from the JPEG compression scheme. The second limitation is the equivalent treatment for each spectral band rather than weighting them by using attention blocks. In fact, DC and some AC spectral bands convey more important information for removing compression artifacts. The third limitation is that we try to map a compressed image to a corresponding uncompressed image. That is to say, we use deep networks as a black box in a reasonable harmonic transform domain. Instead of this, we design modular network structures using the proposed transform domains. Another limitation is not making use of any statistical or natural image priors including low rank and nonlocal self-similarity when designing deep networks. Some other limitations are simple network topology and robust decompression capability for the scenarios where one has no knowledge about quality factor.

Possible future work includes the following:

- Extension of our harmonic nets to color images and video signals with larger and more diverse datasets, including those with various image complexities, resolutions, and content.
- Incorporating other image models, such as nonlocal self-similarity and low-rank priors, into our network.
- Introducing attention mechanisms for weighting each harmonic subband image. They can help models focus on more critical regions of the image during restoration. This can be especially useful for models with fewer parameters, ensuring they focus on essential details.

NOTE

1. In this chapter, we refer the DCT, DST, and DHT as harmonic transforms.

REFERENCES

1. Wallace, G. K. (1992). The JPEG Still Picture Compression Standard. IEEE Transactions on Consumer Electronics, 38(1):18–34.
2. Liu, J., Liu, D., Yang, W., Xia, S., Zhang, X., and Dai, Y. (2020). A Comprehensive Benchmark for Single Image Compression Artifact Reduction. IEEE Transactions on Image Processing, 29:7845–7860.
3. Chen, H., He, X., Qing, L., Xiong, S., and Nguyen, T. Q. (2018). DPW-SDNet: Dual Pixel-Wavelet Domain Deep CNNs for Soft Decoding of JPEG-Compressed Images. In Proceedings of the IEEE Conference on Computer Vision and Pattern Recognition Workshops, pages 711–720.
4. Liu, P., Zhang, H., Zhang, K., Lin, L., and Zuo, W. (2018). Multi-level Wavelet-CNN for Image Restoration. In Proceedings of the IEEE Conference on Computer Vision and Pattern Recognition Workshops, pages 773–782.
5. Zhang, X., Yang, W., Hu, Y., and Liu, J. (2018). DMCNN: Dual-domain Multi-scale Convolutional Neural Network for Compression Artifacts Removal. In 25th IEEE International Conference on Image Processing (ICIP), pages 390–394.
6. Zheng, B., Chen, Y., Tian, X., Zhou, F., and Liu, X. (2019). Implicit Dual-domain Convolutional Network for Robust Color Image Compression Artifact Reduction. IEEE Transactions on Circuits and Systems for Video Technology, 30(11):3982–3994.
7. Reeve III, H. C. and Lim, J. S. (1984). Reduction of Blocking Effects in Image Coding. Optical Engineering, 23(1):34–37.
8. Ramamurthi, B. and Gersho, A. (1986). Nonlinear Space-variant Postprocessing of Block Coded Images. IEEE Transactions on Acoustics, Speech, and Signal Processing, 34(5):1258–1268.

9. Minami, S. and Zakhor, A. (1995). An Optimization Approach for Removing Blocking Effects in Transform Coding. IEEE Transactions on Circuits and Systems for Video Technology, 5(2):74–82.
10. Yang, Y., Galatsanos, N. P., and Katsaggelos, A. K. (1993). Regularized Reconstruction to Reduce Blocking Artifacts of Block Discrete Cosine Transform Compressed Images. IEEE Transactions on Circuits and Systems for Video Technology, 3(6):421–432.
11. Yang, Y., Galatsanos, N. P., and Katsaggelos, A. K. (1995). Projection-Based Spatially Adaptive Reconstruction of Block-Transform Compressed Images. IEEE Transactions on Image Processing, 4(7):896–908.
12. Zhang, X., Xiong, R., Ma, S., and Gao, W. (2012). Reducing Blocking Artifacts in Compressed Images via Transform-domain Non-Local Coefficients Estimation. In IEEE International Conference on Multimedia and Expo, pages 836–841.
13. Ren, J., Liu, J., Li, M., Bai, W., and Guo, Z. (2013). Image Blocking Artifacts Reduction via Patch Clustering and Low-Rank Minimization. In IEEE Data Compression Conference, pages 516.
14. Zhang, X., Lin, W., Xiong, R., Liu, X., Ma, S., and Gao, W. (2016). Low-Rank Decomposition-Based Restoration of Compressed Images via Adaptive Noise Estimation. IEEE Transactions on Image Processing, 25(9):4158–4171.
15. Chang, H., Ng, M. K., and Zeng, T. (2013). Reducing Artifacts in JPEG Decompression via a Learned Dictionary. IEEE Transactions on Signal Processing, 62(3):718–728.
16. Foi, A., Katkovnik, V., and Egiazarian, K. (2007). Pointwise Shape-Adaptive DCT for High-Quality Denoising and Deblocking of Grayscale and Color Images. IEEE Transactions on Image Processing, 16(5):1395–1411.
17. Liu, X., Wu, X., Zhou, J., and Zhao, D. (2015). Data-Driven Sparsity-based Restoration of JPEG-Compressed Images in Dual Transform-Pixel Domain. In Proceedings of the IEEE Conference on Computer Vision and Pattern Recognition, pages 5171–5178.
18. Dong, C., Deng, Y., Loy, C. C., and Tang, X. (2015). Compression Artifacts Reduction by A Deep Convolutional Network. In Proceedings of the IEEE International Conference on Computer Vision, pages 576–584.
19. Zhang, K., Zuo, W., Chen, Y., Meng, D., and Zhang, L. (2017). Beyond a Gaussian Denoiser: Residual Learning of Deep CNN for Image Denoising. IEEE Transactions on Image Processing, 26(7):3142–3155.
20. Tai, Y., Yang, J., Liu, X., and Xu, C. (2017). MemNet: A Persistent Memory Network for Image Restoration. In Proceedings of the IEEE International Conference on Computer Vision, pages 4539–4547.
21. Zhang, K., Li, Y., Zuo, W., Zhang, L., Van Gool, L., and Timofte, R. (2021). Plug-and-play Image Restoration with Deep Denoiser Prior. IEEE Transactions on Pattern Analysis and Machine Intelligence, 44(10):6360–6376.
22. Chen, Y. and Pock, T. (2016). Trainable Nonlinear Reaction Diffusion: A Flexible Frame-for Fast and Effective Image Restoration. IEEE Transactions on Pattern Analysis and Machine Intelligence, 39(6):1256–1272.
23. Fu, X., Wang, M., Cao, X., Ding, X., and Zha, Z.-J. (2022). A Model-Driven Deep Unfolding Method for JPEG Artifacts Removal. IEEE Transactions on Neural Networks and Learning Systems, 33(11):6802–6816.
24. Gonzalez, R. C. and Woods, R. E. (2008). Digital Image Processing. Prentice Hall, Upper Saddle River, N.J.
25. Hel-Or, Y. and Ben-Artzi, G. (2021). The Role of Redundant Bases and Shrinkage Functions in Image Denoising. IEEE Transactions on Image Processing, 30:3778–3792.
26. Jain, A. K. (1989). Fundamentals of Digital Image Processing. Prentice-Hall, Inc.
27. Karaoglu, H. H. and Eksioglu, E. M. (2023). DCTNet: Deep Shrinkage Denoising via DCT Filter banks. Signal, Image and Video Processing, 17(7):3665–3676.

28. Maharjan, P., Xu, N., Xu, X., Song, Y., and Li, Z. (2021). DCTResNet: Transform Domain Image Deblocking for Motion Blur Images. In 2021 International Conference on Visual Communications and Image Processing, pages 1–5.
29. Martin, D., Fowlkes, C., Tal, D., and Malik, J. (2001). A Database of Human Segmented Natural Images and its Application to Evaluating Segmentation Algorithms and Measuring Ecological Statistics. In Proceedings 8th IEEE International Conference on Computer Vision, volume 2, pages 416–423.
30. Vedaldi, A. and Lenc, K. (2015). MatConvNet: Convolutional Neural Networks for Matlab. In Proceedings of the 23rd ACM International Conference on Multimedia, pages 689–692.
31. Kingma, D. P. and Ba, J. (2015). Adam: A Method for Stochastic Optimization. In 3rd International Conference on Learning Representations, San Diego, CA, USA.
32. He, K., Zhang, X., Ren, S., and Sun, J. (2015). Delving Deep into Rectifers: Surpassing Human-level Performance on ImageNet Classification. In Proceedings of the IEEE International Conference on Computer Vision, pages 1026–1034.
33. Ehrlich, M., Davis, L., Lim, S.-N., and Shrivastava, A. (2020). Quantization Guided JPEG Artifact Correction. In 16th European Conference on Computer Vision, pages 293–309. Springer.
34. Kim, Y., Soh, J. W., Park, J., Ahn, B., Lee, H.-S., Moon, Y.-S., and Cho, N. I. (2019). A Pseudo-blind Convolutional Neural Network for the Reduction of Compression Artifacts. IEEE Transactions on Circuits and Systems for Video Technology, 30(4):1121–1135.

4 Modeling Face Emotion Perception from Naturalistic Face Viewing

Insights from Fixational Events and Gaze Strategies

Meisam J. Seikavanidi, Maria J. Barrett, and Paolo Burelli

4.1 INTRODUCTION

Facial emotion recognition (FER) is fundamental to human social interactions, enabling individuals to decipher emotions from facial expressions [1, 2]. While methodologies like brain imaging and physiological signals have been employed to probe this complex process [3–5], eye tracking emerges as a noninvasive yet insightful tool, offering deep insights into visual attention and emotional processing [6].

DOI: 10.1201/9781003570882-5

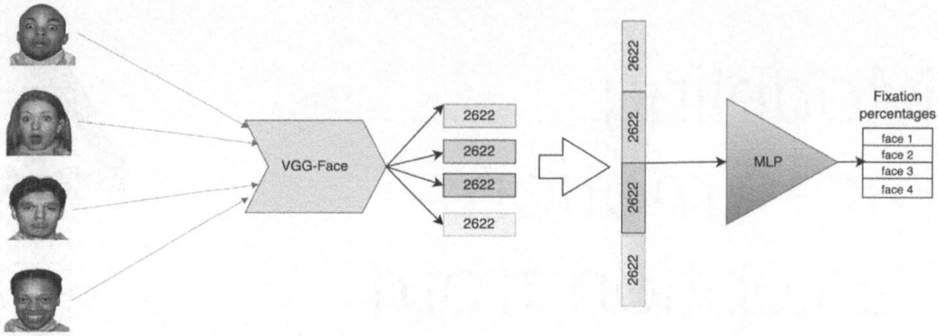

FIGURE 4.1 The network structure designed for predicting fixations in task 2. The model employs a three-layer MLP to analyze spatial, temporal, and spatiotemporal features to predict the dwell time percentage for each face.

Previous research has leveraged eye movements to understand emotion perception (EP) in adults [7, 8], shedding light on atypical EP associated with conditions such as autism, attention-deficit/hyperactivity disorder (ADHD), schizophrenia, and certain types of dementia [9, 10]. Diagnostic eye-tracking tasks, such as the antisaccade task, play a crucial role in identifying diseases like dementia and Alzheimer's [11, 12].

The shift toward naturalistic tasks, mirroring real-life scenarios, gains attraction due to their relaxed environment, making them suitable for lightweight EP assessments. Eye-tracking research extends its applications to clinical diagnosis and human-robot interaction (HRI), with potential implications for designing socially intelligent robots [13].

In this chapter, we extend the work we have done in [14] and delve more into FER using an instructionless paradigm, exploring two FER processes: free viewing and grounded FER. We aim to unravel the intricate relationship between eye movements and emotion perception by scrutinizing fixational, pupillary, and microsaccadic events extracted from eye movement data. By integrating these gaze features with deep learning models, our study uniquely combines the analysis of microsaccadic events and advanced computational techniques to enhance the understanding of emotional processing.

By delineating regions of interest in the face and harnessing deep learning models for face recognition, we investigate the role of eye-gaze strategies in face processing and their link to presented emotions and emotion perception performance. Leveraging features extracted from pretrained deep learning models, we analyze attention during free viewing, enhancing scalability and comparability across datasets and populations. Furthermore, we employ a sequential model with bidirectional Long Short-Term Memory (LSTM) layers to capture the temporal aspects of fixation between regions of interest, providing insights into the dynamic nature of gaze behavior during face viewing.

Our study contributes to the efficiency of EP assessments by predicting FER success based on gaze features during face viewing. This approach not only enhances the precision and ecological validity of emotion recognition systems but also has significant implications for clinical diagnostics and human-computer interaction. By extending the understanding of the eye movements–emotion perception relationship, our work impacts psychology, human–computer interaction, and affective computing domains, offering potential advancements in FER research and applications.

4.2 BACKGROUND

4.2.1 Instructionless FER Task

This section outlines the instructionless FER task initially developed by Russell et al. [10] and our modifications to enhance the understanding of the FER process.

Russell et al. designed the task to identify early-stage frontotemporal dementia, comprising:

1. Displaying four faces with distinct emotions for 10 seconds
2. Presenting an emotion word for 2 seconds
3. Showing the word and faces together for 3 seconds

This setup aimed to intuitively direct participants' gaze to the face matching the emotion if they remembered its position. The constant positions of the faces simulated a memory retrieval task, which helped analyze free-viewing and retrieval phases and considered working memory as a potential confounder.

4.2.2 Modifications to the Instructionless FER Task

To deepen our insight into the FER process and mitigate the effect of working memory, we adjusted Russell et al.'s design by randomizing face positions in Step 3. This required participants to recognize rather than recall face positions, effectively isolating the role of memory. This alteration helped distinguish between the free-viewing (Step 1) and grounded FER (Step 3) phases.

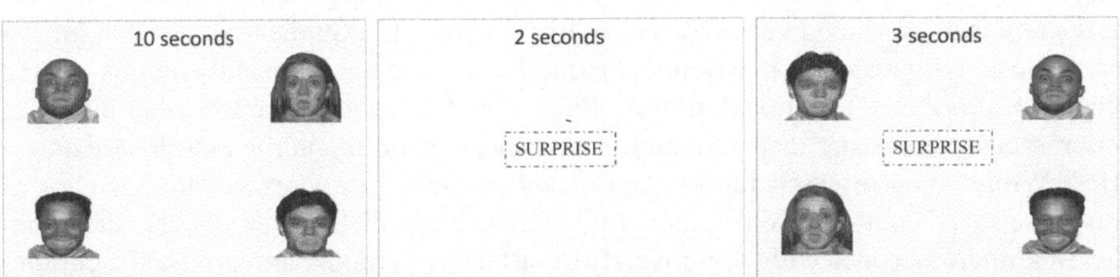

FIGURE 4.2 The FER task unfolds in three steps: (1) overlaying areas of interest, (2) presenting faces with emotions, and (3) randomizing locations to study gaze patterns.

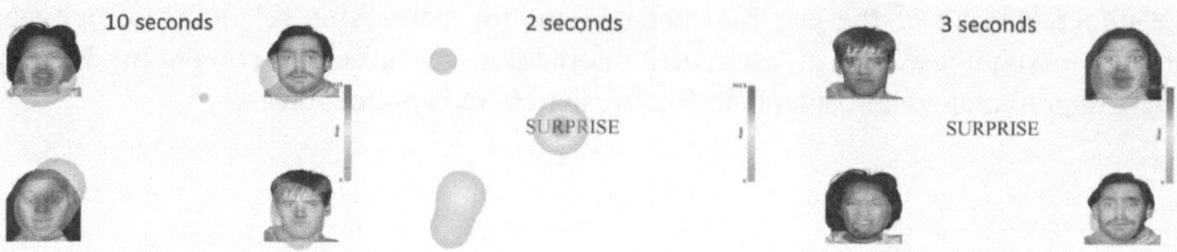

FIGURE 4.3 The heat maps display an FER trial with no instructions. Different emotions have distinct fixation distributions, showcasing varied attention patterns across emotions.

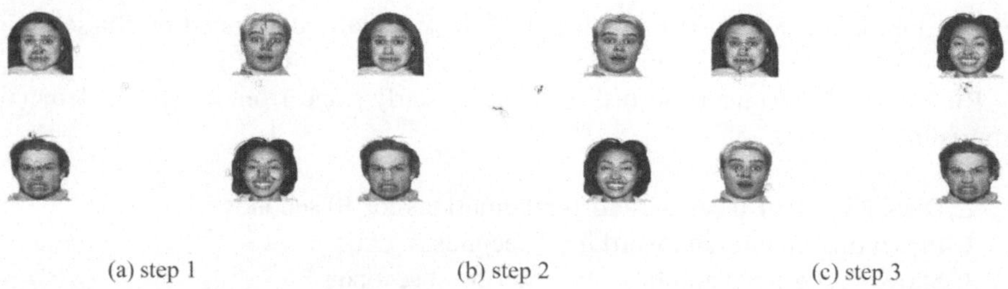

FIGURE 4.4 Microsaccades across different steps. (a) Step 1 shows baseline microsaccadic activity, (b) Step 2 indicates changes in response to emotion words, and (c) Step 3 highlights microsaccades during emotion recognition.

Through these changes, we analyzed gaze behavior and performance variations between these phases, generating quantitative data on cognitive processes in FER. These adjustments better simulate real-world scenarios, enhancing our study of the links among eye movements, emotion perception, and attention in FER tasks.

4.2.3 Microsaccades

Microsaccades are small, involuntary eye movements that occur during visual fixation, typically lasting 6–30 milliseconds and with amplitudes under 0.1 degree of visual angle. Interspersed with slow drifts, they prevent retinal image fading and are vital for tasks requiring sustained visual attention, like reading [15]. Studies suggest that microsaccadic activity responds to emotional stimuli, affecting attention and emotion-related cognitive processes. Emotional arousal, fatigue, and saccade preparation can influence microsaccades, altering their rate and magnitude, depending on the emotional context [16]. While some research reports significant changes in microsaccadic behavior in response to different emotions, others find no noticeable differences [17]. Further studies link microsaccades with cognitive effort, affective priming, and arousal, highlighting their sensitivity to these factors and their role in emotional and cognitive processing [17]. This underlines the importance of microsaccades in understanding visual attention and emotion perception.

TABLE 4.1 Participant characteristics

	AGE	*DRIFT ERROR*
Count	20	20
Range	23–44	0.01–1.21
Mean	29.3	0.41
SD	5.3	0.28

4.2.4 Eye Gaze Strategies

Studies indicate that eye movements during emotion recognition in faces follow both stimulus-driven and goal-driven perceptual strategies [18]. Different facial regions contain varying levels of useful information for distinguishing emotions. For example, joyful faces may draw attention to the lips, while sad faces may attract attention to the eyes. These fixation patterns are influenced by attention to the most diagnostic regions of the face for each emotion, indicating a goal-driven influence on gaze patterns [19]. Furthermore, gaze direction can modulate emotion perception in facial expressions, influencing how emotions are perceived based on gaze direction. A study found that faces with averted gaze were rated higher overall in terms of perceived likelihood of experiencing emotion compared to direct gaze faces, demonstrating an interaction between gaze direction and perceived emotional disposition [20].

4.3 DATA COLLECTION

4.3.1 Participants

We recruited 21 volunteers with normal or corrected vision using glasses or lenses who reported no history of attention deficits or cognitive impairments. One participant was excluded from the analysis due to missing information in some trials. The remaining 20 participants had completed education ranging from high school to Ph.D., with MSc being the mode. Table 4.1 provides an overview of other demographics. The information statement form, which was approved by the legal department of our institution, was signed by all participants.

4.3.2 Apparatus

We utilized the Eyelink 1000 Plus eye tracker from SR Research for recording eye movements during FER tasks. Participants were seated in a dark room with their chin stabilized on a chin rest. An 18-inch high-resolution display screen with a resolution of

1024 × 768 pixels was placed 70 cm from the participants. Before each experiment, we performed a nine-point calibration for the eye tracker, followed by a drift correction between trial rounds to ensure accuracy. Recalibration was done whenever the accuracy dropped below the desired threshold.

4.3.3 Stimuli

The NimStim face emotion dataset [21] was employed, generating 60 trials involving four emotive facial images and one emotion word per trial, aiming for a balanced representation of target emotions and diversity in facial identities. We randomized the positions of the faces and the target face to promote effective FER. Each face, sized 200 × 200 pixels, was placed toward the screen corners. The emotion word appeared in a 40-point Times New Roman font, centrally positioned. A fixation cross was shown for 200 ms at the start of each trial to center participant attention.

4.3.4 Areas of Interest

Areas of interest included the four facial images and their corresponding words per trial. Within each face, we defined subareas for the eyes, nose, and mouth based on a convolutional neural network-based landmark detection model [22]. These landmarks helped segment the faces into seven regions: mouth, both eyebrows, both eyes, nose, and jaw, further grouped into three categories: eye, nose, and mouth regions (see Figure 4.2).

4.3.5 Experiment Protocol

Participants naturally viewed the screen without specific instructions. After six preliminary trials for acclimation, they completed two rounds of 27 trials each, with a short break in between. The total duration was about 15 minutes. Post experiment, participants could opt to complete the Reading Minds in the Eyes test online to assess their emotion decoding ability. Fifteen participants took this test, with their performance detailed in Table 4.1.

4.3.6 Preprocessing and Cleaning

Data from both eyes were collected, but fixation detection relied on the eye with the highest accuracy, confirmed through drift checks. For other analyses, we used binocular data or eye averages. The first six trials were excluded to eliminate initial bias. Fixations were assigned to the nearest area of interest, ensuring clear and consistent data interpretation across trials, simplifying analysis and enhancing data reliability.

4.3.7 Microsaccade Extraction Algorithm

To extract microsaccades from eye-tracking data, we implemented an algorithm that processes velocity and acceleration thresholds to identify significant eye movements. The algorithm preprocesses the data for both left and right eyes, extracts relevant features, and iterates through the samples to detect microsaccades based on predefined velocity and acceleration thresholds. Detected microsaccades were validated by their duration and spatial displacement criteria. Figure 4.4 illustrates the extracted microsaccades across the three key steps of a trial. The complete algorithm is outlined in Algorithm 4.1.

4.4 STATISTICAL ANALYSIS

4.4.1 Overview

Following the guidelines suggested by Skaramagkas et al. [6], we employed the dwell time percentage (dwell time %) as our primary measure for assessing visual attention. This metric calculates the focus duration on a specific area of interest (AOI) as a percentage of the total time spent on a particular step. The change in dwell time for target faces, as proposed by Russell et al. [10], was used to gauge EP performance:

$$\textit{dwell time change} = \textit{dwell time \% step } 3 - \textit{dwell time \% step } 1$$

Algorithm 4.1: Microsaccade Extraction

Input: Eye-tracking data (df), fixation index (value)

Output: Microsaccade count (ms_count), left eye gaze lists (gaze_l_list_all), right eye gaze lists (gaze_r_list_all)

Parameters: velocity_threshold = 15, acc_threshold = 5000, min_duration = 10, max_duration = 100

1. Preprocess data for left and right eyes: cleaned_data_l, cleaned_data_r
2. Extract velocity, acceleration, and gaze data for both eyes
3. Initialize variables: in_ms ← False, ms_count ← 0, ms_duration ← 1
4. Initialize sums: v_l_sum ← [0.0, 0.0], v_r_sum ← [0.0, 0.0], a_l_sum ← [0.0, 0.0], a_r_sum ← [0.0, 0.0], gaze lists
5. Set sample counters: count ← 100, samples_len, samples_count ← 0
6. For each sample of vel_l, acc_l, vel_r, acc_r, gaze_r, gaze_l:
7.   samples_count ← samples_count + 1
8.   If samples_count + 100 > samples_len, then break
9.   While count > 1:
10.   count ← count - 1

```
11.         continue
12.     Accumulate v_l, v_r, a_l, a_r into sums
13.     Append gaze_r, gaze_l to respective lists
14.     If mean velocities and accelerations exceed thresholds:
15.         If in_ms, then:
16.             ms_duration ← ms_duration + 1
17.             continue
18.         Else:
19.             in_ms ← True
20.     Else:
21.         If min_duration ≤ ms_duration ≤ max_duration:
22.             If distance between start and end gaze positions is within range:
23.                 ms_count ← ms_count + 1
24.                 Reset count
25.                 Append gaze lists to gaze_list_all
26.         Reset variables and lists
27. Return ms_count, gaze_l_list_all, gaze_r_list_all
```

Results indicate a pattern where participants significantly focus more on the target face after the emotion word is presented, aligning with findings from previous studies [10, 21, 23]. The specific dwell time percentages and changes across emotions and steps are detailed in Table 4.2.

4.4.2 Analysis Methods

Chi-Square and ANOVA tests were used to analyze the relationship between categorical and numerical variables, respectively, with Bonferroni correction applied to control for the risk of Type I errors across multiple comparisons [24, 25]. The results of these statistical tests are detailed in Tables 4.3 and 4.4, respectively.

TABLE 4.2 Dwell time % per step in the main areas of interest (target face, nontarget face, and word) and D(well) T(ime) C(hange) across emotions

	STEP 1	*STEP 2*			*STEP 3*			*DTC*
		TARGET		*WORD*	*TARGET*		*WORD*	
		NO	*YES*		*NO*	*YES*		
angry	23.8	7.3	10.9	68.6	15.8	42.2	11.9	27.8
disgust	23.5	7.7	11.1	69.8	14.9	45.3	11.4	31.7
fear	25.3	7.5	8.6	69.0	17.8	34.9	10.8	16.0
happy	21.4	6.7	11.0	68.2	14.4	43.8	12.1	34.2
sad	23.0	6.2	11.4	65.8	14.8	38.5	11.3	25.1
surprise	23.9	7.3	8.4	68.9	16.8	40.9	11.5	26.1
average	23.5	7.1	10.2	68.4	15.7	40.9	11.5	26.8

4.4.3 Significant Findings

- **Target Emotion:** Variations in fixation duration and microsaccade activity highlighted differences influenced by the emotion depicted on the target faces.
- **Face Regions:** Significant disparities in microsaccade rates across different facial regions underscore their importance in analyzing facial emotions.
- **Interest Periods:** Observable differences in microsaccades and pupil size among different steps suggest varying cognitive demands, which can inform more nuanced analyses in emotion perception studies.
- **Participants:** The diversity in participant responses to identical stimuli illustrates the challenge in crafting a universally applicable model for emotion perception but also highlights the potential of personalized data analysis.

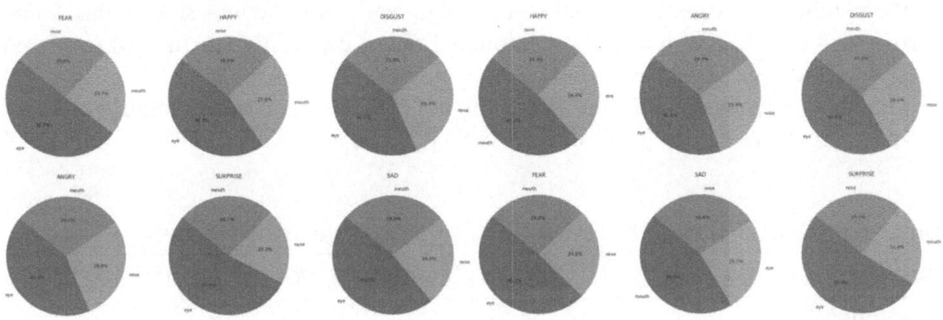

(a) Fixation distribution on step 1 (b) Fixation distribution on step 3

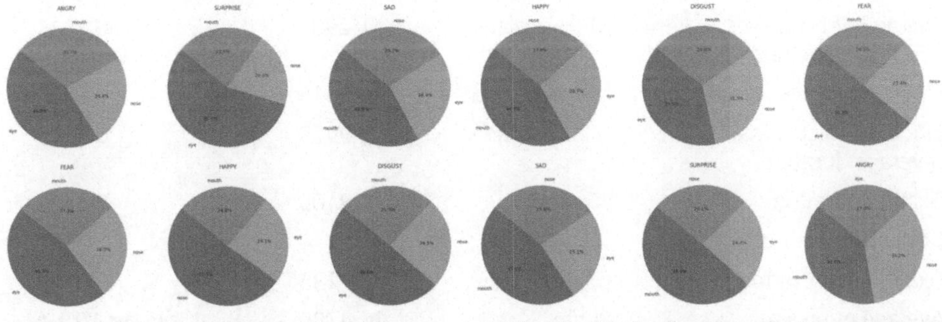

(c) Fixation distribution on step 3 looked at targetwhere participant (d) Fixation distribution on step 3 where participant looked at non-target faces

FIGURE 4.5 Fixation distributions over different face regions (eye, nose, and mouth) across the different steps. These distributions help in understanding the attention strategies employed by participants during the FER tasks.

4.4.4 Performance Analysis

The analysis revealed that fear generally resulted in lower performance scores, whereas happiness was associated with higher scores. This variability demonstrates differences in the efficiency of emotion recognition among participants.

4.4.5 Experimental Observations

In Step 2, participants intuitively sought to match the emotion word to the corresponding face. The emotion word and the position of the target face attracted the most attention, indicative of a memory effect. Specifically, the position of the target face received more focus, particularly for emotions like sadness, fear, and surprise, as depicted in Figure 4.3.

In Step 3, we noticed a new FER process where target faces (M = 40.9, SD = 20.2) received significantly more attention than nontarget faces (M = 15.7, SD = 11.8) ($t(4318) = 64.2$, $p < 0.0001$). Nontarget faces showing fear and surprise still attracted more attention than other nontarget faces. Our results align well with previous studies that found participants tend to fixate longer on emotional faces, especially fearful and surprised ones, during daily communication.

Figure 4.5 depicts the relative distribution of eye, nose, and mouth regions of target and nontarget faces in Step 3 and all faces in Step 1. The results reveal distinct attentional

TABLE 4.3 ANOVA test results between numerical and categorical variables

VARIABLE	*CATEGORY*	*F-STATISTIC*	*P-VALUE (ANOVA)*
Emotions			
Fixation index in trial	Fixation	1.6851	0.1348
Average pupil size	Pupil	1.1715	0.3207
Average of both eyes' microsaccade rate	Microsaccade	1.2997	0.2611
Binocular microsaccade rate	Microsaccade	2.0415	0.0700
Binocular microsaccade average duration	Microsaccade	0.9459	0.4500
Fixation duration	Fixation	1.3105	0.2566
RoI Label			
Fixation index in trial	Fixation	2.7235	0.0123
Average pupil size	Pupil	10.4780	1.7351×10^{-11}*
Average of both eyes' microsaccade rate	Microsaccade	2.7766	0.0108
Binocular microsaccade rate	Microsaccade	4.2456	**0.0003***

(Continued)

TABLE 4.3 *(Continued)* ANOVA test results between numerical and categorical variables

VARIABLE	*CATEGORY*	*F-STATISTIC*	*P-VALUE (ANOVA)*
Binocular microsaccade average duration	Microsaccade	0.4616	0.8371
Fixation duration	Fixation	6.0402	2.7859×10^{-6}*
Face Region			
Fixation index in trial	Fixation	2.5302	0.0556
Average pupil size	Pupil	0.1862	0.9058
Average of both eyes' microsaccade rate	Microsaccade	2.4109	0.0651
Binocular microsaccade rate	Microsaccade	7.6508	4.3650×10^{-5}*
Binocular microsaccade average duration	Microsaccade	0.5778	0.6296
Fixation duration	Fixation	4.3748	**0.0045***
Participant ID			
Fixation index in trial	Fixation	2.9196	0.0542
Average pupil size	Pupil	0.1942	0.8235
Average of both eyes' microsaccade rate	Microsaccade	3.4939	0.0306
Binocular microsaccade rate	Microsaccade	10.2293	3.7856×10^{-5}*
Binocular microsaccade average duration	Microsaccade	0.0011	0.9989
Fixation duration	Fixation	3.7091	0.0247
Interest Period Index			
Fixation index in trial	Fixation	10.8056	1.1613×10^{-31}*
Average pupil size	Pupil	286.4408	**0.0***
Average of both eyes' microsaccade rate	Microsaccade	20.3782	1.4308×10^{-64}*
Binocular microsaccade rate	Microsaccade	36.3224	4.0576×10^{-116}*
Binocular microsaccade average duration	Microsaccade	5.4073	3.7311×10^{-13}*
Fixation duration	Fixation	2.8619	3.3758×10^{-5}*

The results highlight significant associations between various eye-tracking metrics and task-related factors.

strategies for different emotions depending on the task requirements, whether free or emotion-guided observation. For instance, consistent with the findings of Polet et al. [23], the eye region of surprised, fearful, and sad faces appears more crucial than that of other emotions in Step 1. Interestingly, this effect is even more pronounced when recognizing emotions in 3. Additionally, the mouth region is more critical for recognizing anger and disgust in 1 and, to a lesser extent, 3, compared to other emotions.

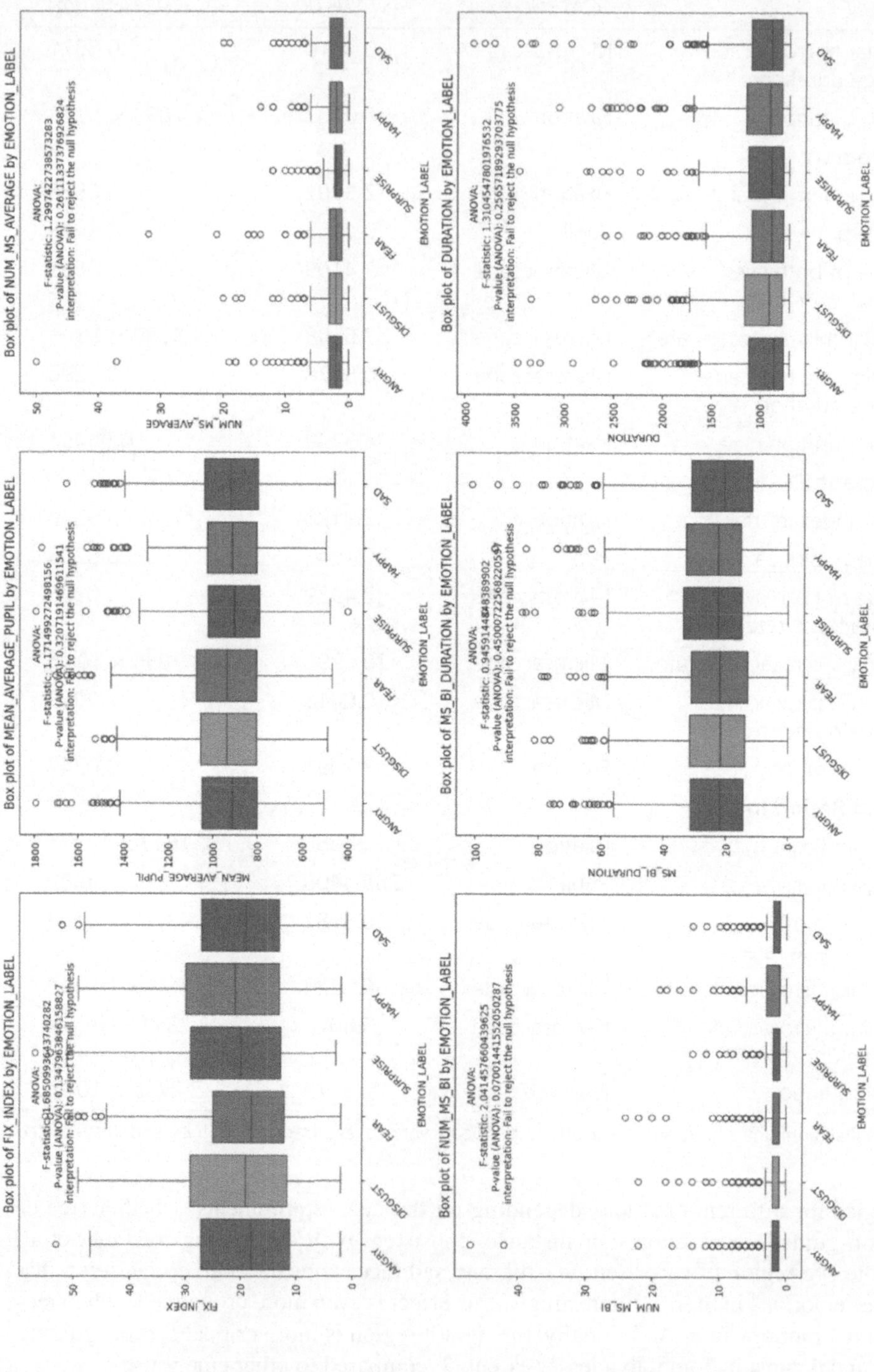

FIGURE 4.6 Distribution of fixational, microsaccadic, and pupillary events across different emotions, showing the variability and potential connections between these events and emotion perception.

This comprehensive analysis highlights the intricate dynamics of eye movements in relation to emotion perception, underscoring the significance of detailed attention to microsaccades and fixation patterns across different stages of the emotion recognition task.

4.5 MODELING

We analyzed data from 20 participants over 54 trials, deriving 54 input-output pairs by averaging fixation events. We employed leave-one-out cross-validation due to the limited dataset size and focused on mean squared error (MSE) as our main performance metric.

4.5.1 Hyperparameter Selection and Justification

Our choice of hyperparameters for deep-learning models was informed by literature, grid search, and practical constraints, aiming to balance complexity and overfitting risk given our small dataset.

4.5.2 Task 1

Task 1 predicted dwell times for each face in Step 3 from Step 1's fixation data. This could improve instruction-free FER tasks reflecting everyday interactions. We predicted using spatial and temporal features of fixation, such as dwell time percentage and fixation counts, and emotion-related features, across four faces into a three-layer MLP with 32-16-4 nodes. The model, trained for 500 epochs at a learning rate of 0.001, emphasized target face predictions by weighting them higher in the loss function.

4.5.3 Task 2

This task forecasted fixation dwell times using only visual features of faces. Employing a pretrained VGG-Face model, we extracted 2622-dimensional embeddings as inputs for a three-layer MLP (100-16-4 nodes). The larger first layer managed the high-dimensional data, reducing it progressively. Figure 4.1 illustrates the structure of the network. The model ran for 1000 epochs with a learning rate of 0.001, focusing separately on data from Steps 1 and 3, prioritizing target face features.

4.5.4 Task 3

Task 3 aimed to predict individual user performance from eye movement events in Step 1. We calculated microsaccade rates and average pupil sizes from seven facial regions of

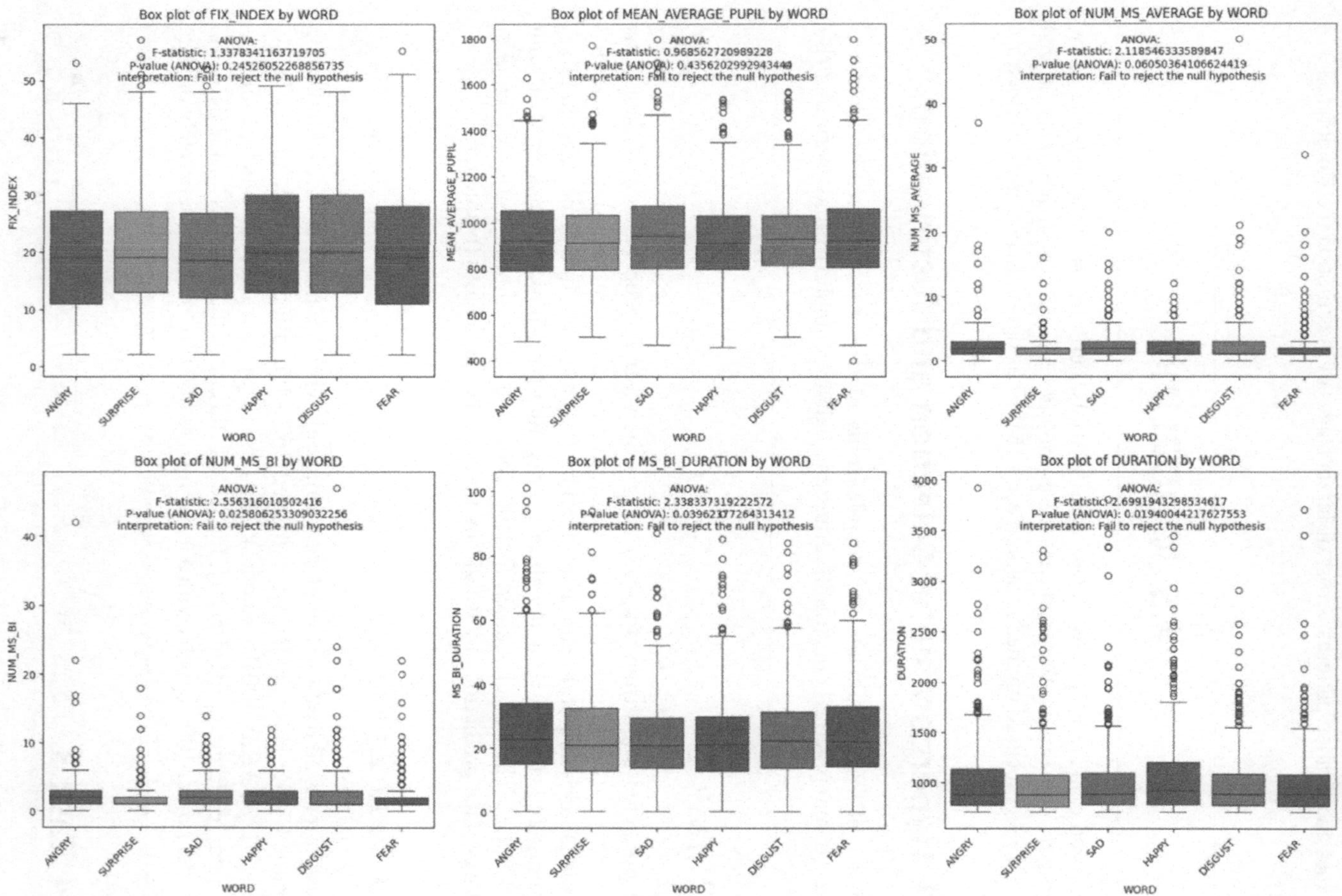

FIGURE 4.7 Comparison of fixation duration, microsaccade rate, and duration across different target emotions. The results suggest emotional differences influence these eye-tracking metrics.

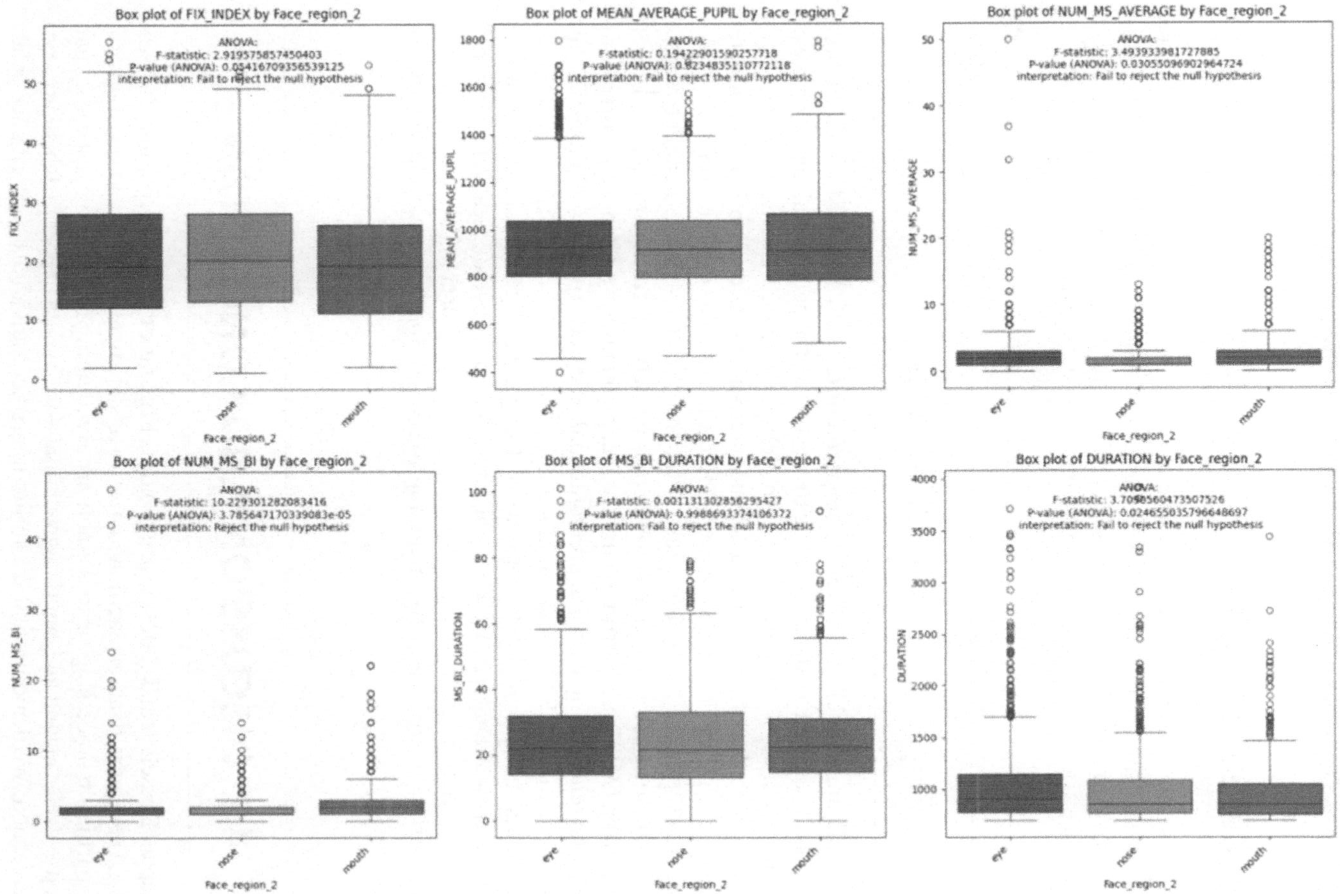

FIGURE 4.8 Microsaccade rate variations across different face regions, highlighting the significance of specific facial areas in emotion perception tasks.

interest for each emotion, using a 64-unit bidirectional LSTM for 100 epochs to capture temporal patterns. Performance prediction was handled by an XGBoost model, chosen for its effectiveness with structured data.

4.5.5 Baseline Model

The baseline model assumed the target face attracted the most attention, allocating the longest viewing time to it and dividing the remainder equally among other faces. This strategy provided a dwell time distribution of 0.50 for the target and 0.1666 for nontarget faces in Tasks 1 and 2's Step 3, and equal times in Task 2's Step 1.

4.6 RESULTS

Results indicated successful dwell time predictions with low MSE rates in both tasks. Task 1 showed that temporal features outperformed spatial ones, suggesting the importance of temporal data in modeling complex FER tasks. Task 2 confirmed the difficulty of emotion-related task predictions, with Step 3 posing greater challenges than Step 1. The average MSE results for these predictions in Tasks 1 and 2 are shown in Table 4.5, illustrating the performance of different feature sets. In evaluating the performance detection models for Task 3, we utilized mean squared error (MSE) as a primary metric, supplemented by Spearman correlation to assess the relationship between predicted and true performances. Contrary to our expectations, the BiLSTM sequential model performed worse than the XGBoost model, though neither model demonstrated strong overall performance. This aligns with the statistical analysis, which highlighted the challenge of creating a universal model for predicting individual performance across all users. The specific performance metrics for these models, including MSE and Spearman correlation values, are presented in Table 4.6.

4.7 DISCUSSION AND CONCLUSION

We adapted Russell et al.'s (2021) instructionless FER task to delve deeper into the FER process. These modifications facilitated extensive statistical analysis and revealed crucial disparities in processing various emotions. By dividing the emotion processing into two steps—free-viewing and emotion grounding—we gained deeper insights into emotion perception. We demonstrated the ability to predict individuals' performance solely from features observed during the free-viewing steps, aligning with eye-gaze strategies for facial emotion perception.

TABLE 4.4 Chi-Square test results between categorical variables

VARIABLE	*CATEGORY*	*CHI-SQUARE*	*P-VALUE*
Emotions			
Emotions	RoI Label	398.9696	4.5144×10^{-66}
Emotions	Target Emotion	11232.5746	**0.0***
Emotions	Face region	286.1583	1.3057×10^{-55}
RoI Label			
RoI Label	Target Emotion	68.7352	**0.0001***
RoI Label	Face region	83284.0	**0.0***
Target Emotion			
Target Emotion	Face region	7.2468	0.7020

The results indicate significant associations between categorical variables such as emotions, RoI labels, and face regions.

TABLE 4.5 Average MSE results for predicting the dwell time of tasks 1 and 2

	TASK 1	
FEATURES	*ALL*	*TARGET*
Baseline	0.0164	0.0060
Spatial	0.0134	0.0066
Temporal	0.0053	0.0030
Spatiotemporal	**0.0046**	**0.0024**
	TASK 2	
FEATURES	*STEP 1*	*STEP 3*
Baseline	0.0152	0.0164
Face embeddings	**0.0065**	**0.0077**

The results indicate that temporal and spatiotemporal features significantly enhance prediction accuracy compared to baseline and spatial features alone.

TABLE 4.6 The performance metrics of models on task 3

MODEL	*MEAN SQUARED ERROR*	*SPEARMAN CORRELATION*
XGBoost	0.0628	0.2496*
LSTM	0.0664	0.1980*

* $p < 0.05$.

Our findings indicate that gaze events, particularly temporal features, can predict FER performance by merely observing faces. We also forecasted the fixation duration of FER tasks based on facial visual features, aiding in the assessment of trial difficulty. Notably, we predicted emotion perception accuracy from free face viewing, marking progress toward lightweight emotion recognition assessments not dependent on language skills.

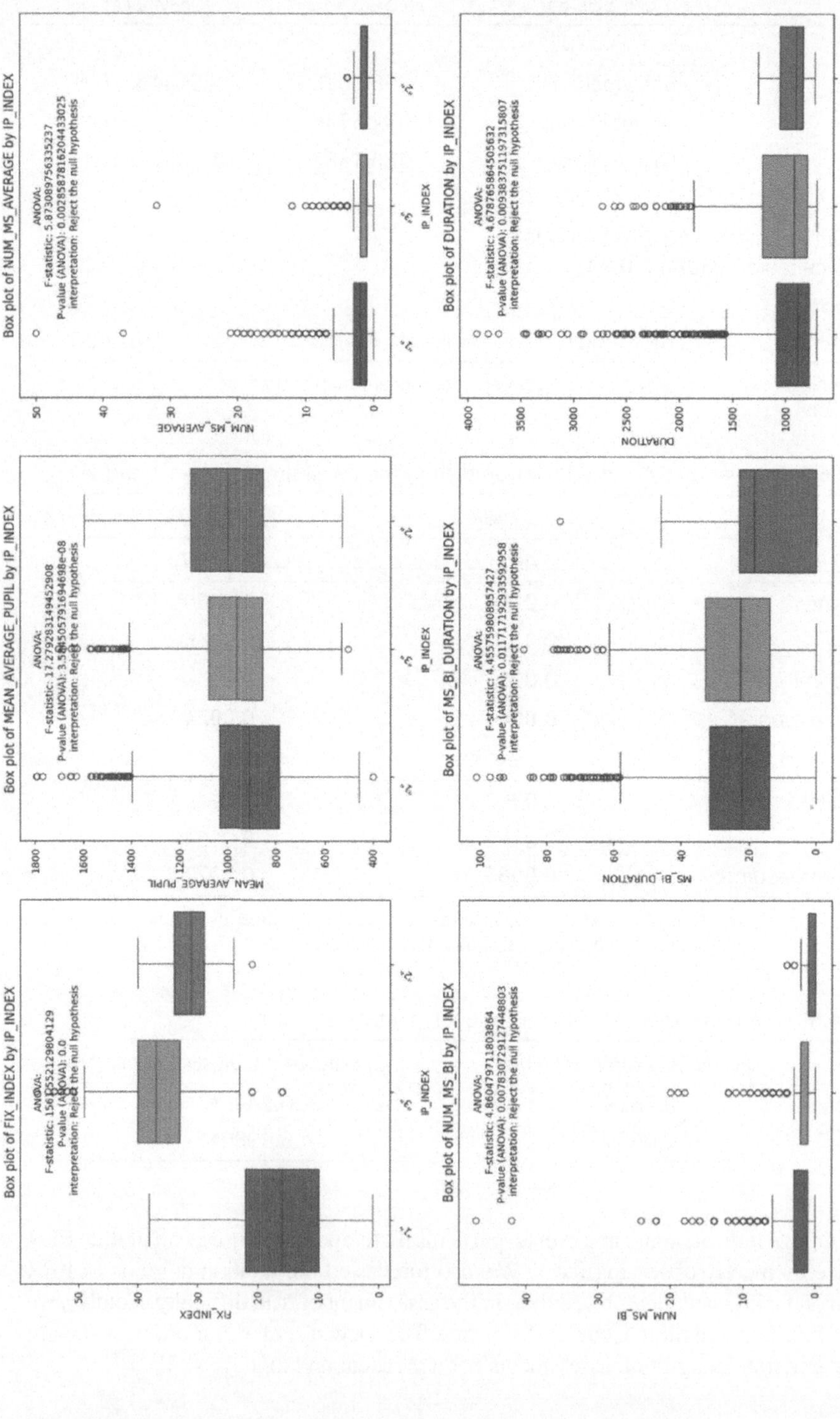

FIGURE 4.9 Comparison of numerical events across different interest periods (steps). Microsaccades and pupil size show significant differences, indicating the varying cognitive load and task complexity in each step.

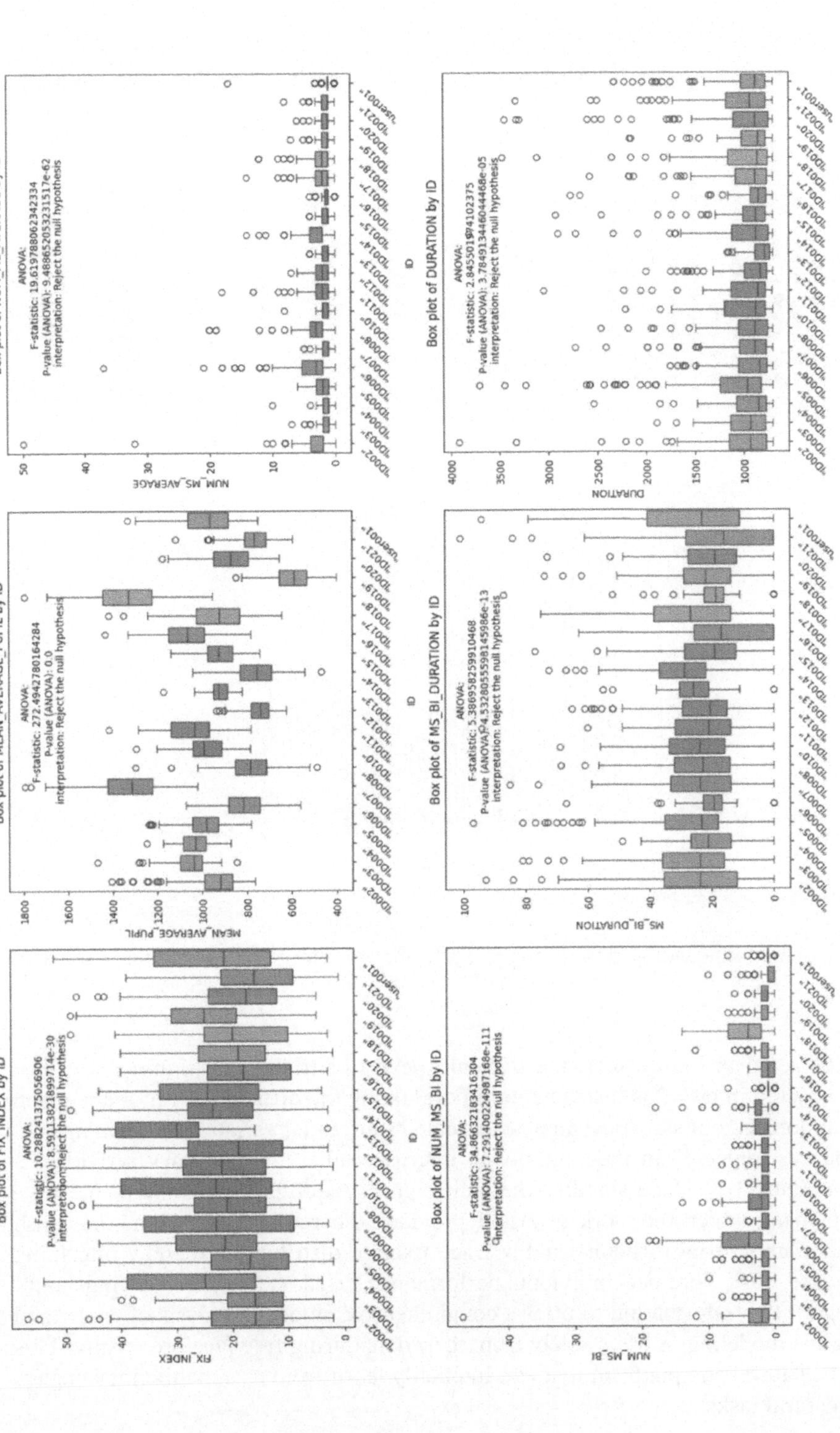

FIGURE 4.10 Variability in pupillary, microsaccadic, and fixational activities among participants. This underscores the individual differences in emotion perception and the challenge of creating a universal model.

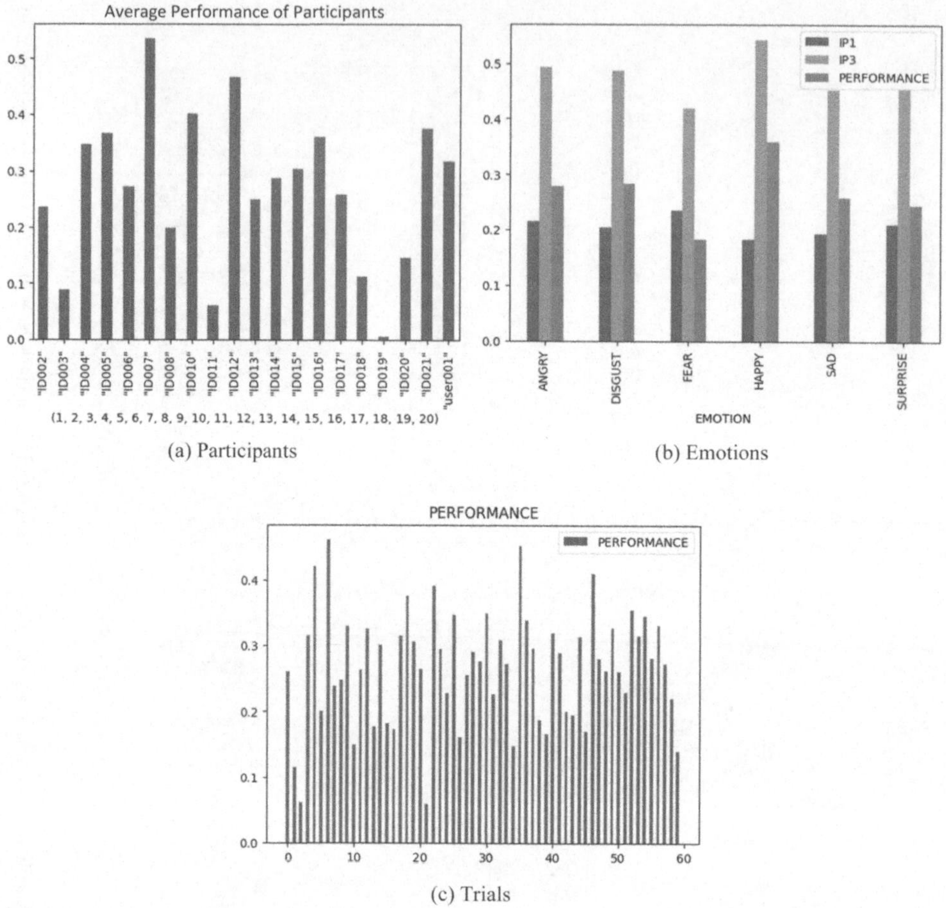

FIGURE 4.11 Performance analysis across (a) participants, (b) trials, and (c) emotions. The analysis highlights individual differences in performance, emotional perception, and trial difficulty.

As depicted in Figure 4.11, the difficulty levels of trials vary significantly. The work conducted in task 2 is instrumental in evaluating the difficulty level of any given trial. Additionally, our statistical analysis revealed substantial variations in participants' performances, reflected in their fixational, microsaccadic, and pupillary activities as shown in Figure 4.10. Understanding these differences requires approximating the average participant, which the work in task 1 provides. Therefore, tasks 1 and 2 furnish information about trial difficulty and average fixation distribution, thereby informing performance. Task 3 predicts individual performance based on free face viewing, amalgamating all this information to offer a comprehensive emotion analysis of a user and enabling the modeling of users solely from their data during free face processing. This provides a naturalistic platform that can implicitly learn from users while they engage in their natural tasks.

Moreover, we introduced a standardized tool for FER datasets, enhancing the comparability of results. Overall, our work offers insights for FER research and could shape the development of more naturalistic emotion recognition assessments.

While the temporal modeling of eye gaze strategies presents a promising avenue, further investigation is warranted. Specifically, the extraction of compatible events and an increase in data collection could significantly enhance performance. Another factor to consider is the challenge posed by having four faces on the screen simultaneously, which can make it difficult to distinguish between regions of interest across different faces. Designing a new setting where each face is presented separately could address this issue and provide more valid temporal data on gaze strategies. Additionally, incorporating participant annotations of perceived emotions could enrich the analysis by providing a more valid ground truth for modeling emotion perception.

In conclusion, our work advances FER by exploring new paradigms and models. Predicting FER performance from free-viewing eye movements offers a pathway for efficient and ecologically valid emotion perception assessments. We anticipate that our work will inspire further research and foster improved tools and methodologies for studying human emotion perception.

REFERENCES

1. A. S. Walker-Andrews, "Emotions and social development: Infants' recognition of emotions in others," *Pediatrics*, vol. 102, no. 5 Suppl E, pp. 1268–1271, 1998.
2. M. L. Smith, G. W. Cottrell, F. Gosselin, and P. G. Schyns, "Transmitting and decoding facial expressions," *Psychological Science*, vol. 16, no. 3, pp. 184–189, 2005.
3. L. Collin, J. Bindra, M. Raju, C. Gillberg, and H. Minnis, "Facial emotion recognition in child psychiatry: A systematic review," *Research in Developmental Disabilities*, vol. 34, no. 5, pp. 1505–1520, 2013.
4. G. Valenza, L. Citi, A. Lanatá, E. P. Scilingo, and R. Barbieri, "Revealing real-time emotional responses: A personalized assessment based on heartbeat dynamics," *Scientific Reports*, vol. 4, no. 1, pp. 1–13, 2014.
5. M. M. Bradley, L. Miccoli, M. A. Escrig, and P. J. Lang, "The pupil as a measure of emotional arousal and autonomic activation," *Psychophysiology*, vol. 45, no. 4, pp. 602–607, 2008.
6. V. Skaramagkas, G. Giannakakis, E. Ktistakis, *et al.*, "Review of eye tracking metrics involved in emotional and cognitive processes," *IEEE Reviews in Biomedical Engineering*, vol. 16, pp. 260–277, 2021.
7. C. Aracena, S. Basterrech, V. Snáel, and J. Velásquez, "Neural networks for emotion recognition based on eye tracking data," in *2015 IEEE International Conference on Systems, Man, and Cybernetics*, Hong Kong: IEEE, 2015, pp. 2632–2637. doi: 10.1109/SMC.2015.460.
8. L. Chaby, I. Hupont, M. Avril, V. Luherne-du Boullay, and M. Chetouani, "Gaze behavior consistency among older and younger adults when looking at emotional faces," *Frontiers in Psychology*, vol. 8, p. 548, 2017.
9. V. Tsang, "Eye-tracking study on facial emotion recognition tasks in individuals with high-functioning autism spectrum disorders," *Autism*, vol. 22, no. 2, pp. 161–170, 2018.

10. L. L. Russell, C. V. Greaves, R. S. Convery, *et al.*, "Novel instructionless eye tracking tasks identify emotion recognition deficits in frontotemporal dementia," *Alzheimer's Research & Therapy*, vol. 13, no. 1, pp. 1–11, 2021.
11. M. R. Readman, M. Polden, M. C. Gibbs, L. Wareing, and T. J. Crawford, "The potential of naturalistic eye movement tasks in the diagnosis of Alzheimer's disease: A review," *Brain Sciences*, vol. 11, no. 11, p. 1503, 2021.
12. L. L. Russell, C. V. Greaves, R. S. Convery, *et al.*, "Eye movements in frontotemporal dementia: Abnormalities of fixation, saccades and anti-saccades," *Alzheimer's & Dementia: Translational Research & Clinical Interventions*, vol. 7, no. 1, p. e12218, 2021.
13. C. Fu, Q. Deng, J. Shen, H. Mahzoon, and H. Ishiguro, "A preliminary study on realizing human–robot mental comforting dialogue via sharing experience emotionally," *Sensors*, vol. 22, no. 3, p. 991, 2022.
14. M. J. Seikavandi, and M. J. Barret, "Gaze reveals emotion perception: Insights from modelling naturalistic face viewing," in *2023 International Conference on Machine Learning and Applications (ICMLA)*, IEEE, 2023, pp. 2022–2025.
15. C. Howard, *Temporal sampling and representation updating*. Academic Press, 2017, vol. 236.
16. K. Kashihara, "Microsaccadic modulation evoked by emotional events," *Journal of Physiological Anthropology*, vol. 39, no. 1, pp. 1–11, 2020.
17. C. Strauch, L. Greiter, and A. Huckauf, "Pupil dilation but not microsaccade rate robustly reveals decision formation," *Scientific Reports*, vol. 8, no. 1, p. 13165, 2018.
18. H. Rodger, N. Sokhn, J. Lao, Y. Liu, and R. Caldara, "Developmental eye movement strategies for decoding facial expressions of emotion," *Journal of Experimental Child Psychology*, vol. 229, p. 105622, 2023.
19. M. Schurgin, J. Nelson, S. Iida, H. Ohira, J. Chiao, and S. Franconeri, "Eye movements during emotion recognition in faces," *Journal of Vision*, vol. 14, no. 13, pp. 14–14, 2014.
20. J. Liang, Y.-Q. Zou, S.-Y. Liang, Y.-W. Wu, and W.-J. Yan, "Emotional gaze: The effects of gaze direction on the perception of facial emotions," *Frontiers in Psychology*, vol. 12, p. 684357, 2021.
21. N. Tottenham, J. W. Tanaka, A. C. Leon, *et al.*, "The nimstim set of facial expressions: Judgments from untrained research participants," *Psychiatry Research*, vol. 168, no. 3, pp. 242–249, 2009.
22. A. Zadeh, Y. C. Lim, T. Baltrusaitis, and L.-P. Morency, "Convolutional experts constrained local model for 3d facial landmark detection," in *Proceedings of the IEEE International Conference on Computer Vision Workshops*, 2017, pp. 2519–2528.
23. K. Polet, S. Hesse, A. Morisot, *et al.*, "Eye-gaze strategies during facial emotion recognition in neurodegenerative diseases and links with neuropsychiatric disorders," *Cognitive and Behavioral Neurology*, vol. 35, no. 1, pp. 14–31, 2022.
24. D. C. Montgomery, E. A. Peck, and G. G. Vining, *Introduction to linear regression analysis*. John Wiley & Sons, 2021.
25. C. Bonferroni, "Teoria statistica delle classi e calcolo delle probabilita," *Pubblicazioni Del R Istituto Superiore Di Scienze Economiche e Commericiali Di Firenze*, vol. 8, pp. 3–62, 1936.

PART TWO

Deep Learning for Natural Language Processing

Large Language Models for Automated Short-Answer Grading and Student Misconception Detection in STEM

5

Indika Kahanda, Nazmul Kazi, and James Becker

5.1 INTRODUCTION

Modern education prioritizes individualized learning for deeper comprehension [1]. Standardized tests, often multiple-choice, focus on memorization over understanding, potentially leading to inaccurate evaluations of cognitive abilities [2]. Multiple-choice questions offer binary feedback, potentially limiting learning from mistakes. Conversely, open-ended questions foster analytical thinking and metacognitive growth, providing a comprehensive assessment of understanding and application of ideas [3]. Written tasks, where students express their understanding, offer valuable insights into their comprehension and perspectives [4]. These tasks document cognitive activities and, when used for evaluation, affirm knowledge and promote positive learning outcomes.

DOI: 10.1201/9781003570882-7

Education aims to equip students with knowledge and skills, but misconceptions can hinder progress. Detecting misconceptions is separate from grading, focusing on understanding rather than correctness [5]. However, manual scoring and misconception identification can be demanding, distracting teachers from instruction [5]. Automatic short-answer grading (ASAG) simplifies this process, allowing for immediate feedback and more instructional time [1]. Automated misconception detection (AMD) goes beyond ASAG, identifying errors in comprehension [6]. Both ASAG and AMD can benefit from large language models (LLMs), which understand and generate human-like text across domains. This chapter examines ASAG (Section 5.2) and AMD (Section 5.3), focusing on enhancing LLMs' transfer learning and identifying misconceptions in a beginner's circuit analysis course.

5.2 AUTOMATED SHORT-ANSWER GRADING

5.2.1 Introduction

Adaptive evaluations, crucial for individualized learning paths, often use multiple-choice questions due to their ease of execution and scoring. However, these questions primarily test fact recall, not comprehensive understanding, and can distort a student's cognitive skills evaluation [1–2]. They offer binary feedback, limiting learning from errors. In contrast, open-ended questions foster critical thinking and meta-cognitive abilities, providing a thorough assessment of concept grasp and application [3]. They require students to articulate their thought processes, offering educators insights into learning trajectories and areas needing improvement. However, manual evaluation of these questions can be laborious [5]. ASAG in intelligent tutoring systems (ITS) autonomously grades brief responses, providing immediate feedback and aiding effective student assessment [1]. This allows educators to focus more on instruction and student support.

Consequently, there is an increasing interest in creating automated systems capable of assessing student answers to open-ended questions across a variety of fields and subjects [1]. ASAG is a complex task, necessitating both semantic understanding and the ability to recognize textual entailment (RTE). Furthermore, ASAG introduces an extra layer of complexity by demanding transfer learning to ensure compatibility across different domains, making it fundamentally a data-driven issue. The SemEval-2013 Task 7 [7] and the SciEntsBank dataset, which are included in this challenge, serve as widely accepted benchmarks for ASAG research.

LLMs have shown potential in various natural language processing tasks, including ASAG, due to their ability to understand complex contextual information [8, 9]. They may improve ASAG's precision and efficiency compared to traditional ML methods. Dzikovska et al. [7] reported a best weighted-average F1 of 0.63 for three-way labeling with classical ML models. Recent studies show LLMs' superior performance in ASAG. Sung et al. [10] achieved a weighted-average F1 of 0.68 using BERT-base [8], while Zhu et al. [1] reported a weighted-average F1 of 0.67 using BERT-base and 0.69 using a BERT-based deep neural network (DNN). Camus and Filighera [11] found that RoBERTa Large fine-tuned over the multigenre natural language inference (MNLI) corpus yielded the best result with a weighted-average F1 of 0.72.

In this section, we explore two primary research questions associated with automated short-answer grading using LLMs. First, we evaluate the capability of various renowned classical machine learning (ML) algorithms to exceed the performance of the lexical baseline set in SemEval-2013 Task 7. Second, we examine whether fine-tuning RoBERTa-Large on a broader and more diverse corpus, like the MNLI corpus [12], could assist the model with semantic inference and transfer learning to enhance the model performance. By addressing these research questions, we aim to contribute to the ongoing efforts to enhance the accuracy and efficiency of ASAG using LLMs.

5.2.2 Datasets

5.2.2.1 SciEntsBank

We utilized a segment of the student response analysis (SRA) corpus [13], referred to as the SciEntsBank dataset.[1] This dataset has been annotated with SRA labels by human annotators [7]. It was released with three distinct labeling versions: five-way, three-way, and two-way. For our experiment, we used the three-way labeling, where each sample is labeled as either *correct*, *contradictory*, or *incorrect*.

The SciEntsBank dataset is composed of four distinct sets: a training set and three test sets. These test sets are designed to evaluate a model's adaptability across various problems and domains. The creation and purpose of these test sets are detailed below: (1) Unseen domains (UDs): The authors reserved the complete set of questions and answers from three science domains during training to create this set. The aim of this set is to assess a model's flexibility and adaptability across different knowledge domains; (2) Unseen questions (UQs): From the 12 domains chosen for training, the authors randomly selected a subset of questions and excluded all responses to these selected questions to create this set. This set evaluates a model's ability to handle new questions within known domains; and (3) Unseen answers (UAs): From the questions chosen for the training set, the authors excluded a subset of randomly selected responses from the training set to create this set. This set of unseen answers is the most common approach to model evaluation. Its purpose is to evaluate the model's skill in grading responses it has not seen before.

The training set consists of samples that are not included in any of the three test sets. Table 5.1 shows the distribution of the three-way labels in the dataset. It is important to note that the dataset is imbalanced. The *Contradictory* class has a significantly lower number of samples (only 10% of the dataset) compared to the other two classes.

5.2.2.2 Natural language inference (NLI) corpora

In this research, we have employed two NLI corpora: a) Stanford natural language inference (SNLI) corpus [14], and b) multigenre natural language inference (MNLI)

TABLE 5.1 The three-way label distribution of SciEntsBank dataset

		TEST		
LABELS	*TRAIN*	*UNSEEN ANSWERS*	*UNSEEN QUESTIONS*	*UNSEEN DOMAINS*
Correct	2,008	233	301	1,917
Contradictory	499	58	64	417
Incorrect	2,462	249	368	2,228
Total	4,969	540	733	4,562
			5,835	
			10,804	

TABLE 5.2 Sample distribution of SNLI and MNLI corpora across train, validation (i.e., eval), and test sets

CORPUS	*TRAIN*	*VALIDATION*	*TEST*	*TOTAL*
SNLI	550,152	10,000	10,000	570,152
MNLI	392,702	20,000	20,000	432,702

corpus [12]. These carefully assembled corpora are key resources in the domain, acting as well-recognized and broadly used benchmark datasets for NLI tasks, especially in the RTE. The SNLI corpus encompasses a wide array of sentence pairs labeled for entailment, contradiction, or neutrality. Conversely, the MNLI corpus, which is designed based on the SNLI corpus, broadens the scope by incorporating various genres, thereby ensuring a more exhaustive evaluation of models across diverse linguistic contexts. It is crucial to mention that there is no overlap in samples between the MNLI and SNLI corpora. The sample distribution of these corpora is illustrated in Table 5.2.

5.2.3 Models

5.2.3.1 Classical Machine Learning models (Baseline)

We set a benchmark for our deep learning algorithms using four well-known traditional ML models. These models necessitate feature engineering, selection, and extraction. We employ the TfidfVectorizer with a word analyzer to create features. To decrease the dimensionality of the feature space, we eliminate all English stop words and apply lemmatization to the remaining words. Our feature space includes unigrams and bigrams, but we restrict it to the top 10,000 features for training.

We conduct experiments with two tree-based models: decision tree (DT) and random forest (RF). DT is a nonparametric model that derives parameters from the given features and constructs tree structures for decision-making. Conversely, RF fits multiple DTs on various subsets of the dataset and determines a label by averaging

TABLE 5.3 F1 scores of linear SVM for different *C* values

	MACRO			*WEIGHTED*		
Value of C	0.1	0.5	1.0	0.1	0.5	1.0
Unseen Answers (UAs)	0.43	0.48	0.51	0.54	0.59	0.61
Unseen Questions (UQs)	0.32	0.34	0.35	0.42	0.45	0.46
Unseen Domains (UDs)	0.34	0.36	0.39	0.37	0.45	0.47

the results of all DTs. We employ an RF model composed of a hundred DTs to predict the labels.

We also examine the support vector machines (SVMs), a supervised learning model that aims to establish a hyperplane in the feature space that distinctly separates the data points. SVM has shown encouraging results in numerous NLP tasks. We utilize the linear SVM classifier from Sci-Kit Learn, and we test with three different *C* values, finding the optimal performance for $C = 1$ (refer to Table 5.3).

Furthermore, we assess the performance of an artificial neural network (ANN) model, specifically, the multilayer perceptron (MLP) model, a fully connected feedforward neural network. We use ReLU as the activation function and the *Adam* optimizer with two hidden layers consisting of 1,000 and 100 neurons, respectively, given the complexity of the problem.

5.2.3.2 Large language models

Pretrained LLMs like RoBERTa Large have demonstrated impressive results in various natural language processing tasks. RoBERTa Large, a pretrained LLM, has been trained on a substantial volume of unlabeled text data. When a pretrained RoBERTa Large model is fine-tuned for a specific task, it can yield considerable performance enhancements [9].

In our experiment, we utilized the pretrained RoBERTa Large model without any further pretraining. The fine-tuning of the RoBERTa Large model was carried out using the Adam optimizer with the following hyperparameters: a learning rate of 2e-5, an Adam epsilon of 1e-08, a batch size of 5, a warm-up step of 500, and a weight decay of 0.01. The model was fine-tuned over 20 epochs, and its performance was recorded at the conclusion of each epoch. We fine-tuned a RoBERTa Large model exclusively on the SciEntsBank dataset, which we refer to as "RoBERTa Large". In addition, we fine-tuned another RoBERTa Large model on the MNLI corpus, a commonly used benchmark dataset for natural language inference, and then on the SciEntsBank dataset. This model is referred to as "RoBERTa Large MNLI".

Every entry in the SciEntsBank dataset is composed of a question, a reference answer, a student's response, and a corresponding label. The objective is to categorize the student's response in the context of the given question and reference answer. In this study, we approach ASAG as an RTE issue. To accomplish this, we form the premise by joining the question with the reference answer. The student's response then acts as the hypothesis in our modeling strategy. We developed three models, utilizing the pretrained RoBERTa LLM as the fundamental base. The goals and the setup of these three models are detailed below:

1. RoBERTa Large: This model was fine-tuned exclusively on the SciEntsBank dataset. The main goal was to evaluate the intrinsic abilities of RoBERTa Large on the SciEntsBank dataset.
2. RoBERTa Large MNLI: We fine-tuned a RoBERTa Large model initially on the MNLI corpus and then on the SciEntsBank dataset. The objective of this model was to explore the possible performance improvement that could be attained by fine-tuning a pretrained model on a corpus related to the task.
3. RoBERTa Large 2NLI: Remarkably, the RoBERTa Large MNLI model demonstrated a significantly superior performance in comparison to the RoBERTa Large model. Noticing the advantages of fine-tuning on a corpus related to the task, we further investigated whether fine-tuning on multiple corpora could result in additional enhancements in the model's performance. Consequently, we created this model that was fine-tuned first on the MNLI corpus, then on the SNLI corpus, and finally on the SciEntsBank dataset. We designated this model as 2NLI, reflecting its fine-tuning on two NLI datasets.

We established two separate sets of hyperparameters for fine-tuning the models: one set for fine-tuning on the NLI corpus and another set for fine-tuning on the SciEntsBank dataset. The details of both sets of hyperparameters are provided below:

1. MNLI and SNLI: The hyperparameters we used were adapted from the configuration suggested by the creators of the RoBERTa Large model [9]. Our modifications included the use of the Adam optimizer with $\epsilon = 1e - 6$, $\beta_1 = 0.9$, $\beta_1 = 0.98$, a learning rate of $2e - 5$, and a weight decay of 0.1. The learning rate scheduler was set to linear, and a warm-up ratio of 6% was used to ensure a smooth transition into the main training phase. We fine-tuned for a maximum of 10 epochs with early stopping implemented. For the MNLI, we employed a batch size of 32. A batch size of 64 yielded superior results (data not shown) for the SNLI. The performance of the model was recorded after each epoch, with the best model being selected based on the epoch that achieved the highest macro F1 score.
2. SciEntsBank: These hyperparameters were customized to match the unique characteristics of the dataset. We used the Adam optimizer with ϵ = 1e − 8, a learning rate of 2e − 5, and a weight decay of 0.01. A warm-up step of 500 was set up to ensure a smooth start to the training. We fine-tuned for a maximum of 20 epochs with early stopping and a batch size of 5. The model's performance was recorded after each epoch, and the best model was chosen based on the epoch that displayed the highest macro F1 score.

5.2.3.3 Experimental setup

The performance of our models is assessed using the same three metrics as [7]: a) accuracy, b) macro-average F1, and c) weighted-average F1. The macro-average F1 computes the average F1 score for all classes, disregarding the size of each class. It is important to note that the dataset is imbalanced, as illustrated in Table 5.1, with the *Contradictory* class having significantly fewer samples than the other two classes. The weighted-average F1 considers the size of each class in its calculation, providing a balanced score for imbalanced datasets. For traditional ML models and evaluation metrics, we utilized the Sci-Kit Learn library. The LLMs were implemented using the PyTorch and Hugging Face Transformers libraries.

5.2.4 Results and Discussion

Table 5.4 showcases the performance of four traditional ML models: DT, RF, SVM, and MLP across three test sets. The table also includes the lexical baseline from SemEval-2013 Task 7 [7]. Among all classifiers, MLP outshines on the Unseen Answers test set in all metrics. On the Unseen Questions test set, RF scores the highest in all metrics, matching the macro-averaged F1 of the lexical baseline. On the Unseen Domains test set, while RF achieves the highest accuracy and weighted-average F1, MLP delivers the best macro-averaged F1 among our models, but none surpass the lexical baseline. Overall, MLP exhibits superior performance over other ML models for intra-domain classification, whereas RF demonstrates high potential for transfer learning. We establish a new baseline for the LLMs by taking the maximum score per test set and metric—a method chosen to highlight the best potential performance of classical ML across different aspects—which is denoted as "Baseline" in Table 5.5.

Table 5.5 displays the performance of LLMs on all test sets. The baseline indicates that the LLMs have significantly outperformed. The SemEval 2013 Task 7 results are also presented in the table, which shows the maximum scores achieved by any model/algorithm in that competition. These scores serve as another point of comparison for

TABLE 5.4 Performance of classical ML models

	UNSEEN ANSWERS			*UNSEEN QUESTIONS*			*UNSEEN DOMAINS*		
CLASSIFIER	*ACC*	*M-F1*	*W-F1*	*ACC*	*M-F1*	*W-F1*	*ACC*	*M-F1*	*W-F1*
Lexical Baseline [7]	0.56	0.41	0.52	0.54	0.39	0.52	0.58	0.42	0.55
Decision Tree (DT)	0.65	0.57	0.64	0.48	0.33	0.44	0.48	0.34	0.43
Random Forest (RF)	0.63	0.54	0.62	0.56	0.39	0.53	0.55	0.37	0.52
Support Vector Machines (SVM)	0.63	0.52	0.61	0.47	0.34	0.46	0.50	0.38	0.47
Multilayer Perceptron (MLP)	0.67	0.63	0.67	0.38	0.33	0.38	0.47	0.39	0.47

Acc: Accuracy, M-F1: Macro-Average F1, W-F1: Weighted-Average F1.

TABLE 5.5 Performance of large language models (LLMs)

	UNSEEN ANSWERS			*UNSEEN QUESTIONS*			*UNSEEN DOMAINS*		
CLASSIFIER	*ACC*	*M-F1*	*W-F1*	*ACC*	*M-F1*	*W-F1*	*ACC*	*M-F1*	*W-F1*
Baseline	0.67	0.63	67	0.56	0.39	0.53	0.58	0.42	0.55
SemEval 2013 Task 7 (the best score per set and metric) [7]	0.72	0.65	0.71	0.66	0.47	0.63	0.64	0.49	0.62
BERT Base by Sung et al. [10]	0.76	0.72	0.76	0.65	0.58	0.65	0.64	0.58	0.63
BERT Base by Zhu et al. [1]	0.74	0.69	0.73	0.66	0.55	0.65	0.66	0.56	0.64
BERT-based DNN by Zhu et al. [1]	0.77	0.71	0.76	0.69	0.58	0.67	0.66	0.56	0.65
RoBERTa (Large + MNLI) by Camus and Filighera [11]	0.79	0.78	0.79	0.66	0.66	0.66	0.72	0.71	0.72
RoBERTa Large	0.76	0.71	0.75	0.65	0.54	0.66	0.67	0.60	0.67
RoBERTa Large MNLI	0.77	0.74	0.77	0.72	0.67	0.72	0.72	0.70	0.72
RoBERTa Large 2NLI	0.78	0.75	0.78	0.72	0.68	0.72	0.72	0.67	0.72

Acc: Accuracy, M-F1: Macro-Average F1, W-F1: Weighted-Average F1.

the LLMs. The BERT-base fine-tuned by Sung et al. [10] surpasses the baseline in all test sets and metrics except UQs and UDs on the accuracy, whereas the BERT-base fine-tuned by Zhu et al. [1] surpasses the baseline in each test set and metric. The BERT-base fine-tuned by Sung et al. [10] performs slightly better than the BERT-base fine-tuned by Zhu et al. [1] based on weighted-average F1 (0.68 versus 0.67). BERT-based DNN performs equally or better than both BERT-base models except for underperformance for UA and UD on macro-averaged F1 compared to the BERT-base fine-tuned by Sung et al. [10] (0.71 versus 0.72 and 0.56 versus 0.58, respectively). With a weighted average F1 of 0.69, BERT-based DNN outperforms both BERT-base models. RoBERTa Large underperforms compared to BERT-based DNN on UAs and UQs but outperforms the model on UD in all metrics. RoBERTa Large MNLI outperforms all models on all test sets and all metrics. RoBERTa Large MNLI significantly improves upon the UQ and UD sets compared to any other models on any metrics. Notably, RoBERTa Large MNLI shows closely similar performance on the UQ and UD sets portraying little to no difference between unseen questions and domains while narrowing down the performance gap with UAs.

The RoBERTa Large model fine-tuned by Camus and Filighera [11] on the MNLI corpus demonstrates similar performance to our own. While their model achieved

TABLE 5.6 Improvement in F1 scores illustrating the impact of MNLI corpus and transfer learning in model performance on three-way labeling scheme

	UNSEEN ANSWERS			*UNSEEN QUESTIONS*			*UNSEEN DOMAINS*		
CLASSIFIER	*CR*	*CN*	*IN*	*CR*	*CN*	*IN*	*CR*	*CN*	*IN*
RoBERTa Large	0.77	0.58	0.78	0.68	0.24	0.70	0.67	0.40	0.72
RoBERTa Large MNLI	0.78	0.65	0.78	0.70	0.55	0.76	0.71	0.64	0.75
Improvement	0.01	0.07	0.00	0.02	0.31	0.06	0.04	0.24	0.03

CR: correct, CN: contradictory, and IN: incorrect.

higher performance for the UA set on all metrics, our model excels on the UQ set. Notably, both models exhibit almost identical performance on the UD set. We strongly attribute the performance differences to our choice of hyperparameters, emphasizing that batch size can significantly impact the scores as well. Importantly, our results not only validate their model performance but also reciprocally affirm our findings. It is crucial to emphasize that the objective of this experiment is not solely to validate their model's performance but rather to investigate the effect of fine-tuning on the MNLI corpus, shedding light on its influence on model performance and its utility in transfer learning—an aspect unexplored by Camus and Filighera [11].

Table 5.6 provides a performance comparison between RoBERTa Large and RoBERTa Large MNLI, based on F1 scores. It is clear that fine-tuning the model on the MNLI corpus has significantly boosted its semantic inference capabilities, leading to a noticeable improvement in performance, particularly within the contradictory class. This enhancement is most evident in the UQ and UD sets, with increases of 0.31 and 0.24, respectively. Additionally, the UA set shows a commendable increase of 0.07.

It is worth noting that the SciEntsBank dataset contains a significantly smaller number of training samples within the contradictory class, making up only 10% of the dataset (see Table 5.1). As a result, RoBERTa Large initially shows relatively lower performance for *contradictory* compared to both *correct* and *incorrect* across all test sets. However, fine-tuning the model on the MNLI dataset has been instrumental in improving its performance within this minority class through effective transfer learning.

5.2.5 Summary and Open Problems

Understanding and accurately classifying student responses is key in automated scoring systems. However, the complexity of student answers poses challenges. LLMs have shown superior performance over traditional rule-based methods in this domain. Our experiments found that fine-tuning LLMs on the MNLI corpus significantly improves model performance, highlighting the effectiveness of transfer learning for tasks like ASAG. Future research could include the SRA corpus, investigate optimal hyperparameter configurations, and evaluate the performance of newer models like Megatron-Turing NLG or PaLM 2 on the SRA corpus. In the context of ASAG, recall is arguably more important than precision, making it crucial to minimize false negatives [6]. Future evaluations could consider the F-beta score for a more nuanced evaluation, adjusting the balance between precision and recall. However, in multiclass classification, increasing recall for one label can lower the recall of another label(s).

5.3 AUTOMATED MISCONCEPTION DETECTION

5.3.1 Introduction

Education is structured to equip students with essential knowledge and skills. However, misconceptions can obstruct educational progression [15]. Stemming from misunderstood concepts or incomplete information, misconceptions form cognitive barriers, affecting academic performance and real-world application of knowledge if unaddressed [16, 17]. Writing exercises, promoting critical thinking and metacognitive skills, serve as an invaluable tool for academic evaluation and misconception detection [6]. They encourage deep exploration of concepts, offering insights into students' thought processes and learning progress. However, identifying misconceptions in student writing is demanding and time-consuming [19]. It requires a thorough review of each student's work, pinpointing subtle hints of misconceptions, and providing constructive feedback [20]. As class size increases, this task becomes more difficult, compromising prompt and comprehensive feedback [21]. The field of automated misconception detection (AMD) systems, integrating semantic inference, textual entailment recognition, and transfer learning techniques, is gaining interest [6]. AMD's complexity lies in its capacity to detect subtle misconceptions, requiring a deep understanding of the subject matter and common misconceptions within that domain.

Several studies have addressed educational challenges through various methodologies. Schmidt et al. [16] identified misconceptions in chemical terms and developed test questions for teachers. Danielsiek et al. [22] identified error-prone topics in algorithms and data structures, emphasizing a dual approach to data collection. Britos et al. [23] used data mining tools and decision trees to detect misconceptions in programming. Michalenko et al. [24] introduced an NLP framework using probabilistic classical ML models for misconception detection. Yang et al. [25] used concept maps and lists to diagnose misconceptions in circuit courses. Arbogast et al. [26] explored linguistic indicators to correlate linguistic changes with conceptual understanding in engineering. Elmadani et al. [27] investigated data-driven techniques to uncover misconceptions within ITS interactions. Goncher et al. [28] focused on thematic analysis and concept extraction from open-ended responses to assess understanding.

While the AMD community has made valuable contributions, it has not received as much attention as other educational assessment fields. Previous methods' limitations include restricted ability to handle nuanced misconceptions and time-consuming manual rule-based coding. However, LLMs, with their pretrained knowledge and context understanding, can identify complex linguistic patterns indicating misconceptions.

The capabilities of LLMs for AMD remain unexplored, presenting an opportunity to advance misconception detection in education. In this section, our objective is to present a framework that utilizes the capabilities of LLMs to detect common misconceptions among students taking introductory STEM courses. We focus on an introductory circuit analysis course titled EELE 201: Circuits I for Engineering [6] at Montana State University. To the best of our knowledge, this is the first study that explores the feasibility of using LLMs for automated misconception detection for writing assignments.

5.3.2 Writing Exercise and Student Response Annotation

Figure 5.1 depicts the quiz question used to collect student responses for this project. This quiz was developed by Becker et al. [17] with a dual purpose: to identify students at risk and to detect common misconceptions. Notably, this quiz is given on the fifth day of the class.

Below are two examples of student responses. In these cases, the bold text indicates a *sequential misconception*. The sequential misconception in electric circuits is the belief that elements located further downstream from a source, such as R_2 and R_3 in Figure 5.1, receive current later than elements positioned closer to the source, like R_1.

> *To start, if the resistance of R2 decreases, the voltage source will remain unchanged, still giving out a constant voltage.* ***Since R1 comes before R2 (decreasing) and R3, R1 will remain unchanged.*** *For R2 and R3, when they have equal resistance, power runs through them equally, but when R2 decreases its resistance, compared to R3, R2 will have more power going through it, because R2 would be the path of least resistance. R3 will still have some power going through it just not as much compared to R2. This is allowed to happen because R2 and R3 are in parallel and the Vs, and the Vs and R1 are in series.*

> ***The power in R1 remains the same because R2 has a lower voltage drop but is after R1 so the resistance across R2 doesn't affect R1.*** *Furthermore, the power across R3 drops because R3 and R2 are in parallel so the KCL says that as more current flows through R2 less flows through R3. The power across Vs increases because there is less resistance in the circuit as R2 decreases so there is more power because P=VI and the current across R2 increases V=IR.*

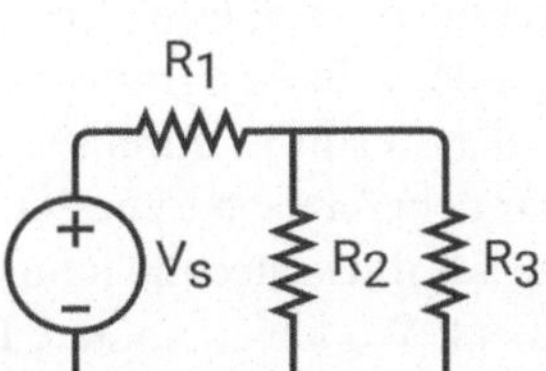

Consider the four-element circuit depicted below and argue what will happen to the power (increase, decrease, or remain the same) of each of the circuit's four elements when the resistance of resistor R_2 decreases. Treat all elements as ideal, including the independent voltage source and thoroughly justify your response.

FIGURE 5.1 The question of the writing quiz employed to collect student responses [6].

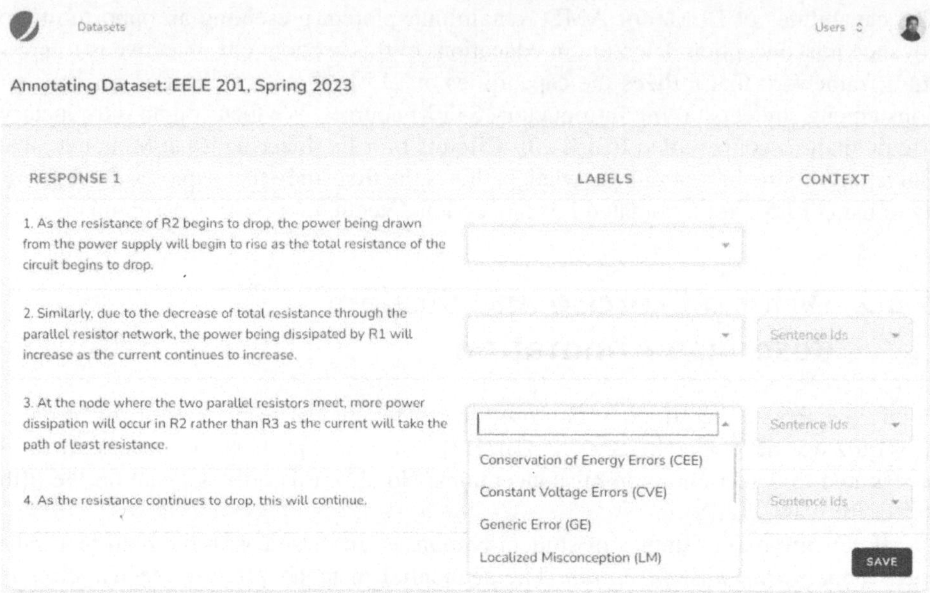

FIGURE 5.2 An overview of the annotation page featuring a response with four sentences. The label menu for the third sentence is open, displaying the available labels.

5.3.2.1 Annotation web app

As we commenced the annotation of new data, the need for an annotation tool to standardize the process became evident. With no other viable options available through the community, we dedicated our efforts to developing an in-house annotation web app (see Figure 5.2).

5.3.3 Datasets

Our curated dataset comprised over 300 responses annotated with seven misconceptions (see Table 5.7). These seven misconception label types are described elsewhere [6]. Each response underwent a meticulous review, addressing misspellings and potential typos, as misspelled words can introduce noise for most LLMs in the BERT series, impacting tokenization. Grammatical errors, if present, were retained, as altering them might compromise the originality of the responses. Training a model with data from one semester and testing it on responses from different semesters may result in unreliable outcomes. To address this, we store and annotate responses separately for each semester.

Figure 5.3 depicts the overall class label distribution. Table 5.8 illustrates the number of responses per label and their distribution across semesters. We have annotated the responses at the sentence level, and the distribution of sentences across labels and semesters is presented in Table 5.9. It is important to acknowledge that certain sentences carry multiple labels, with the count reflected under each applicable label. Notably, in

TABLE 5.7 Abbreviations and full titles of the misconception labels

ABBREVIATION	*FULL NAME*
CEE	Conservation of Energy Errors
CVE	Constant Voltage Errors
IIVSM	Ideal Independent Voltage Source Misconception
LM	Localized Misconception
PCM	Precedence of Current Misconception
RCE	Resistor Combination Errors
SM	Sequential Misconception

The full definitions of these seven misconception label types are provided elsewhere [6].

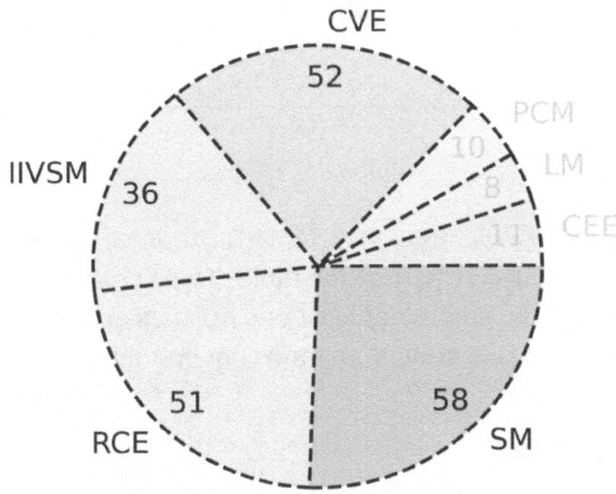

FIGURE 5.3 Overall class label distribution.

TABLE 5.8 Distribution of collected responses across eight semesters and seven misconceptions

MISCONCEPTIONS	*F19*	*S20*	*S21*	*F21*	*S22*	*F22*	*S23*	*F23*	*TOTAL*
Number of responses	57	46	26	42	27	47	26	42	313
Shows misconception	31	19	10	28	17	23	13	35	176
No misconception	26	27	16	14	10	24	13	7	137
CEE	0	0	2	2	2	2	0	3	11
CVE	9	8	4	8	7	6	3	7	52
IIVSM	8	6	1	6	1	6	3	5	36
LM	1	0	0	1	2	2	0	2	8
PCM	1	0	2	0	2	4	1	0	10
RCE	14	5	0	10	5	2	5	10	51
SM	7	4	4	9	4	11	5	14	58

Please note that the three rows (e.g., *Number of responses*) in the top section of the table are not sums of any numbers in that column since the responses are annotated at the sentence level with multiple labels.

TABLE 5.9 Distribution of sentences, in the collected responses, across eight semesters and seven misconceptions

MISCONCEPTIONS	*F19*	*S20*	*S21*	*F21*	*S22*	*F22*	*S23*	*F23*	*TOTAL*
Number of sentences	317	212	143	257	186	284	165	227	1,791
Shows misconception	44	29	16	41	28	37	19	76	290
No misconception	273	183	127	216	158	247	146	151	1,501
CEE	0	0	2	2	2	2	0	3	11
CVE	10	9	5	11	10	6	3	12	66
IIVSM	8	6	1	7	1	7	3	6	39
LM	1	0	0	1	2	2	0	3	9
PCM	1	0	2	0	3	4	1	0	11
RCE	17	9	0	11	5	2	6	11	61
SM	8	5	7	10	5	14	7	24	80

both tables, the *Number of responses* and *Shows misconception* rows do not represent any sums derived from the lower part of the table. The *Shows misconception* row indicates the count of samples presenting at least one misconception, while the *No misconception* row signifies samples devoid of any misconceptions.

5.3.3.1 Filtering data

Sentences from merely six responses across five semesters have been annotated with multiple labels. While detecting misconceptions inherently involves a multilabel classification, our dataset lacks the comprehensive information needed to address it as such. Given this limitation, our primary emphasis lies in modeling the data for multiclass classification. Recognizing the presence of a few responses with multiple labels, we have opted to exclude these six responses entirely from our training data.

Figure 5.4 illustrates a histogram of response distribution based on the number of sentences, with the majority containing three to five sentences. Figure 5.5 presents a histogram based on the number of tokens, revealing that the majority of responses contain around 105 tokens. To address the challenge of short responses while minimizing data loss, we set the token threshold at 60 and remove 25 responses. A few responses with less than three sentences met the token threshold and remain in the training data.

Upon finalizing the structure of the inputs (see Section 5.3.4.1), we were restricted to only 349 tokens for each response. Fortunately, only one response exceeded this limit with 384 tokens. We exclude it from our training data.

Out of the seven misconceptions used to annotate the dataset, there are only a handful of responses under CEE, LM, and PCM (refer to Table 5.8). These labels lack sufficient responses to be distributed across train, validation, and test sets. Therefore, we opt to exclude these labels from fine-tuning, retaining the responses without these labels.

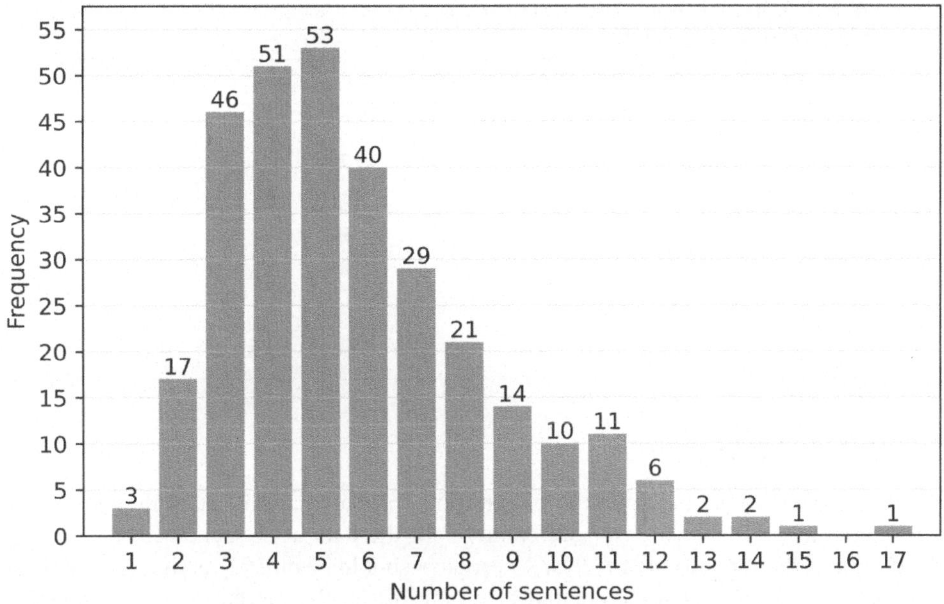

FIGURE 5.4 Distribution of responses based on the number of sentences.

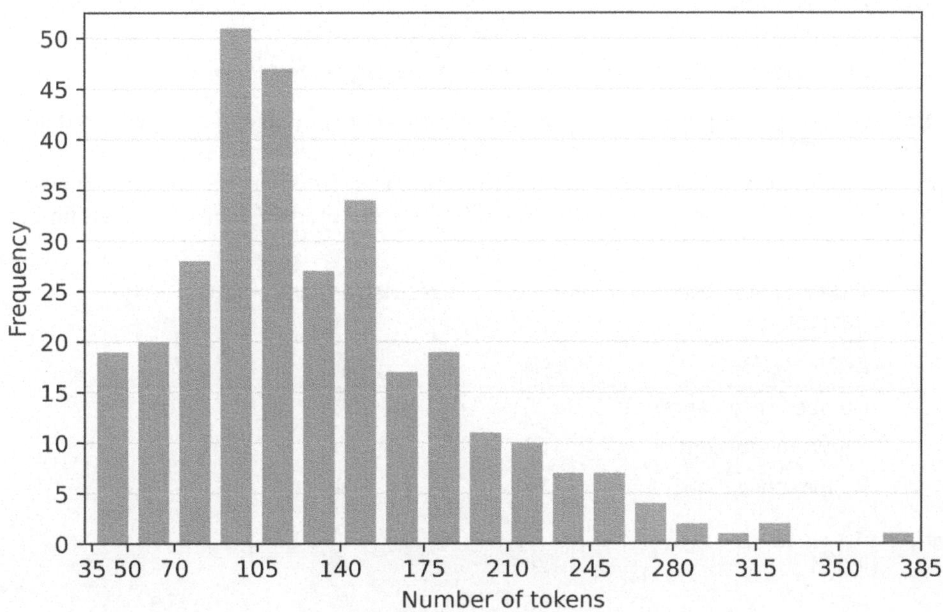

FIGURE 5.5 Distribution of responses based on the number of tokens.

TABLE 5.10 Distribution of responses across labels and semesters in the train, validation, and test sets

DATA SPLITS	*TRAIN*	*VALID*	*TEST*
Number of responses	211	35	35
Shows misconception	102	17	17
No misconception	109	18	18
CVE	39	6	5
IIVSM	22	5	4
RCE	31	6	6
SM	39	6	8

5.3.3.2 Splitting data

We divide the training data into three sets: train, validation, and test. Although we are fine-tuning at the sentence level, we cannot split the data at the sentence level due to the design of the input to the models. A response should be entirely assigned to at most one set. Since responses are annotated with multiple labels, we must consider the label distribution while splitting the data. The distribution of responses in train, validation, and test sets is depicted in Table 5.10. The distribution of sentences in train, validation, and test sets is depicted in Table 5.11.

5.3.4 Models

5.3.4.1 Fine-tuning approach and design decisions

Drawing from our experience with the ASAG, we conceptualize misconception detection as an RTE problem. In this framework, we map the response as a hypothesis and provide the requisite knowledge through the premise. In our case, the quiz question contains valuable information for the model, given the models' lack of background or

TABLE 5.11 Distribution of sentences across labels and semesters in the train set

DATA SPLITS	*TRAIN*	*VALID*	*TEST*
Number of sentences	1275	204	208
Shows misconception	159	30	30
No misconception	1116	174	178
CVE	48	8	7
IIVSM	24	5	4
RCE	36	7	8
SM	51	10	11

subject knowledge. However, the quiz question includes a circuit diagram, and we cannot input images into a text model. We extend the quiz question by describing the circuit diagram using words to address this issue. We augment the premise with a reference answer, following our established practice from the ASAG project. We concatenate the quiz question and the reference answer with a single space. The same premise is used for all inputs to the model.

The input size of LLMs is restricted by a predetermined number of tokens, established during pretraining. For this task, we utilize RoBERTa Large LLMs [9], which have a token limit of 512. Consequently, the combined length of the premise and hypothesis cannot exceed 509 tokens. This poses a challenge since the reference answer provided by the instructor is 352 tokens long. To overcome this limitation and ensure there is sufficient space for the hypothesis, we focus on condensing the token count in the premise. We address this challenge by manually summarizing both the question and the reference answer, resulting in 62 tokens (43 words) for the question and 98 tokens (76 words) for the reference answer, while preserving essential information. In summarizing the reference answer, we aimed to succinctly convey the changes in power and the reasons behind them using as few words as possible. This yields a premise of 160 tokens, leaving 349 tokens for the hypothesis. The summarized question and reference answer are provided below:

Question: Resistors R_1 and R_2 are connected in series with a voltage source Vs and resistor R_3 is connected in parallel to R_2. When the resistance of R_2 decreases, will the power on R_1, R_2, R_3, and Vs increase, decrease, or remain the same?

Reference Answer: As the resistance of R_2 decreases, the power of R_1 will increase since $P = I\^{}2R$ and the current flow increases. The power consumed by the Vs will increase, as it carries more current through the decreasing equivalent resistance. The power of R_3 will decrease because less voltage drops across the parallel combination of R_2 and R_3. As R_2 approaches zero resistance, the power of R_2 will eventually decrease to zero, as it will sustain no voltage drop.

Upon reviewing the annotations, we observed that in many cases, a sentence presents a misconception only in the context of a preceding sentence. Therefore, it is crucial to include the preceding sentences in the input. To keep the preceding sentences, separate from the classifying sentence, we concatenate them with a newline character. Since all the sentences in a response are collapsed into a single paragraph and have been segmented into sentences, a response cannot contain a newline character. Thus, we chose the newline character as glue in the hypothesis and hypothesized (due to the black-box nature of the models) that it would be recognized by the model as a separator. In the case of the first sentence of a response, we employed an empty string in place of the preceding sentences. In other words, the hypothesis begins with a newline character.

Due to the design of our input structure, any form of truncation would lead to the loss of information. Truncating the premise would jeopardize the reference answer.

Placing the classifying sentence at the beginning would compromise context and dependencies from the preceding sentences. Positioning the classifying sentence at the end would compromise the sentence itself. Consequently, we have disabled truncation.

5.3.4.2 Hyperparameters and tokenizer

We fine-tune RoBERTa Large MNLI using the Adam optimizer with $\epsilon = 1e - 8$, a learning rate of $1e - 5$, a weight decay of 0.01, and cross entropy loss function. The training is warmed up linearly over the first 10% of the data. The fine-tuning process lasts for a maximum of 20 epochs with early stopping. A batch size of five is utilized. Model performance is recorded after each epoch, and the best model is selected based on the epoch displaying the highest macro F1 score on the validation set.

The responses contain domain-specific words that may be unfamiliar to the model's tokenizer. Such words are segmented into smaller sub-word units recognizable by the tokenizer. However, when segmentation is not possible, the tokenizer marks these words as *unknown,* and they are subsequently ignored by the model. Given the substantial weight and semantic context carried by domain-specific words in this task, it is crucial to incorporate them into the tokenizer's vocabulary and the model's embeddings. To achieve this, we generate a list of all unique words present in the dataset. Subsequently, we compare this list with the tokenizer's vocabulary (case-sensitive) to identify words unknown to the model. We have identified and added the following 27 words to the model's vocabulary:

- G_x (denotes the conductance of resistor x): G_1, G_2, G_3
- I_x (denotes the current of Rx): I_1, I_2, I_3
- Ieq (unclear what this denotes, but assumed to be "equivalent current" or "total current")
- Kirchhoff
- KCL, KVL (Kirchhoff's current/voltage laws)
- P_x (denotes power associated with resistor x): P_1, P_2, P_3
- R_x (denotes resistor x or associated resistance): R_1, R_2, R_3
- R_{eff} (assumed to denote effective resistance)
- R_{eq} (assumed to denote equivalent resistance)
- R_n (unclear what this denotes)
- R_{tot}, Rtotal (denotes total resistance)
- V_x (denotes voltage associated with resistor x): V_1, V_2, V_3
- V_{out} (denotes output voltage)
- amperage, wattage

Remarkably, we discovered that many domain-specific words are already present in the model's vocabulary. Examples include R, V_s, W, amps, ohms, and current. R denotes resistor and Vs refers to a voltage source (i.e., battery) in the electrical engineering domain. Although these terms exist in the model's vocabulary, the specific meanings of words like R and V_s to the model remain unknown to us.

5.3.5 Results and Discussion

The comprehensive performance of our models is illustrated in Table 5.12. Initially, we fine-tune a RoBERTa Large MNLI model with input tokenized using the stock tokenizer (i.e., the tokenizer with the original vocabulary), codenamed ST (stock tokenizer). Subsequently, we fine-tune another RoBERTa Large MNLI model using our tokenizer with an extended vocabulary, referred to as ET (extended tokenizer). The ET model significantly outperforms the ST model in terms of recall (0.60 versus 0.55) and F1 score (0.60 versus 0.54). However, the ST model exhibits marginally better precision (0.64 versus 0.63) than the ET model. To validate the importance of preceding sentences as context, we fine-tune a RoBERTa Large MNLI model, codenamed LC (limited context), at the sentence level without incorporating the preceding sentences in the inputs. The ET model surpasses the LC model across all metrics, showcasing the significance of preceding sentences in providing valuable context and improving both precision (0.58 versus 0.63) and recall (0.55 versus 0.60). The epoch column denotes the training epoch at which the model achieves the highest F1 on the validation set. Notably, the ET model has a lower epoch than both ST and LC models and verifies the effectiveness of our design.

The performance of the ET model for each label is detailed in Table 5.13 and Figure 5.6. The ET model excels in detecting SM, boasting perfect precision (1.00) and a recall of 0.64. It achieves the highest recall on RCE (0.75) with the second-best precision (0.67) and F1 score (0.71). However, the model faces challenges in detecting CVE and IIVSM, attaining the same F1 score (0.29) for these labels with a higher recall on CVE (0.43 versus 0.23). A notable observation is that the model, despite having twice the sample size for CVE (48) compared to IIVSM (24) and only a few samples less than SM (51), achieves the lowest F1 on CVE. This suggests that detecting CVE in responses is inherently more complex than the other labels. Remarkably, the model achieves an F1 of 0.94 in detecting sentences with no misconceptions.

To gain valuable insights into the challenges and complexities of detecting misconceptions in general and specific instances, we fine-tune six binary models using RoBERTa Large MNLI, prefixed with BIN for binary, and the scores are presented in

TABLE 5.12 Performance of RoBERTa large MNLI multiclass models fine-tuned with three distinct configurations

		VALID			*TEST*		
MODEL	*EPOCH*	*P*	*R*	*F1*	*P*	*R*	*F1*
ST (Stock Tokenizer)	15	0.45	0.42	0.42	0.64	0.55	0.54
LC (Limited Context)	19	0.60	0.56	0.55	0.58	0.55	0.56
ET (Extended Tokenizer)	12	0.61	0.56	0.57	0.63	0.60	0.60

TABLE 5.13 Performance of the RoBERTa Large MNLI model with extended tokenizer (ET), the best multiclass model, with scores detailed per misconception label

	VALID			*TEST*		
LABEL	*P*	*R*	*F1*	*P*	*R*	*F1*
CVE	0.19	0.38	0.25	0.21	0.43	0.29
IIVSM	0.40	0.40	0.40	0.33	0.25	0.29
RCE	0.83	0.71	0.77	0.67	0.75	0.71
SM	0.67	0.40	0.50	1.00	0.64	0.78
No Misconception	0.94	0.93	0.93	0.95	0.93	0.94
Macro Average	0.61	0.56	0.57	0.63	0.60	0.60
Weighted Average	0.88	0.86	0.87	0.90	0.88	0.89

Table 5.14 and Figure 5.6. First, we fine-tune a binary model, codenamed BIN RES, to detect the presence of any misconceptions at the response level. The BIN RES model exhibits higher scores on each metric, notably a significantly higher recall (0.71 versus 0.60) than ET. This suggests that detecting misconceptions, in general, might share common patterns and supports our choice of approaching misconception detection as an RTE problem. Second, we fine-tune another binary model, codenamed BIN SENT, to investigate how the challenge of detecting the presence of misconceptions changes when transitioning from the response to the sentence level. The BIN SENT model achieves a lower recall (0.67 versus 0.71) compared to the BIN RES model while maintaining the same precision. This indicates that detecting misconceptions at the sentence level

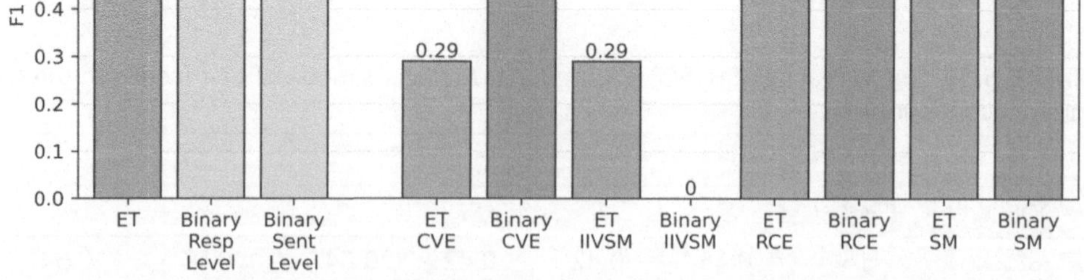

FIGURE 5.6 Performance of the RoBERTa Large MNLI model with ET, with scores detailed per misconception label (ET CVE, ET IIVSM, ET RCE, ET SM) compared to the performance of two RoBERTa Large MNLI binary models fine-tuned at the response (Binary Resp Level) and sentence levels (Binary Sent Level), and four models fine-tuned for each misconception label (Binary CVE, Binary IIVSM, Binary RCE, Binary SM).

TABLE 5.14 Performance of two RoBERTa Large MNLI binary models fine-tuned at the response and sentence levels, and four models fine-tuned for each misconception label

		VALID			*TEST*		
MODEL	*EPOCH*	*P*	*R*	*F1*	*P*	*R*	*F1*
BIN RES	3	0.86	0.71	0.78	0.67	0.71	0.69
BIN SENT	5	0.65	0.73	0.69	0.67	0.67	0.67
BIN CVE	7	0.43	0.38	0.40	0.50	0.57	0.53
BIN IIVSM	10	0.00	0.00	0.00	0.00	0.00	0.00
BIN RCE	5	1.00	0.71	0.83	0.86	0.75	0.80
BIN SM	4	0.73	0.80	0.76	0.73	0.73	0.73

is harder than at the response level. However, the BIN SENT model outperforms the ET model in all metrics, particularly in F1 (0.67 versus 0.60). This result further supports our hypothesis that common patterns across different misconception types make it easier to detect misconceptions in general.

Third, we fine-tune a binary model for each misconception label at the sentence level. These misconception-specific models are codenamed with the prefix BIN, followed by their corresponding label name. The BIN CVE model significantly outperforms the ET model in all metrics, particularly in precision (0.50 versus 0.21) and F1 (0.53 versus 0.29), for detecting CVE. Similarly, the BIN RCE model exceeds the ET model in precision (0.86 versus 0.67) and F1 (0.80 versus 0.71), while maintaining the same recall. However, the ET model surpasses the binary models in F1 for detecting IIVSM (0.29 versus 0.00) and SM (0.78 versus 0.73). Notably, the BIN IIVSM model labels all samples as misconception-free even after 10 epochs. Although extensive hyperparameter tuning may enhance the performance of BIN IIVSM, it remains unexplored given the low performance of the ET model on IIVSM. Interestingly, the binary models excel for two labels while underperforming for the other two, and this performance trend does not correlate with the sample size. The fact that the epoch of BIN CVE is higher than all other binary models and nearly double that of the BIN SM model aligns with the earlier observation suggesting that detecting CVE is inherently more challenging, pointing to its complexity as a misconception.

Recognizing the strengths and weaknesses of the models, we develop an ensemble model by combining the ET model with multiple binary models. BIN CVE, BIN RCE, and BIN SM outperform the ET model in F1 on the validation set. Consequently, we use these three binary models to predict their corresponding labels, while the ET model is used for IIVSM. We apply a misconception label to a sample if the sample is predicted for misconception by the BIN SENT model. The performance of the ensemble model is outlined in Table 5.15 and Figure 5.7. The ensemble model achieves higher F1 scores for all labels, with a notable increase for CVE (0.50 versus 0.29). The precision for SM decreased (0.89 versus 1.00), whereas the recall has increased (0.73 versus 0.64). Notably, the precision for RCE (1.00 versus 0.67) and IIVSM (0.50 versus 0.33) have also increased significantly. Overall, the recall remains unchanged while the precision increases significantly (0.78 versus 0.63) contributing to a higher F1 score (0.67 versus 0.60).

TABLE 5.15 Performance of RoBERTa Large MNLI ensemble model with scores outlined per misconception label

	VALID			*TEST*		
LABEL	*P*	*R*	*F1*	*P*	*R*	*F1*
CVE	0.50	0.12	0.20	0.60	0.43	0.50
	0.31 ↑	0.26 ↓	0.05 ↓	0.39 ↑		0.21 ↑
IIVSM	0.40	0.40	0.40	0.50	0.25	0.33
				0.17 ↑		0.04 ↑
RCE	1.00	0.71	0.83	1.00	0.62	0.77
	0.17 ↑		0.06 ↑	0.33 ↑	0.13 ↓	0.06 ↑
SM	0.86	0.60	0.71	0.89	0.73	0.80
	0.19 ↑	0.20 ↑	0.21 ↑	0.11 ↓	0.09 ↑	0.02 ↑
No Misconception	0.92	0.98	0.95	0.94	0.98	0.96
	0.02 ↓	0.05 ↑	0.02 ↑	0.01 ↓	0.05 ↑	0.02 ↑
Macro Average	0.74	0.56	0.62	0.78	0.60	0.67
	0.13 ↑		0.05 ↑	0.15 ↑		0.07 ↑
Weighted Average	0.89	0.91	0.89	0.92	0.92	0.92
	0.01 ↑	0.05 ↑	0.02 ↑	0.02 ↑	0.04 ↑	0.03 ↑

The second row within each label indicates the difference in performance compared to the ET model. The up arrows signify an increase in performance, while the down arrows indicate a decrease in performance.

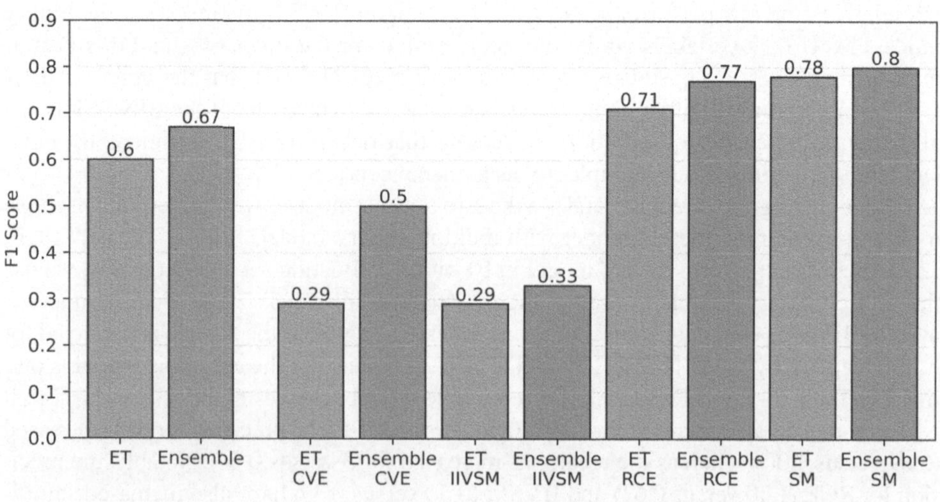

FIGURE 5.7 Performance of RoBERTa Large MNLI ensemble model compared with the ET mode. In addition, scores per misconception label is depicted for both models.

5.3.6 Summary and Open Problems

Our study dives into the task of misconceptions detection in student responses, employing state-of-the-art language models like RoBERTa Large MNLI. Formulating the task as recognizing textual entailment by incorporating the question along with a reference answer as premise and the student response as hypothesis demonstrates notable advancements in the detection of specific misconceptions. While the ensemble model exhibits moderate precision and F1 scores with considerable recall, indicating its proficiency in capturing misconceptions, it is crucial to acknowledge the challenges posed by certain complex misconceptions.

Our exploration extends beyond the binary classification approach that is often regarded easier, embracing a multiclass classification framework. The results underscore the intricacies involved in developing the dataset and fine-tuning models for diverse misconceptions, with varying levels of complexity and prevalence. Our study lays the groundwork for investigating the impact of sample size on model performance. Future work could explore deeper into understanding how sample distribution and size influence the efficacy of misconception detection, potentially leading to insights that inform data collection strategies.

As we navigate the evolving landscape of misconception detection, several other avenues for future research emerge. First, to adhere to the token limitation of the model, we had to minimize the size of both the quiz question and the reference answer, sacrificing valuable information. Exploring LLMs, such as T5, PaLM, and Megatron-Turing NLG, that have a higher token limit and surpass RoBERTa on RTE could be a promising avenue for future research. T5 boasts a token limit of 4,096 tokens, while both PaLM and Megatron-Turing NLG further extend the limit to 8,000 tokens, providing an excellent opportunity to enrich the context through premise. Second, the observed challenges in detecting complex misconceptions highlight the need for more sophisticated strategies and thorough model design. Third, we perform shallow hyperparameter tuning on RoBERTa Large MNLI and extensive hyperparameter tuning may further improve the overall performance or on specific misconceptions. Fourth, misconception detection is by default a multilabel classification problem that is not explored due to the lack of sufficient data. Fifth, we have excluded three misconceptions from our training data owing to their low sample counts, which can be a part of future research. In essence, our study opens the door to a realm of possibilities in the pursuit of enhancing misconceptions detection in student responses.

ACKNOWLEDGMENTS

We are grateful to the National Research Platform for providing access to the Nautilus HyperCluster, which was used for the computations presented in this chapter. A portion of this chapter is based upon work supported by the National Science Foundation under Grant IUSE-2120466. AI-based writing assistance tools were used for improving the content that was originally generated by the authors.

NOTE

1. https://huggingface.co/datasets/nkazi/SciEntsBank

REFERENCES

1. X. Zhu, H. Wu, and L. Zhang, "Automatic short-answer grading via BERT-based deep neural networks," *IEEE Trans. Learn. Technol.*, vol. 15, no. 3, pp. 364–375, 2022.
2. O. L. Liu, H.-S. Lee, C. Hofstetter, and M. C. Linn, "Assessing knowledge integration in science: Construct, measures, and evidence," *Educ. Assess.*, vol. 13, no. 1, pp. 33–55, 2008.
3. S. Bonthu, S. R. Sree, and M. K. Prasad, "Automated short answer grading using deep learning: A survey," in *Machine Learning and Knowledge Extraction: 5th IFIP TC 5, TC 12, WG 8.4, WG 8.9, WG 12.9 International Cross-Domain Conference, CD-MAKE 2021, Virtual Event, August 17–20, 2021, Proceedings 5*, pp. 61–78, 2021. Springer.
4. D. K. Pugalee, *Writing to develop mathematical understanding.* Christopher-Gordon, 2005.
5. S. Kumar, S. Chakrabarti, and S. Roy, "Earth mover's distance pooling over Siamese LSTMs for automatic short answer grading," in *IJCAI*, pp. 2046–2052, 2017.
6. J. P. Becker, I. Kahanda, and N. H. Kazi, "WIP: Detection of student misconceptions of electrical circuit concepts in a short answer question using NLP," in *2021 ASEE Virtual Annual Conference Content Access*, 2021.
7. M. O. Dzikovska, R. D. Nielsen, C. Brew, C. Leacock, D. Giampiccolo, L. Bentivogli, P. Clark, I. Dagan, and H. T. Dang, "Semeval-2013 task 7: The joint student response analysis and 8th recognizing textual entailment challenge," *Technical report, University of North Texas*, 2013.
8. J. Devlin, M.-W. Chang, K. Lee, and K. Toutanova, "BERT: Pre-training of deep bidirectional transformers for language understanding," in *Proceedings of the 2019 Conference of the North American Chapter of the Association for Computational Linguistics: Human Language Technologies, Volume 1*, pp. 4171–4186, Minneapolis, Minnesota, June 2019. Association for Computational Linguistics.
9. Y. Liu, M. Ott, N. Goyal, J. Du, M. Joshi, D. Chen, O. Levy, M. Lewis, L. Zettlemoyer, and V. Stoyanov, "RoBERTa: A robustly optimized BERT pretraining approach," *arXiv*, 2019.
10. C. Sung, T. I. Dhamecha, and N. Mukhi, "Improving short answer grading using Transformer-based pre-training," in *Artificial Intelligence in Education: 20th International Conference, AIED 2019, Chicago, IL, USA, June 25-29, 2019*, Proceedings, *Part* I *20*, pp. 469–481, 2019. Springer.
11. L. Camus, and A. Filighera, "Investigating transformers for automatic short answer grading," in *Artificial Intelligence in Education: 21st International Conference, AIED 2020, Ifrane, Morocco, July 6–10, 2020, Proceedings, Part II 21*, pp. 43–48, 2020. Springer.
12. A. Williams, N. Nangia, and S. Bowman, "A broad-coverage challenge corpus for sentence understanding through inference," in *Proceedings of the 2018 Conference of the North American Chapter of the Association for Computational Linguistics: Human Language Technologies, Volume 1 (Long Papers)*, pp. 1112–1122. Association for Computational Linguistics, 2018.

13. M. O. Dzikovska, R. Nielsen, and C. Brew, "Towards effective tutorial feedback for explanation questions: A dataset and baselines," in *Proceedings of the 2012 Conference of the North American Chapter of the Association for Computational Linguistics: Human Language Technologies*, pp. 200–210, 2012.
14. S. R. Bowman, G. Angeli, C. Potts, and C. D. Manning, "A large annotated corpus for learning natural language inference," in *Proceedings of the 2015 Conference on Empirical Methods in Natural Language Processing*, pp. 632–642, Lisbon, Portugal, September 2015. Association for Computational Linguistics.
15. J. P. Becker, and C. Plumb, "Identifying at-risk students in a basic electric circuits course using instruments to probe students' conceptual understanding," in *2018 ASEE Annual Conference Proceedings*, 2018.
16. H.-J. Schmidt, "Students' misconceptions—Looking for a pattern," *Sci. Educ.*, vol. 81, no. 2, pp. 123–135, 1997.
17. J. P. Becker, E. Sior, J. Hoy, and I. Kahanda "Board 11: Predicting at-risk students in a circuit analysis course using supervised machine learning," in *2019 ASEE Annual Conference & Exposition*, 2019.
18. J. Michael, "Misconceptions—What students think they know," *Adv. Physiol. Educ.*, vol. 26, no. 1, pp. 5–6, 2002.
19. A. Stevens, A. Collins, and S. E. Goldin, "Misconceptions in student's understanding," *Int. J. Man-Mach. Stud.*, vol. 11, no. 1, pp. 145–156, 1979.
20. A. J. Rossman, B. Chance, and C. P. S. L. Obispo, "Anticipating and addressing student misconceptions," in *ARTIST Conference on Assessment in Statistics, Lawrence University*, pp. 1–4, 2004.
21. C. Mulryan-Kyne, "Teaching large classes at college and university level: Challenges and opportunities," *Teach. High. Educ.*, vol. 15, no. 2, pp. 175–185, 2010.
22. H. Danielsiek, W. Paul, and J. Vahrenhold, "Detecting and understanding students' misconceptions related to algorithms and data structures," in *Proceedings of the 43rd ACM technical symposium on Computer Science Education*, pp. 21–26, 2012.
23. P. Britos, E. J. Rey, D. Rodríguez, and R. García-Martínez, "Work in progress-programming misunderstandings discovering process based on intelligent data mining tools," in *2008 38th Annual Frontiers in Education Conference*, pp. F4H–1, 2008. IEEE.
24. J. J. Michalenko, A. S. Lan, and R. G. Baraniuk, "Data-mining textual responses to uncover misconception patterns," in *Proceedings of the Fourth (2017) ACM Conference on Learning @ Scale, L@S '17*, pp. 245–248, New York, NY, USA, 2017. Association for Computing Machinery.
25. Y. Yang, H. Ye, J. Deng, C. You, W. Chen, Y. Zhang, and X. Xiong, "Diagnosis of misconception in electronic circuits engineering courses: A case study at UESTC, China," in *2019 IEEE International Conference on Engineering, Technology and Education (TALE)*, pp. 1–10, 2019. IEEE.
26. C. A. Arbogast, and D. Montfort, "Applying natural language processing techniques to an assessment of student conceptual understanding," in *2016 ASEE Annual Conference & Exposition*, 2016.
27. M. Elmadani, M. Mathews, and A. Mitrovic, "Data-driven misconception discovery in constraint-based intelligent tutoring systems," in *20th International Conference on Computers in Education (ICCE)*, 2012.
28. A. Goncher, W. Boles, and D. Jayalath, "Using automated text analysis to evaluate students' conceptual understanding," in *Proceedings of the 2014 Annual Conference of the Australasian Association for Engineering Education (AAEE2014)*, pp. 1–9. Massey University, 2014.

Word Class and Syntax Rule Representations Spontaneously Emerge in Recurrent Language Models

6

Patrick Krauss, Kishore Surendra, Paul Stoewer, Andreas Maier, Claus Metzner, and Achim Schilling

6.1 INTRODUCTION

The question of how humans come to language is one of the oldest scientific problems [1]. According to the Greek historian Herodotus, already 2,500 years ago the Egyptian pharaoh Psamtik sought to discover the origin of language. Therefore, he conducted an experiment with two children whom he gave as newborn babies to a shepherd who should feed and care for them, but had the instruction not to speak to them. Psamtik hypothesized that the infants' first word would be uttered in the root language of all people. Consequently, as one of the children cried "bekos," which was the sound of the Phrygian word for "bread," Psamtik concluded that Phrygian was the root language

DOI: 10.1201/9781003570882-8

of all humans [2]. Obviously, the assumption behind this cruel language deprivation experiment was that humans are born with innate words and their meanings, and that this root language is somehow "overruled" during individual development and first language learning.

Nowadays, it is of course clear that words and meanings are not innate but rather learned during language acquisition [3], and that there is no causal relation between the signifier (sound pattern) and the signified (meaning) [4]. However, it is still highly debated to what extent language capacities are innate or must be learned.

According to Chomsky's theory of universal grammar, humans have an innate, genetically determined language faculty that distinguishes between different word classes such as nouns and verbs, making it easier and faster for children to learn to speak [5–7]. In contrast, in cognitive linguistics and usage-based approaches, a profound relationship between language structure and language use is assumed [8–11]. In particular, contextual mental processing and mental representations are assumed to have the cognitive capacity to capture the complexity of actual language use at all levels [12–17]. According to Diessel, grammar is a "dynamic system of emergent structures" and it needs to be explained "how linguistic structures evolve" during language acquisition [18].

Predictive coding and processing are thought to be canonical computations of the human brain [19–22], in particular during speech and language processing, which involves the prediction of which words come next [23–29]. In previous studies, it has already been demonstrated that efficient successor representations to form cognitive maps of space and language can be learned by artificial neural networks [30–33]. In particular, it has been demonstrated how a neural network model can infer the underlying word classes of a simplified artificial language just by observing sequences of words (i.e., sentences) and without any prior knowledge about actual word classes or grammar. The emerging representations share important properties with network-like cognitive maps, enabling navigation in arbitrary abstract and conceptual spaces, and thereby broadly supporting domain- general cognition, as proposed by Bellmund et al. [34]. In a further study, we addressed the question if abstract linguistic categories and structures (i.e., representations of word classes) can be learned from experienced language alone in a more complex and naturalistic linguistic task (i.e., word prediction) in a natural language scenario [35].

The present study constitutes an extension of the aforementioned study. In particular, we trained the recurrent neural networks (RNNs) on three different tasks: First, prediction of the next word given a sequence of nine consecutive words as input from a German novel (see [35]). Secondly, in order to test if the results are independent from the particular language, we repeated the experiments using an English novel. Finally, in a third experiment, we predicted the next two words, using the German novel again. Subsequently, we analyzed the emerging activation patterns in the hidden layers of the deep RNN. Strikingly, we find that the internal representations of nine-word input sequences cluster according to the word class of the tenth word to be predicted as output in both languages, even though the neural network did not receive any explicit information about word classes during training. Furthermore, in the RNN trained on prediction of the next two words, we find that the internal representations of nine-word input sequences cluster according to the word class combinations of the tenth and eleventh word to be predicted as output.

These surprising results suggests that also in the human brain abstract representational categories such as word classes or syntax rules (word class combinations) may naturally emerge as a consequence of predictive coding and processing of language input. Based on these findings, we hypothesize that during language acquisition – which at least partly corresponds to learn to predict which words or utterances come next – word classes and grammatical rules spontaneously emerge as clusters of successor representations of perceived utterances. We conclude that word classes and syntax rules need not be innate to enable efficient language acquisition. Thus, our findings contradict the idea of the necessity of an innate "universal grammar."

6.2 METHODS

6.2.1 Data Preprocessing

The German novel *Gut gegen Nordwind* by Danial Glattauer and the English novel *Hitchhiker's Guide to the Galaxy* by Douglas Adams served as natural language text data for training and testing our RNNs. Both text data consist of a total number of approximately 40,000 tokens and 6,000 types. Prior to further analysis, punctuation and special characters have been removed from the text corpus. Furthermore, repetitive words and extra white spaces have been removed, and all numbers have been replaced by a single word (cf. Table 6.1). All words are converted to lowercase to maintain uniformity, so that the same word occurring in a different case is considered as two tokens of the same type, instead of two different types. All words have been encoded as 384-dimensional word vectors using the word2vec embedding function from the Python library *spaCy* [36]. Sequences of nine consecutive word vectors served as input, while one (the tenth) or two (tenth and eleventh) word vectors served as corresponding outputs. Finally, all word vector sequences were split into a training (80%) and a test dataset (20%).

TABLE 6.1 Data cleaning

CHARACTER/WORD	*OPERATION*
Repetitive words: RE:,AW:,Eine,Zwei,..,Stunden,Sekunden, Stunden...,später,Am nächsten,Kein Betreff,Betreff	Remove completely
Punctuation and other characters:.,',.,?,%,&,',',!	Remove completely
Numbers: 18,1,500,...	Replace with "nummer"
Extra whitespaces	Replace with single space
E-mail	Replace with "email"

Words, characters, and their replacements during data cleaning.

6.2.2 Recurrent Neural Network Architecture and Training Procedure

For the tasks at hand (i.e., to predict the tenth word [or the tenth and the eleventh word], given a prior sequence of nine words occurring in the corpus), RNNs are perfectly suited. Here, we implemented a neural network consisting of four bidirectional LSTM (long short-term memories) layers (with 128, 128, 64, and 64 neurons) followed by a flatten layer and a dense output layer (384 neurons). The input consisted of sequences of nine 384-dimensional word embedding vectors generated as described above. The expected output is a single (or a sequence of two) 384-dimensional word embedding vectors. Weights were initialized using the Glorot uniform initialization, which is Keras's default initializer. As an optimizer, we used Adam, with a learning rate of 0.001; as loss function, we used the mean-squared error and trained the network for 100 epochs.

6.2.3 Word Classes

Word classes were analyzed by applying *part-of-speech (POS) tagging* [37–39], as implemented in the Python library *spaCy* [36]. The used POS tags comprised the following 13 default word classes: "NUM", "VERB", "ADJ", "X", "PART", "NOUN", "SCONJ", "ADP", "DET", "PRON", "CONJ", "AUX", and "ADV". Their exact definitions can be found in [40]. Note that during training we did not provide any information about word classes as input, but exclusively provided sequences of word2vec embeddings, which only comprise semantic features.

6.2.4 Multidimensional Scaling

A frequently used method to generate low-dimensional embeddings of high-dimensional data is t-distributed stochastic neighbor embedding (t-SNE) [41]. However, in t-SNE the resulting low- dimensional projections can be highly dependent on the detailed parameter settings [42], sensitive to noise, and may not preserve, but rather often scramble the global structure in data [43, 44]. In contrast to that, multidimensional scaling (MDS) [45–51] is an efficient embedding technique to visualize high-dimensional point clouds by projecting them onto a two-dimensional plane. Furthermore, MDS has the decisive advantage that it is parameter-free and all mutual distances of the points are preserved, thereby conserving both the global and local structure of the underlying data.

When interpreting patterns as points in high-dimensional space and dissimilarities between patterns as distances between corresponding points, MDS is an elegant method to visualize high-dimensional data. By color-coding each projected data point of a dataset according to its label, the representation of the data can be visualized as a set of point clusters. For instance, MDS has already been applied to visualize for instance word class distributions of different linguistic corpora [23], hidden layer representations (embeddings) of artificial neural networks [52, 53], structure and dynamics of highly recurrent neural networks [54–57], or brain activity patterns assessed during pure tone or speech perception [23, 58] or even during sleep [50, 51, 59, 60]. In all these cases, the apparent compactness and mutual overlap of the point clusters permit a qualitative assessment of how well the different classes separate.

6.2.5 Generalized Discrimination Value

We used the generalized discrimination value (GDV) to calculate cluster separability, as published and explained in detail in [52]. Briefly, we consider N points $\mathbf{x}_{\mathbf{n=1.N}} = (x_{n,1}, \cdots, x_{n,D})$, distributed within D-dimensional space. A label l_n assigns each point to one of L distinct classes $C_{l=1.L}$. In order to become invariant against scaling and translation, each dimension is separately z-scored and, for later convenience, multiplied with ½:

$$s_{n,d} = \frac{1}{2} \cdot \frac{x_{n,d} - \mu_d}{\sigma_d}. \tag{6.1}$$

Here, $\mu_d = \frac{1}{N} \sum_{n=1}^{N} x_{n,d}$ denotes the mean

and $\sigma_d = \sqrt{\frac{1}{N} \sum_{n=1}^{N} (x_{n,d} - \mu_d)^2}$ the standard deviation of dimension d.

Based on the rescaled data points $\mathbf{s_n} = (s_{n,1}, \cdots, s_{n,D})$, we calculate the *mean intra-class dis- tances* for each class C_l:

$$\bar{d}(C_l) = \frac{2}{N_l(N_l - 1)} \sum_{i=1}^{N_l - 1} \sum_{j=i+1}^{N_l} d(s_i^{(l)}, s_j^{(l)}), \tag{6.2}$$

and the *mean inter-class distances* for each pair of classes C_l and C_m:

$$\bar{d}(C_l, C_m) = \frac{1}{N_l N_m} \sum_{i=1}^{N_l} \sum_{j=1}^{N_m} d(s_i^{(l)}, s_j^{(m)}). \tag{6.3}$$

Here, N_k is the number of points in class k, and $\mathbf{s}^{(k)}$ is the ith point of class k. The quantity $d(\mathbf{a}, \mathbf{b})$ is the Euclidean distance between $\mathbf{a}$ and $\mathbf{b}$. Finally, the GDV is calculated from the mean intra-class and inter-class distances as follows:

$$\mathrm{GDV} = \frac{1}{\sqrt{D}} \left[\frac{1}{L} \sum_{l=1}^{L} \bar{d}(C_l) - \frac{2}{L(L-1)} \sum_{l=1}^{L-1} \sum_{m=l+1}^{L} \bar{d}(C_l, C_m) \right] \tag{6.4}$$

whereas the factor $\frac{1}{\sqrt{D}}$ is introduced for dimensionality invariance of the GDV with D as the number of dimensions.

Note that the GDV is invariant with respect to a global scaling or shifting of the data (due to the z-scoring), and also invariant with respect to a permutation of the components in the N-dimensional data vectors (because the Euclidean distance measure has this symmetry). The GDV is zero for completely overlapping, nonseparated clusters, and it becomes more negative as the separation increases. A GDV of –1 signifies already a very strong separation.

6.2.6 Code Implementation

The models were coded in Python. The neural networks were designed using the Keras [61] and Keras-RL [62] libraries. Mathematical operations were performed with numpy [63] and scikit-learn [64] libraries. Visualizations were realized with matplotlib [65] and networkX [66]. For natural language processing (NLP), we used SpaCy [36].

6.3 RESULTS

6.3.1 Next Word Prediction

We trained a recurrent neural network on next word prediction using sequences of nine consecutive word vectors as input from a German novel, as described in detail in the Methods section. The trained network was tested with sequences of nine words not used for training. The resulting neural activation of each layer was read out and the corresponding activation vectors were projected onto a two-dimensional plane using MDS. All projected points were then color-coded according to the word class of the subsequent word of the corresponding input sequence. Word classes were assessed after training using POS tagging and did not serve as input during training. While layer 1 shows a random distribution of the data points (Figure 6.1), we find a remarkably strong clustering according to world classes in the last layer of the neural network (Figure 6.2). This is also confirmed by the corresponding GDV curve across the layers (Figure 6.3). This means that the neural network organizes its internal representations of input word sequences according to the word class of the next word to be predicted (see also [35]). For the English novel, we find very similar results; again, there is a remarkably strong clustering in the last layer according to the word class of the word to be predicted (Figure 6.4). This finding indicates that the described findings are a universal phenomenon independent from any particular language.

6.3.2 Next Two Words Prediction

In a follow-up experiment, we trained a recurrent neural network on next two words prediction using sequences of nine consecutive word vectors as input. The trained network was tested with sequences of nine words not used for training. The resulting neural activation pattern of each layer was read out and the corresponding activation vectors were projected onto a two-dimensional plane using MDS, analogously to the previous experiments. Since there would be $13 \times 13 = 169$ different word class combinations, and the difficulty of displaying 169 labels with unique colors, we restricted our analysis on the top 10 frequently occurring word class combinations. In particular, all projected points were color-coded according to the top 10 most frequently occurring word class combinations of the two subsequent words of the corresponding input sequence. Word classes were assessed after training using POS tagging and did not serve as input during training. As in

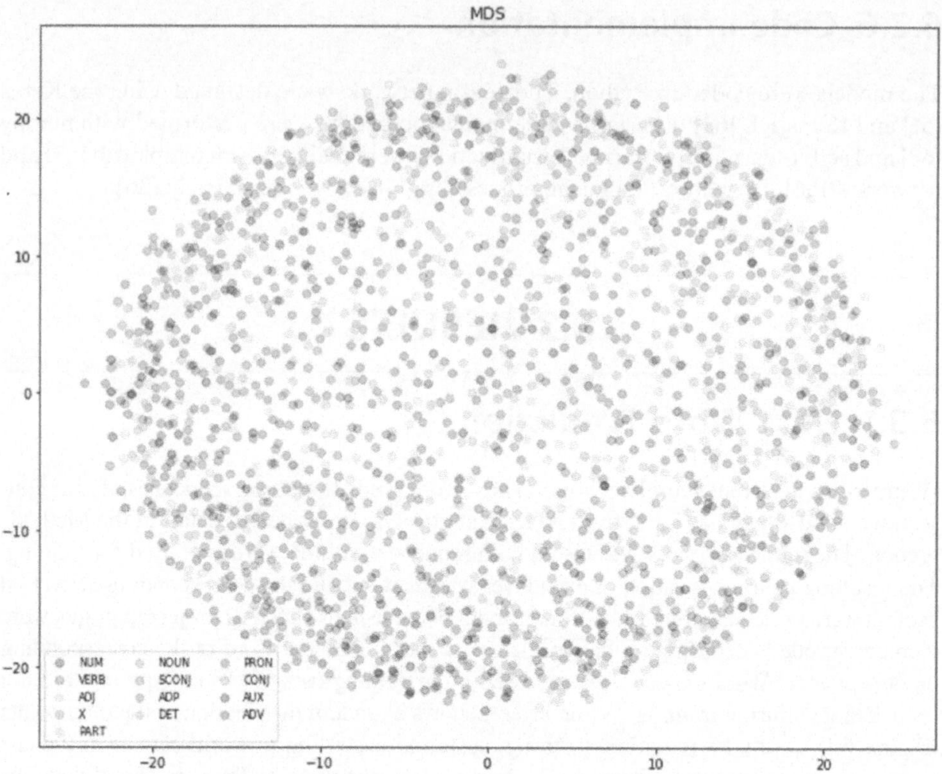

FIGURE 6.1 German novel. Layer 1 results of recurrent neural network testing and projection onto a two-dimensional plane using MDS, with color-coding according to subsequent word class. The points are randomly distributed. The POS tags comprised the following 13 default word classes: "NUM", "VERB", "ADJ", "X", "PART", "NOUN", "SCONJ", "ADP", "DET", "PRON", "CONJ", "AUX", and "ADV". (Note that the two axes and their scaling have no particular meaning other than illustrating the mutual distances between points, i.e., their dissimilarities.)

the single-word prediction task, we find a remarkably strong clustering according to world class combinations in the last layer of the neural network (Figure 6.5). This is again confirmed by the corresponding GDV curve across the layers (Figure 6.6). This means that the neural network organizes its internal representations of input word sequences according to the word class combinations (syntax rules) of the next two words to be predicted.

6.4 DISCUSSION

The results of our study provide evidence that abstract linguistic categories, such as word classes and syntax or grammar rules, can emerge spontaneously in neural

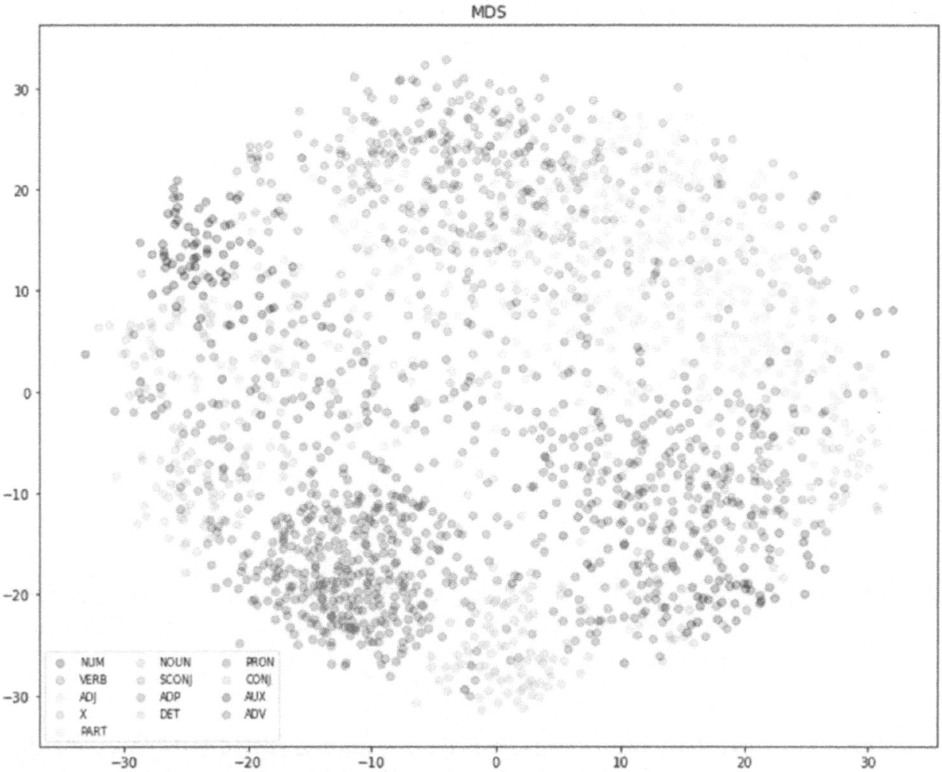

FIGURE 6.2 German novel. Last layer results of recurrent neural network testing and projection onto a two-dimensional plane using MDS, with color-coding according to subsequent word class. The final layer shows strong clustering by word class, indicating that the neural network organizes internal representations based on the predicted subsequent word's class. The used POS tags comprised the following 13 default word classes: "NUM", "VERB", "ADJ", "X", "PART", "NOUN", "SCONJ", "ADP", "DET", "PRON", "CONJ", "AUX", and "ADV". (Note that the two axes and their scaling have no particular meaning other than illustrating the mutual distances between points, i.e., their dissimilarities.)

representations of linguistic input. This finding challenges the notion that the ability to recognize and categorize words by their grammatical function is innate and hardwired in the human brain, as proposed by Chomsky's theory of universal grammar [5, 6, 67]. Our results suggest that language acquisition involves, at least in part, the learning of predictive structures and categories based on statistical regularities in the input, rather than relying solely on innate linguistic knowledge. This is consistent with the view that language is a complex adaptive system shaped by both biological and environmental factors [17, 68].

It is interesting to note that the clustering of input sequences by word classes of the successor word is evident only in the last layer of the neural network, suggesting that the network may gradually learn and refine more abstract and complex features of language

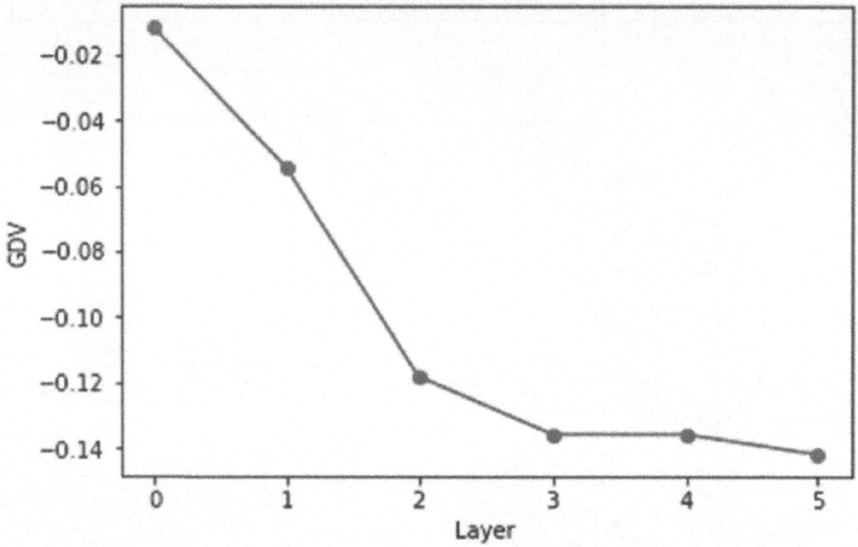

FIGURE 6.3 GDV curve across layers of the RNN. The decline of the GDV indicates that the RNN has learned to cluster internal representations according to the subsequent word's class with increasingly strong clustering from input to output layer.

as information flows through its layers. Firstly, this finding provides evidence that the clustering according to word classes and syntax rules is not an intrinsic effect already caused by producing word embeddings but rather emerges during the training process of the recurrent neural networks. Thus, word embeddings mainly cluster according to semantic similarities (see [69–72]) instead of syntactic structures. Secondly, this finding is consistent with the hierarchical NLP, in which higher-level representations build on lower-level representations [73]. The fact that we found clustering according to syntactic structures in later layers of the RNNs using a German as well as an English novel and extended the experiments by predicting the next two words emphasizes the universality of these findings.

One potential application of our findings is in NLP, where understanding the organization of neural representations of language input can help improve language modeling, machine translation, and other related tasks. In addition, our study provides a starting point for further investigations into the neural mechanisms underlying language acquisition and processing. In conclusion, our study provides compelling evidence that neural networks can spontaneously learn to organize their internal representations of language input according to abstract linguistic categories such as word classes. Our results support the view that language acquisition is a complex and dynamic process that relies on both innate mechanisms and statistical learning from environmental input.

Large language models (LLMs), such as ChatGPT, GPT4, Bard, and others, use statistical learning to predict language patterns, mirroring the study's findings that linguistic categorization can emerge from exposure to language input rather than solely

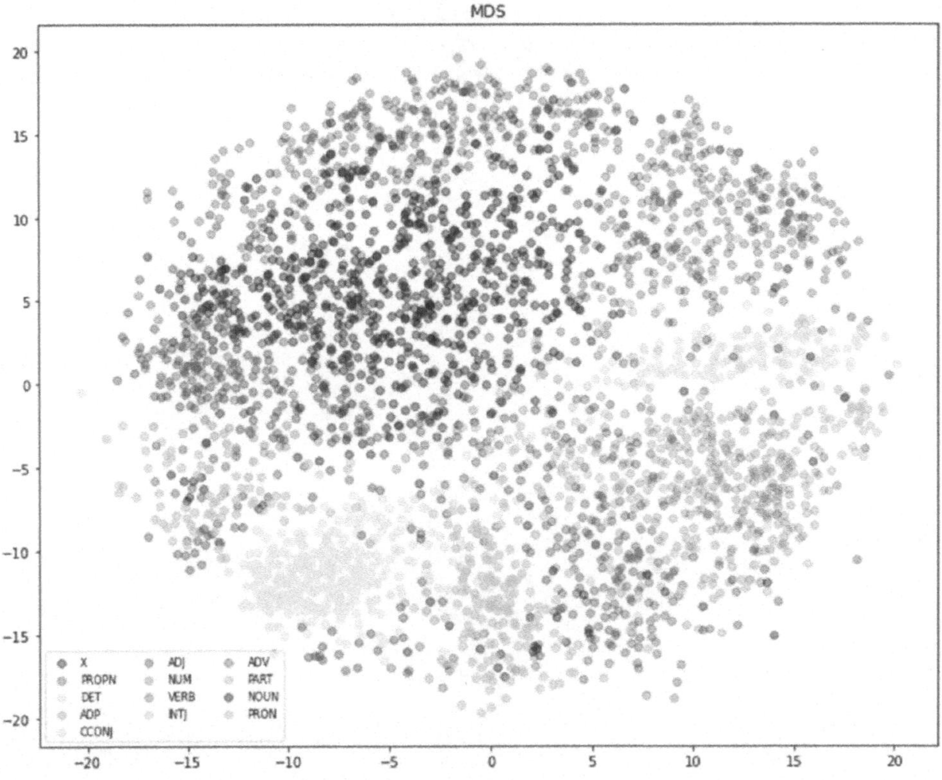

FIGURE 6.4 English novel. Last layer results of recurrent neural network testing and projection onto a two-dimensional plane using MDS, with color coding according to subsequent word class. The final layer shows strong clustering by word class, indicating that the neural network organizes internal representations based on the predicted subsequent word's class. The used POS tags comprised the following 13 default word classes: "NUM", "VERB", "ADJ", "X", "PART", "NOUN", "SCONJ", "ADP", "DET", "PRON", "CONJ", "AUX", and "ADV". (Note that the two axes and their scaling have no particular meaning other than illustrating the mutual distances between points, i.e., their dissimilarities.)

from innate abilities [74–76]. These models' multilayered architecture reflects the hierarchical nature of language processing, with lower-level syntax and higher-level semantic features. Insights into these models' operations can enhance fields like NLP and machine translation, and they can also serve as tools for exploring language acquisition and processing mechanisms in the human brain [77]. Therefore, representations of the LLM networks have to be compared to neuroimaging data recorded during the presentation of continuous speech/language [23–27, 78]. We argue that using artificial neural networks as model systems and tools for neuroscience [53, 57, 79–83] can boost research on information processing in the brain on the one hand [84–86], and that a deeper understanding of brain mechanisms can help to further improve artificial intelligence (AI) systems on the other hand [87–88].

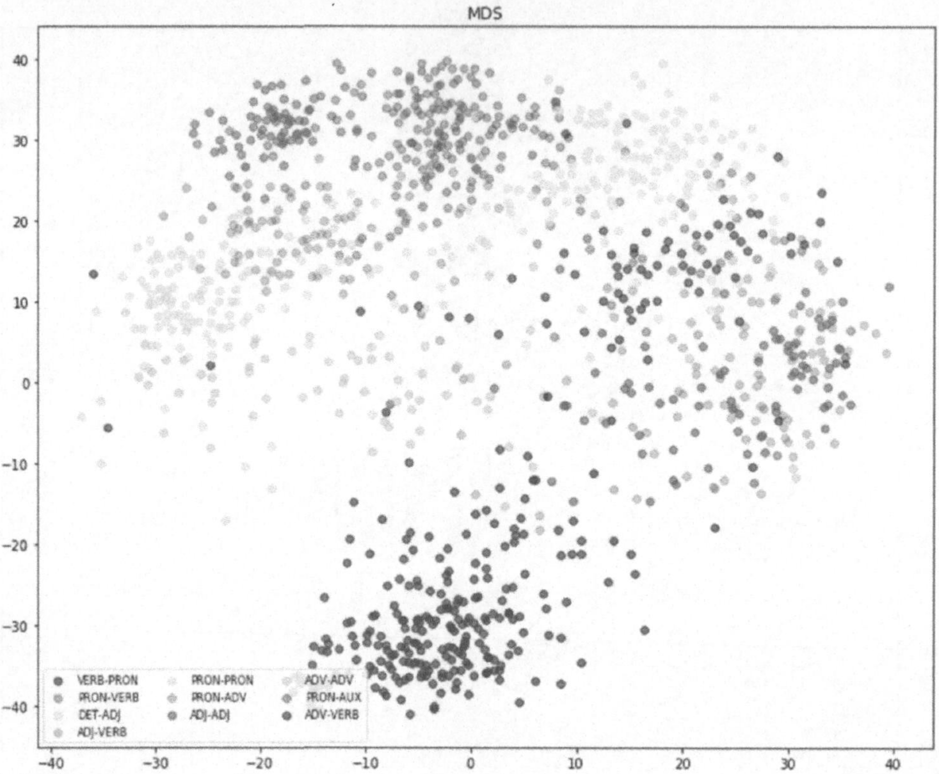

FIGURE 6.5 Last layer results of recurrent neural network testing and projection onto a two-dimensional plane using MDS, with color coding according to subsequent word class combinations. The final layer shows strong clustering by word class combination, indicating that the RNN organizes internal representations based on the predicted subsequent word combinations. The used POS tag combinations comprised the following 10 most frequently occurring word class combinations: "VERB-PRON", "PRON-VERB", "DET- ADJ", "ADJ-VERB", "PRON-PRON", "PRON-ADV", "ADJ-ADJ", "ADV-ADV", "PRON-AUX", and "ADV-VERB". (Note that the two axes and their scaling have no particular meaning other than illustrating the mutual distances between points, i.e., their dissimilarities.)

6.5 STATEMENT

This invited book chapter constitutes a substantial extension of our recently presented study at ICMLA 2023 [35]. The new results (i.e., two-word prediction with the German novel and next word prediction with an English novel) demonstrate the universality of the findings, previously described in Surendra et al. [34].

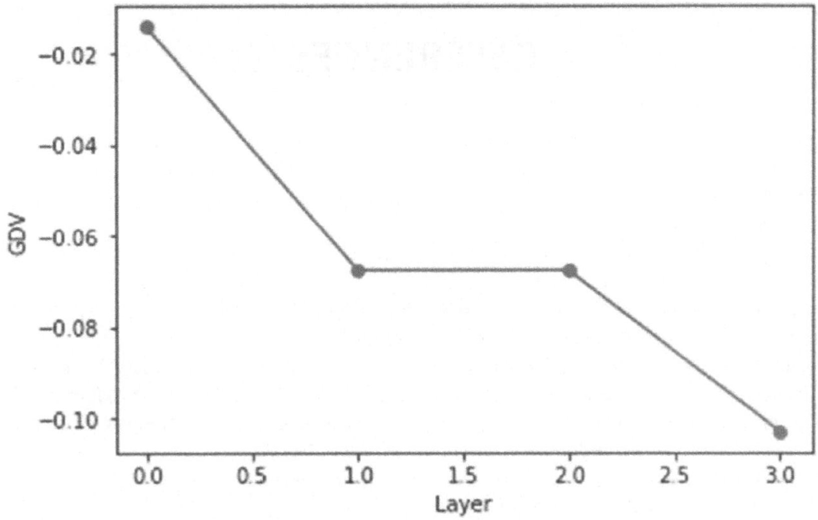

FIGURE 6.6 GDV curve across layers of the RNN. The decline of the GDV indicates that the RNN has learned to cluster internal representations according to the subsequent word class combinations with increasingly strong clustering from input to output layer.

ACKNOWLEDGMENTS

This work was funded by the Deutsche Forschungsgemeinschaft (DFG, German Research Foundation): grants KR 5148/2-1 (project number 436456810), KR 5148/3-1 (project number 510395418), KR 5148/5-1 (project number 542747151), and GRK 2839 (project number 468527017) to PK, and grant SCHI 1482/3-1 (project number 451810794) to AS, and by the Emerging Talents Initiative (ETI) of the University Erlangen-Nuremberg (grant 2019/2-Phil-01 to PK).

AUTHOR CONTRIBUTIONS

All authors contributed to the study and wrote the manuscript.

COMPETING INTERESTS

The authors declare no competing financial interests.

REFERENCES

1. Dieter E Zimmer. *So kommt der Mensch zur Sprache: über Spracherwerb, Sprachentstehung und Sprache & Denken*, volume 16. Heyne TB, 1986.
2. Henry Creswicke Rawlinson, John Gardner Wilkinson, et al. *The history of Herodotus*, J. Murray, volume 1. 1861.
3. Helen Goodluck. *Language acquisition: A linguistic introduction*. Basil Blackwell, 1991.
4. Ferdinand De Saussure. *Course in general linguistics*. Columbia University Press, 2011.
5. Noam Chomsky. On the nature, use and acquisition of language. In *Language and meaning in cognitive science*. pages 13–32. Routledge, 2012.
6. Noam Chomsky. *Aspects of the theory of syntax*, volume 11. MIT Press, 2014.
7. Charles Yang, Stephen Crain, Robert C Berwick, Noam Chomsky, and Johan J Bolhuis. The growth of language: universal grammar, experience, and principles of computation. *Neuro- Science & Biobehavioral Reviews*, 81:103–119, 2017.
8. Adele E Goldberg. *Constructions: a construction grammar approach to argument structure*. University of Chicago Press, 1995.
9. Adele E Goldberg. Constructions: a new theoretical approach to language. *Trends in Cognitive Sciences*, 7(5):219–224, 2003.
10. Michael Tomasello. *Constructing a language: a usage-based theory of language acquisition*. Harvard University Press, 2005.
11. Ronald W Langacker. Cognitive grammar. *Basic Readings*, page 29, 2008. Joan Bybee, Revere Perkins, William Pagliuca, et al. *The evolution of grammar: Tense, aspect, and modality in the languages of the world*. University of Chicago Press, 1994.
12. Paul J Hopper, and Joan L Bybee. Frequency and the emergence of linguistic structure. In *Frequency and the emergence of linguistic structure*. Torrossa, pages 1–502, 2001.
13. Joan L Bybee. Usage-based theory and exemplar representations of constructions. Oxford Handbooks Online. 2013.
14. Holger Diessel, Ewa Dabrowska, and Dagmar Divjak. Usage-based construction grammar. *Cognitive Linguistics*, 2:50–80, 2019.
15. Adele Goldberg, and Adele E Goldberg. *Explain me this*. Princeton University Press, 2019.
16. Hans-Jörg Schmid. *The dynamics of the linguistic system: usage, conventionalization, and entrenchment*. Oxford University Press, 2020.
17. Holger Diessel. Usage-based construction grammar. In *Handbook of cognitive linguistics*. pages 296–322. De Gruyter Mouton, 2015.
18. James M Kilner, Karl J Friston, and Chris D Frith. Predictive coding: an account of the mirror neuron system. *Cognitive Processing*, 8(3):159–166, 2007.
19. Andre M Bastos, W Martin Usrey, Rick A Adams, George R Mangun, Pascal Fries, and Karl J Friston. Canonical microcircuits for predictive coding. *Neuron*, 76(4):695–711, 2012.
20. B Keller, and Thomas D Mrsic-Flogel. Predictive processing: a canonical cortical computation. *Neuron*, 100(2):424–435, 2018.
21. Achim Schilling, William Sedley, Richard Gerum, Claus Metzner, Konstantin Tziridis, Andreas Maier, Holger Schulze, Fan-Gang Zeng, Karl J Friston, and Patrick Krauss. Predictive coding and stochastic resonance as fundamental principles of auditory phantom perception. Brain, awad255, 146(12), 4809–4825. 2023.
22. Achim Schilling, Rosario Tomasello, Malte R Henningsen-Schomers, Alexandra Zankl, Kishore Surendra, Martin Haller, Valerie Karl, Peter Uhrig, Andreas Maier, and Patrick Krauss. Analysis of continuous neuronal activity evoked by natural speech with computational corpus linguistics methods. *Language, Cognition and Neuroscience*, 36(2):167–186, 2021.

23. Armine Garibyan, Achim Schilling, Claudia Boehm, Alexandra Zankl, and Patrick Krauss. Neural correlates of linguistic collocations during continuous speech perception. *Frontiers in Psychology*, 13:1076339, 2022.
24. Nikola Koelbl, Achim Schilling, and Patrick Krauss. Adaptive ICA for speech EEG artifact removal. In *2023 5th International Conference on Bio-engineering for Smart Technologies (BioS- MART)*, pages 1–4. IEEE, 2023.
25. Schüller, A., Schilling, A., Krauss, P., & Reichenbach, T. (2024). The early subcortical response at the fundamental frequency of speech is temporally separated from later cortical contributions. Journal of Cognitive Neuroscience, 36(3), 475–491.
26. Schüller, A., Schilling, A., Krauss, P., Rampp, S., & Reichenbach, T. (2023). Attentional modulation of the cortical contribution to the frequency-following response evoked by continuous speech. Journal of Neuroscience, 43(44), 7429–7440.
27. Alina Schüller, Achim Schilling, Patrick Krauss, and Tobias Reichenbach. The early subcortical response at the fundamental frequency of speech is temporally separated from later cortical contributions. *Journal of Cognitive Neuroscience*, 36(3):475–491, 2024.
28. Achim Schilling, and Patrick Krauss. Tinnitus is associated with improved cognitive performance and speech perception–can stochastic resonance explain? *Frontiers in Aging Neuroscience*, 14:1073149, 2022.
29. Paul Stoewer, Christian Schlieker, Achim Schilling, Claus Metzner, Andreas Maier, and Patrick Krauss. Neural network based successor representations to form cognitive maps of space and language. *Scientific Reports*, 12(1):1–13, 2022.
30. Paul Stoewer, Achim Schilling, Andreas Maier, and Patrick Krauss. Neural network based formation of cognitive maps of semantic spaces and the putative emergence of abstract concepts. *Scientific Reports*, 13(1):3644, 2023.
31. Paul Stöwer, Achim Schilling, Andreas Maier, and Patrick Krauss. Conceptual cognitive maps formation with neural successor networks and word embeddings. In *2023 IEEE International Conference on Development and Learning (ICDL)*, pages 391–395. IEEE, 2023.
32. Paul Stoewer, Achim Schilling, Andreas Maier, and Patrick Krauss. Multi-modal cognitive maps based on neural networks trained on successor representations. *arXiv preprint arXiv:2401.01364*, 2023.
33. Jacob LS Bellmund, Peter G ardenfors, Edvard I Moser, and Christian F Doeller. Navigating cognition: spatial codes for human thinking. *Science*, 362(6415), 2018.
34. Kishore Surendra, Achim Schilling, Paul Stoewer, Andreas Maier, and Patrick Krauss. Word class representations spontaneously emerge in a deep neural network trained on next word prediction. In *2023 International Conference on Machine Learning and Applications (ICMLA)*, pages 1481–1486, 2023.
35. URL: https://spacy. io, 2017.
36. Adwait Ratnaparkhi. A maximum entropy model for part-of-speech tagging. In *Conference on Empirical Methods in Natural Language Processing*, 1996.
37. Horacio Rodríguez, and Lluís Màrquez. Part-of-speech tagging using decision trees. In *European Conference on Machine Learning*, pages 25–36. Springer, 1998.
38. Dan Jurafsky, and James H Martin. *Speech and language processing.* vol. 3, 2014.
39. Slav Petrov, Dipanjan Das, and Ryan McDonald. A universal part-of-speech tagset. *arXiv preprint arXiv:1104.2086*, Pearson: London. 2011.
40. Laurens Van der Maaten, and Geoffrey Hinton. Visualizing data using t-SNE. *Journal of Machine Learning Research*, 9(11), 2008.
41. Martin Wattenberg, Fernanda Viégas, and Ian Johnson. How to use t-SNE effectively. *Distill*, 1(10):e2, 2016.
42. Catalina A Vallejos. Exploring a world of a thousand dimensions. *Nature Biotechnology*, 37(12):1423–1424, 2019.

43. Kevin R Moon, David van Dijk, Zheng Wang, Scott Gigante, Daniel B Burkhardt, William S Chen, Kristina Yim, Antonia van den Elzen, Matthew J Hirn, Ronald R Coifman, et al. Visualizing structure and transitions in high-dimensional biological data. *Nature Biotechnology*, 37(12):1482–1492, 2019.
44. Warren S Torgerson. Multidimensional scaling: I. Theory and method. *Psychometrika*, 17(4):401–419, 1952.
45. Joseph B Kruskal. Nonmetric multidimensional scaling: a numerical method. *Psychometrika*, 29(2):115–129, 1964.
46. Joseph B Kruskal, and Myron Wish. *Multidimensional scaling*, volume 11. SAGE, 1978.
47. Michael AA Cox, and Trevor F Cox. Multidimensional scaling. In *Handbook of data visualization*. pages 315–347. Springer, 2008.
48. Claus Metzner, Achim Schilling, Maximilian Traxdorf, Holger Schulze, and Patrick Krauss. Sleep as a random walk: a super-statistical analysis of EEG data across sleep stages. *Communications Biology*, 4(1):1385, 2021.
49. Claus Metzner, Achim Schilling, Maximilian Traxdorf, Holger Schulze, Konstantin Tziridis, and Patrick Krauss. Extracting continuous sleep depth from EEG data without machine learning. *Neurobiology of Sleep and Circadian Rhythms*, 14:100097, 2023.
50. Claus Metzner, Achim Schilling, Maximilian Traxdorf, Konstantin Tziridis, Andreas Maier, Holger Schulze, and Patrick Krauss. Classification at the accuracy limit: facing the problem of data ambiguity. *Scientific Reports*, 12(1):22121, 2022.
51. Achim Schilling, Andreas Maier, Richard Gerum, Claus Metzner, and Patrick Krauss. Quantifying the separability of data classes in neural networks. *Neural Networks*, 139:278–293, 2021.
52. Patrick Krauss, Claus Metzner, Nidhi Joshi, Holger Schulze, Maximilian Traxdorf, Andreas Maier, and Achim Schilling. Analysis and visualization of sleep stages based on deep neural networks. *Neurobiology of Sleep and Circadian Rhythms*, 10:100064, 2021.
53. Patrick Krauss, Alexandra Zankl, Achim Schilling, Holger Schulze, and Claus Metzner. Analysis of structure and dynamics in three-neuron motifs. *Frontiers in Computational Neuroscience*, 13(5), 2019.
54. Krauss, P., Prebeck, K., Schilling, A., & Metzner, C. (2019). Recurrence resonance" in three-neuron motifs. Frontiers in computational neuroscience, 13, 64.
55. Patrick Krauss, Marc Schuster, Verena Dietrich, Achim Schilling, Holger Schulze, and Claus Metzner. Weight statistics controls dynamics in recurrent neural networks. *PLoS One*, 14(4):e0214541, 2019.
56. Claus Metzner, Marius E Yamakou, Dennis Voelkl, Achim Schilling, and Patrick Krauss. Quantifying and maximizing the information flux in recurrent neural networks. *arXiv preprint arXiv:2301.12892*, 2023.
57. Patrick Krauss, Claus Metzner, Achim Schilling, Konstantin Tziridis, Maximilian Traxdorf, Andreas Wollbrink, Stefan Rampp, Christo Pantev, and Holger Schulze. A statistical method for analyzing and comparing spatiotemporal cortical activation patterns. *Scientific Reports*, 8(1):1–9, 2018.
58. Patrick Krauss, Achim Schilling, Judith Bauer, Konstantin Tziridis, Claus Metzner, Holger Schulze, and Maximilian Traxdorf. Analysis of multichannel EEG patterns during human sleep: a novel approach. *Frontiers in Human Neuroscience*, 12:121, 2018.
59. Maximilian Traxdorf, Patrick Krauss, Achim Schilling, Holger Schulze, and Konstantin Tziridis. Microstructure of cortical activity during sleep reflects respiratory events and state of daytime vigilance. *Somnologie*, 23(2):72–79, 2019.
60. Cois Chollet Fran et al. Keras. https://keras.io, 2015. Last visited: July 15, 2024.
61. Matthias Plappert. keras-rl. https://github.com/keras-rl/keras-rl, 2016. Last visited: July 15, 2024.
62. Charles R. Harris, K. Jarrod Millman, St'efan J. van der Walt, Ralf Gommers, Pauli Vir- tanen, David Cournapeau, Eric Wieser, Julian Taylor, Sebastian Berg, Nathaniel J. Smith, Robert Kern, Matti Picus, Stephan Hoyer, Marten H. van Kerkwijk, Matthew Brett,

Allan Haldane, Jaime Fernàndez del Rìo, Mark Wiebe, Pearu Peterson, Pierre G'erard-Marchant, Kevin Sheppard, Tyler Reddy, Warren Weckesser, Hameer Abbasi, Christoph Gohlke, and Travis E. Oliphant. Array programming with NumPy. *Nature*, 585(7825):357–362, September 2020.

63. F. Pedregosa, G. Varoquaux, A. Gramfort, V. Michel, B. Thirion, O. Grisel, M. Blondel, P. Prettenhofer, R. Weiss, V. Dubourg, J. Vanderplas, A. Passos, D. Cournapeau, M. Brucher, M. Perrot, and E. Duchesnay. Scikit-learn: machine learning in python. *Journal of Machine Learning Research*, 12:2825–2830, 2011.
64. J. D. Hunter. Matplotlib: a 2d graphics environment. *Computing in Science & Engineering*, 9(3):90–95, 2007.
65. Aric A. Hagberg, Daniel A. Schult, and Pieter J. Swart. Exploring network structure, dynamics, and function using networkX. In Gäel Varoquaux, Travis Vaught, and Jarrod Millman, editors, *Proceedings of the 7th Python in Science Conference*, pages 11–15, Pasadena, CA USA, 2008.
66. Noam Chomsky. *Aspects of the theory of syntax*. MIT Press, Cambridge, Mass., 1965.
67. Diessel, H. (2020). A dynamic network approach to the study of syntax. Frontiers in psychology, 11, 604853.
68. Jun Li Li Zhang, and Chao Wang. Automatic synonym extraction using word2vec and spectral clustering. In *2017 36th Chinese Control Conference (CCC)*, pages 5629–5632. IEEE, 2017.
69. Zhiwei Chen, Zhe He, Xiuwen Liu, and Jiang Bian. Evaluating semantic relations in neural word embeddings with biomedical and general domain knowledge bases. *BMC Medical Informatics and Decision Making*, 18:53–68, 2018.
70. Bofang Li, Aleksandr Drozd, Yuhe Guo, Tao Liu, Satoshi Matsuoka, and Xiaoyong Du. Scaling word2vec on big corpus. *Data Science and Engineering*, 4:157–175, 2019.
71. Neguine Rezaii, Elaine Walker, and Phillip Wolff. A machine learning approach to predicting psychosis using semantic density and latent content analysis. *NPJ Schizophrenia*, 5(1):9, 2019.
72. Holger Diessel. *The grammar network*. Cambridge University Press, 2019.
73. Ashish Vaswani, Noam Shazeer, Niki Parmar, Jakob Uszkoreit, Llion Jones, Aidan N Gomez, L- ukasz Kaiser, and Illia Polosukhin. Attention is all you need. *Advances in Neural Information Processing Systems*, 30, 2017.
74. Md Saidur Rahaman, MM Tahmid Ahsan, Nishath Anjum, Harold Jan R Terano, and Md Miza- nur Rahman. From ChatGPT-3 to GPT-4: a significant advancement in AI-driven NLP tools. *Journal of Engineering and Emerging Technologies*, 2(1):1–11, 2023.
75. SIAD, S. M. (2023). The Promise and Perils of Google's Bard for Scientific Research. https://doi.org/10.17613/yb4n-mc79
76. Patrick Krauss, Jannik Hösch, Claus Metzner, Andreas Maier, Peter Uhrig, and Achim Schilling. Analyzing narrative processing in large language models (llms): Using GPT4 to test bert. *arXiv preprint arXiv:2405.02024*, 2024.
77. Achim Schilling, Richard Gerum, Claudia Boehm, Jwan Rasheed, Claus Metzner, Andreas Maier, Caroline Reindl, Hajo Hamer, and Patrick Krauss. Deep learning based decoding of single local field potential events. NeuroImage, 120696, 297, 2024.
78. Nikolaus Kriegeskorte, and Pamela K Douglas. Cognitive computational neuroscience. *Nature Neuroscience*, 21(9):1148–1160, 2018.
79. Schilling, A., Gerum, R., Boehm, C., Rasheed, J., Metzner, C., Maier, A., ... & Krauss, P. (2024). Deep learning based decoding of single local field potential events. NeuroImage, 297, 120696.
80. Richard C Gerum, and Achim Schilling. Integration of leaky-integrate-and-fire neurons in standard machine learning architectures to generate hybrid networks: a surrogate gradient approach. *Neural Computation*, 33(10):2827–2852, 2021.
81. Holger Schulze, Achim Schilling, Patrick Krauss, and Konstantin Tziridis. Erlanger modell der tinnitusentstehung–perspektivwechsel und neue behandlungsstrategie. HNO, 1–7, 71(10), 662–668, 2023.

82. Richard Gerum, André Erpenbeck, Patrick Krauss, and Achim Schilling. Leaky-integrate-and- fire neuron-like long-short-term-memory units as model system in computational biology. In *2023 International Joint Conference on Neural Networks (IJCNN)*, pages 1–9. IEEE, 2023.
83. Achim Schilling, Richard Gerum, Claus Metzner, Andreas Maier, and Patrick Krauss. Intrinsic noise improves speech recognition in a computational model of the auditory pathway. *Frontiers in Neuroscience*, 16:908330, 2022.
84. Andreas Stoll, Andreas Maier, Patrick Krauss, Richard Gerum, and Achim Schilling. Coincidence detection and integration behavior in spiking neural networks. Cognitive Neurodynamics, 18(4), 1753–1765, 2023.
85. Claus Metzner, Marius E Yamakou, Dennis Voelkl, Achim Schilling, and Patrick Krauss. Quantifying and maximizing the information flux in recurrent neural networks. *Neural Computation*, 36(3):351–384, 2024.
86. Zijin Yang, Achim Schilling, Andreas Maier, and Patrick Krauss. Neural networks with fixed binary random projections improve accuracy in classifying noisy data. In *Bildverarbeitung für die Medizin 2021*. pages 211–216. Springer, 2021.
87. Richard C Gerum, André Erpenbeck, Patrick Krauss, and Achim Schilling. Sparsity through evolutionary pruning prevents neuronal networks from overfitting. *Neural Networks*, 128:305–312, 2020.
88. Andreas Maier, Harald Köstler, Marco Heisig, Patrick Krauss, and Seung Hee Yang. Known operator learning and hybrid machine learning in medical imaging—a review of the past, the present, and the future. *Progress in Biomedical Engineering*, 4(2):022002, 2022.

Detection of Emerging Cyberthreats Through Active Learning

7

Joel Brynielsson, Amanda Carp, and Agnes Tegen

7.1 INTRODUCTION

Machine learning holds great promise for detecting and responding to increasingly sophisticated cyberthreats [1]. The large volume of available data can, however, present challenges for effective data management. Annotating data typically requires a significant amount of time and resources, particularly if the task necessitates specialized knowledge. This raises the question of whether there are methods to reduce the manual labeling effort, while still achieving satisfactory performance. Active learning (AL) can reduce the necessary amount of labeled data by introducing human interaction in the training process. Through different query strategies, AL investigates the selection of data points by identifying the most informative samples to be labeled [2, 3].

This chapter, extending previous work [4], presents a study of the potential and application of AL to increase model performance for a binary text classification task. The aim is to fine-tune a transformer model for the purpose of classifying tweets to determine if an advanced persistent threat (APT) is mentioned. Incorporating AL in the training process seeks to avoid the laborious process of labeling data points that do not contribute further to spanning the outcome space. The main objective is to study which

DOI: 10.1201/9781003570882-9

AL approaches and strategies that are suitable for continuous improvement of identification of APTs in tweets. Hence, the research question studied is the following:

- What active learning approaches are effective for continuous improvement of classification of advanced persistent threats in tweets?

The remainder of this chapter is structured as follows. In Section 7.2, background to active learning is provided, along with how the technique can be related to the work of detecting cyberthreat actors. Section 7.3 then describes related work, followed by Section 7.4 discussing various strategies used to select data points for labeling, which is the central issue in active learning. Section 7.5 then outlines the method for the example of detecting cyberthreat actors in text fragments, which is explored in this chapter to illustrate the use of active learning for threat analysis. Section 7.6 presents the results of the conducted experiments in terms of how different strategies and parameters affect the machine learning performance. Section 7.7 includes a discussion of the results and experimental limitations to consider, followed by conclusions and recommendations for future work in Section 7.8.

7.2 BACKGROUND

This study focuses on the application of AL approaches for classifying tweets to identify potential threat actors within a cybersecurity context. Understanding the cyber perspective, the importance of identifying threat actors, and how it relates to continuous updating of a machine learning classifier, is therefore crucial for the purpose of this work.

7.2.1 Active Learning

AL has been used successfully in a wide variety of applications, such as computer vision [5, 6], activity recognition [7, 8], and natural language processing (NLP) [9]. AL aims to reduce the labeling effort required to train a model. Rather than labeling the entire dataset, only a small subset is labeled by querying an oracle. The oracle can be a human or a computer software. AL strategies choose which data points to label based on some type of informativeness criterion [10]. The goal is to select a representative set of labeled instances that capture the underlying distribution of the entire dataset. The performance of the model trained on this smaller set of labeled data can, with an optimal selection of data points, be comparable to a model trained on a much larger labeled dataset [2]. In most work on AL, it is assumed that the oracle is always correct and acts in accordance with what is expected, but this is not always the case [11, 12]. It is important to consider whether the assumptions about the oracle are reasonable, given the application at hand, for example contemplating whether the oracle is an expert or not.

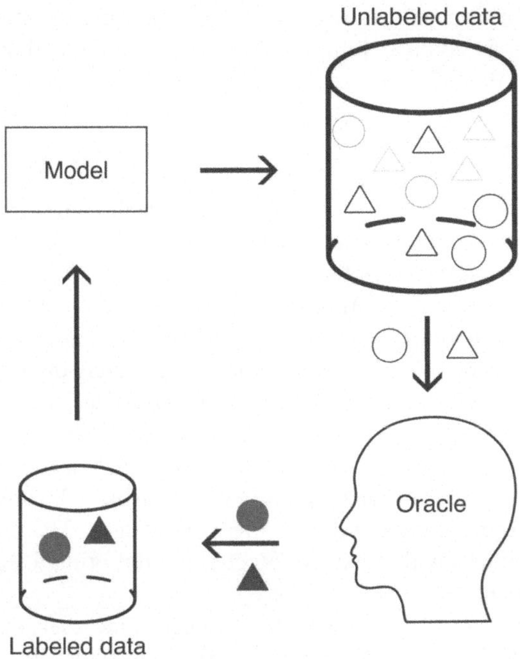

FIGURE 7.1 Active learning cycle.

Figure 7.1 illustrates an iterative training process of active learning, which is commonly employed. The training process is repeated until the model performs robustly, the labeling budget is used, or certain criteria are met. The labeling budget is often a percentage of the total amount of data [13]. The choice of AL strategy is dependent on the problem at hand. It can be difficult to estimate whether one strategy performs better than another before they have been put to the test. Settles [14] states that random sampling, which typically is used as a baseline, might be advisable if the problem is not well understood.

7.2.2 The Cyberthreat

Traditional cyberthreat detection methods rely on preventive work and network monitoring [15]. However, identifying and remediating cyberattacks take time. As threat actors grow more complex, new complementary technologies and methods are needed to identify and counter cyberattacks [16]. Cyber actors use social media, open forums, and the darknet to plan attacks, and the results of attacks are often sold or exposed online. Analyzing unstructured data from open sources can thus assist in predicting cyberattacks and cyberthreats.

APT is a term that is used to label a specific type of threat actor. An APT is usually a particularly well-resourced, stealthy adversary who is able to target specific information, and also eventually acquire it through persistent efforts [17]. The APT will typically succeed even if the target is a competent high-profile company or even a government. APTs are typically conducting long-term campaigns that involve multiple stages, utilizing the full range of their capabilities. According to the U.S. National Institute of Standards and Technology [18], APTs demonstrate a high level of expertise while they also possess large amounts of resources, enabling them to leverage multiple attack vectors, such as cyber, physical, and deceptive tactics. These attacks primarily involve infiltrating the targeted entity's information technology infrastructure to gain confidential information, disrupt vital aspects of a mission or organization, or position themselves to achieve similar objectives in the future.

APTs are typically given a name or a number by the first organization that discovers and publishes findings about them. However, these organizations, often antivirus and other types of cybersecurity companies, normally use their own naming conventions for an APT, regardless of who named it first [19]. This can lead to serious confusion. APT28, for example, has multiple aliases, such as Fancy Bear, Strontium, Pawn Storm, Sofacy, Sednit, and Tsar Team [19]. APT28, mentioned here as an example of an active APT, is a Russian-associated group that has been extensively documented and analyzed due to its involvement in multiple high-profile cyberattacks. The group has a long history of performing attacks with the common goal of promoting the political interests of the Russian government.

Cyber intelligence analysts have various roles. Some seek to assess the various APTs' capabilities to make threat assessments by analyzing and evaluating computer networks and systems [15]. They typically use various tools and techniques to monitor network traffic and activity, to detect patterns or anomalies that may indicate a cyberattack or a security breach. Actions to prevent or mitigate cyberattacks can then take place at different levels. At the strategic level, long-term measures are required, for example, replacing an entire system, or overhauling an architecture, due to an excessive number of security risks [20]. At the tactical level, responses are often more time-sensitive. Associated necessary measures should be implemented more swiftly, which may include, for example, updates of firewall rules or changes in routing tables.

Intelligence analysts possess considerable expertise in identifying and recognizing APTs. As such, they are potential users of the outcome of the study presented in this chapter, where the intelligence analysts fulfill the role of labeling data points. Through this process, the analysts can make valuable contributions to the training of the system through AL approaches, without necessarily having to share secret data with a system designer. This, in turn, secures that the system continues to stay pertinent, while accommodating additional data.

7.3 RELATED WORK

Several surveys explore AL within NLP applications. Olsson [9] presents an overview of the area, especially focusing on the theory and methodology that different AL approaches use for data selection. Much of the content can be generalized to AL in other applications as well, and is not specific to NLP. Miller et al. [21] present an overview, as well as simulation studies to investigate performance, efficiency, and practical applicability. They use support vector machines (SVMs) with data from Twitter, Wikipedia talk pages, and news articles in their experiments. Margin sampling, that is, uncertainty sampling based on the distance from the data points to the SVM hyperplanes, performs the best in their experiments. They also find that the length and style of the text data affect the results. In Wang et al.'s [22] study, human-in-the-loop NLP frameworks are discussed from both the machine learning perspective and the human-computer interaction perspective. They classify the surveyed papers in terms of task, goal, human interaction, and feedback learning method. Zhang et al. [23] showcase how the number of publications focusing on AL in the ACL Anthology[1] has increased over the last 15 years, indicating an increased interest in the subject. They discuss the current status of AL in NLP, as well as suggested future directions. Stiennon et al. [24] introduce a human feedback model for producing summaries of text data. In experiments with data from Reddit, they show that their model improves the quality of summaries, compared to supervised learning. Zhang et al. [25] study how active learning can aid in the fine-tuning of large language models (LLMs). LLMs, like deep learning in general, need a large amount of data to be trained. The paper introduces an approach where AL is combined with existing fine-tuning techniques to improve data efficiency. Experiments show that the new approach is better than baseline models on three complex reasoning tasks. Hu et al. [26] explore how AL together with LLMs handle code-related tasks. They study 11 different sampling strategies with three different LLMs, and find that classification-related tasks yield good performance, while nonclassification tasks do not yield as good results. In their experiments, they also find that clustering-based strategies outperform uncertainty-based strategies.

In the work mentioned above, AL within different NLP applications is studied, but not explicitly with regard to the cybersecurity domain. Bhattacharjee et al. [27] study the task of classifying phishing or malicious URLs. They propose an uncertainty-based AL strategy to help with the task, and experiments show an increase in the results. Lin et al. [28] investigate how malicious mislabeling and data poisoning attacks can affect AL of deep neural networks. They propose a clustering-based strategy and perform experiments on an image dataset. The results show that the suggested AL strategy is robust against mislabeling and data poisoning attacks. Moskal and Yang [29] introduce an approach that combines AL, transfer learning, and pseudolabels to aid analysts in interpreting possible intrusion alerts. They use a minimal amount of data, but still yield significant results. Pal et al. [30] take the perspective of the attacker, instead of defending against attacks. They focus on model extraction, where the aim of the attacker is to be able to replicate an unknown model, without access to the training data or knowledge about the employed base model. Their framework, denoted ActiveThief, is able to extract models in a variety of domains, from image to text. They compare different AL strategies within the framework and find that they get better results compared to the baseline random strategy.

Li et al. [31], Srivastava et al. [32], and Xie et al. [33] all study how AL can be used in named entity recognition (NER) tasks. Deep neural network models have resulted in good performance, but are typically dependent on the amount of data, which is why AL can be a possible solution. Li et al. [31] present an adversarial AL framework to pick the most informative samples for annotation. Their AL strategy is based on comparing the similarity between unlabeled samples and the already-labeled samples. Srivastava et al. [32] combine reinforcement learning with AL in their proposed method. Xie et al. [33] focus specifically on Chinese NER, which has not been studied much and is a complex task. They introduce a strategy combining uncertainty, confidence, and diversity.

Compared to previous work as discussed, this study focuses on evaluating the performance of different AL strategies in classifying APTs in tweets. A new dataset is used and the case when annotated data is scarce is studied specifically. The scarcity of data is motivated, as previously mentioned, by the expensive process of annotating data, especially when expert knowledge is needed.

7.4 ACTIVE LEARNING STRATEGIES

The central research question in active learning concerns how data points for annotation are selected. An example of how the selection of data points can be performed using two different strategies is illustrated in Figure 7.2. Four AL strategies are studied in the experiments, including the random strategy. The random query strategy (AL-random) selects data points randomly for labeling, and is used as a baseline for comparison with the other strategies. The other strategies are uncertainty sampling with entropy for uncertainty measurement (AL-entropy), diversity sampling using K-means (AL-kmeans), and cost-effective active learning (CEAL-entropy).

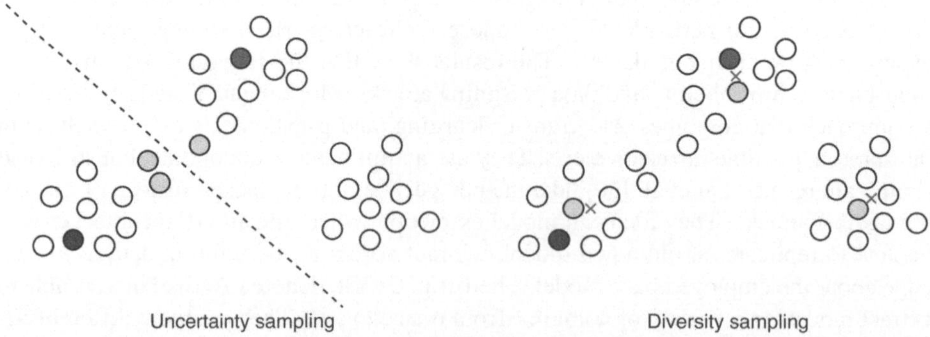

FIGURE 7.2 Data point selection by different AL query strategies. Red and blue points indicate labeled samples for two classes, and gray points represent samples selected for oracle labeling for the respective query strategy. The dashed line marks the calculated dividing line between the two classes, and the red crosses mark the centroids of the clusters.

7.4.1 Uncertainty Sampling

Uncertainty sampling is based on selecting the samples that the model is most uncertain about how to classify. Thus, instances where the model is highly uncertain are supposed to be maximally informative [2]. A common approach to evaluate the predictions made by a model is to assess the probabilistic distribution of the classes. There are several uncertainty-based query strategies to measure this, such as least confidence, margin of confidence, and entropy. Entropy, a measure of impurity of a system [34], is widely used in machine learning as a measure of uncertainty of a model. The higher the entropy value, the more uncertain the model is about which class the data point belongs to. Hence, for binary classification, entropy-based sampling is the same as choosing the data point with posterior closest to 0.5.

7.4.2 Diversity Sampling

Uncertainty sampling is prone to selecting outliers and data that may not accurately represent the dataset [35]. Conversely, diversity sampling mitigates these concerns by identifying a subset of samples that comprehensively cover the entire dataset. Depending on the methodology employed to construct the subset, there exists a variety of diversity sampling techniques. One technique is cluster-based sampling, which is a method used to find structures among the unlabeled data points, where a commonly used strategy is K-means. This involves clustering a set of samples into K clusters, each characterized by a centroid point that minimizes the associated inertia [36].

7.4.3 Cost-Effective Active Learning

In addition to only selecting data points that the model is least confident about, cost-effective active learning (CEAL also considers samples where the model is most confident [37]. For instance, if the model predicts a data point belonging to a class with certainty 0.5, it is a likely candidate for uncertainty sampling. However, if the prediction is 1, it can be inferred that the model is maximally confident in its classification. In each AL cycle, the CEAL technique selects samples at both extremes: those with the highest uncertainty, and those with the lowest. For the latter, CEAL suggests provisionally labeling them based on the model's predictions, creating so-called pseudolabeled samples [37]. Subsequently, both the pseudolabeled samples and the oracle-labeled data points are added to the labeled training dataset, which is used to train a new model. Upon completing the training of the new model, the pseudolabeled samples are eliminated from the training dataset, and a new CEAL cycle is initiated. This process is depicted in Figure 7.3.

The unlabeled samples with an uncertainty measurement below a predetermined threshold δ are considered the most certain. The threshold for high-confidence sample

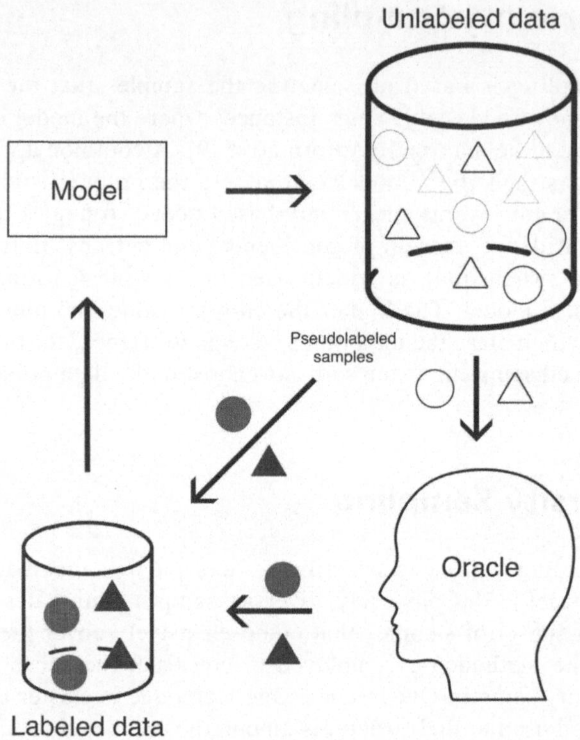

FIGURE 7.3 Cost-effective active learning.

selection is updated at each epoch, according to Equation 7.1. This is to be done to ensure that the labeling process remains dependable [37]. The threshold δ is defined by:

$$\delta = \begin{cases} \delta_0, & \text{for } t = 0, \\ \delta - dr \times t, & \text{for } t > 0, \end{cases} \tag{7.1}$$

where δ_0 is the initial threshold, dr controls the threshold decay rate, and t is the current epoch.

7.5 METHOD

This section describes the model used, the data, and the experimental setup. Figure 7.4 displays the iterative process used to train and evaluate the model. The chosen AL strategy picks data points from the pool of unlabeled data based on the defined selection criteria. The chosen data points are presented to the oracle, who in turn annotates them. The now-annotated data points are added to the annotated

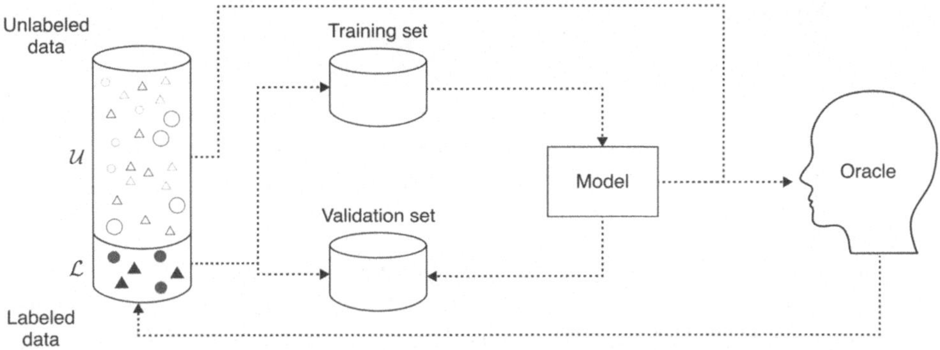

FIGURE 7.4 The iterative AL training and validation process employed in the experiments.

data, which in turn is used as a training set to retrain the model. The updated model is tested on the validation set and a performance score is received. After this final step, the cycle restarts and the AL strategy picks new data points for the oracle to annotate.

7.5.1 DistilBERT for Text Classification

DistilBERT is an open-source NLP framework used for the text classification part of the experiments. DistilBERT is smaller, faster, cheaper, and lighter than its predecessor BERT [38]. By using a small model, the time and resource costs associated with model training can be reduced while still maintaining high performance. The pretrained transformer DistilBERT, as described, was used through the Hugging Face library [39]. The tweets were tokenized and [CLS] and [SEP] tokens, used for classification and sentence separation, were added. The [CLS] token captures the entire context of the input for simple downstream tasks, such as classification. For sentence representations used in classification tasks, the size of the [CLS] token is equal to the number of data points × the number of hidden states. The tokenized input was padded to match the length of the longest tweet in the dataset. An attention mask was also created to distinguish the padded tokens from the nonpadded ones. The stochastic optimizer Adam [40] was utilized. A small search was conducted to identify an optimal learning rate for this classification task. Various learning rates were tested, focusing on values near the suggested learning rates mentioned for the original BERT model [41]. The search resulted in an optimal learning rate of $2e-5$. A single linear layer was added at the output hidden state of the [CLS] token, on top of the DistilBERT model, to perform classification.

The pretrained model and the additional untrained classification layer were trained and updated at every iteration for the specific task. The cross-entropy loss was used to measure the performance of the model, calculated by comparing the divergence between the predicted probability and the actual label.

7.5.2 Dataset

The dataset used in the experiments contains approximately 35,000 labeled tweets [42]. Cyber-related tweets were identified by their association with keywords such as “cyber” and “malware.” An existing infrastructure for data download and rule-based detection of known APTs was leveraged to download large amounts of cyber-related tweets and automatically categorize them into two groups: texts with and without (known) APTs.

The dataset contains a total of 70 different APTs. A language detector from the fastText [43] library was utilized to identify and discard all tweets where English was not the most probable language. To enhance the model in locating APTs, distracting elements in all tweets were eliminated. Links, email addresses, phone numbers, and usernames were replaced with their respective masking tokens (“LINK,” “MAIL,” “PHONE,” and “USER”). Emojis were then converted to descriptive ones (for example, “👍” was changed to “:thumbs up:”) using the demoji Python package.[2] Duplicate tweets were removed, and the tweets were also normalized, for example, replacing “a . m .” and “p . m .” with “a.m.” and “p.m.”

Approximately 19,000 tweets remained after the cleaning. The entire dataset contained twice as many tweets belonging to the negative class as the positive class. To prevent the model from overtraining on a small number of negative examples, a skewed distribution with three times more negative examples was chosen for the training. An even distribution between positive and negative was chosen for the validation set. For the unlabeled pool dataset, the remaining data was added, resulting in 66% negative samples. For clarification, this is presented in Table 7.1.

Figure 7.5 displays the final preprocessed dataset, with the *x*-axis representing the length of tweets, and the *y*-axis representing the number of tweets, starting with the first bar representing one-word length.

7.5.3 Experimental Setup

In Algorithm 1, the main loop of the experiment is documented using pseudocode. The algorithm shows that the total number of data points added to the training dataset $\mathcal{L}$ is $K \times N$, where K is the number of samples in a batch and N is the number of epochs. The F-score and the accuracy were accumulated over all batches and logged at each epoch for the validation dataset. To obtain a fair evaluation and comparison between AL approaches, the training was averaged over three runs with 10

TABLE 7.1 Data distribution per class

DATASET	*POSITIVE*	*NEGATIVE*
Training	25%	75%
Validation	50%	50%
Unlabeled pool	34%	66%

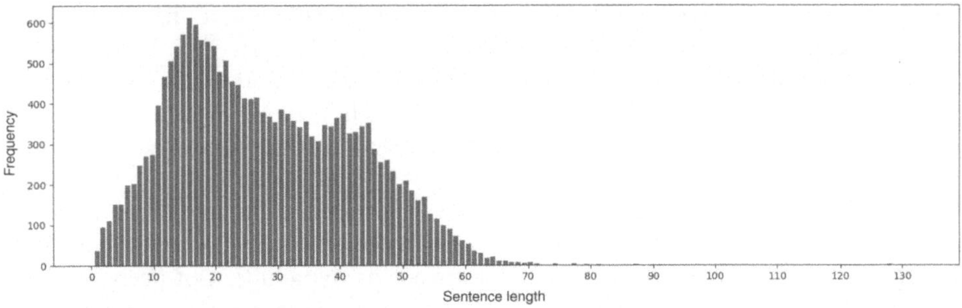

FIGURE 7.5 The frequency of sentence lengths in the final preprocessed dataset.

different seeds (101, 102, ..., 110). The stopping criterion for training was when the maximum number of epochs was achieved. At every iteration, the model $\mathcal{M}$ is fine-tuned and thereby updated. At the end of every epoch, data points chosen according to a query strategy are added to the training dataset. In experiments involving the CEAL approach, the training dataset is augmented with pseudosamples during the fine-tuning of the model, and after the fine-tuning these pseudosamples are then returned to the unlabeled data pool before the next iteration, as described in Section 7.4.3.

Algorithm 1 Active learning experiment design

Inputs: Pretrained transformer model $\mathcal{M}$, unlabeled pool dataset $\mathcal{U}$, initially labeled dataset $\mathcal{L}$, validation dataset $\mathcal{V}$, acquisition size K for AL sampling, threshold δ for CEAL pseudosample inclusion (0 if CEAL is not to be used), maximum number of epochs N.

Output: Fine-tuned model $\mathcal{M}$.

for $i = 0, 1, \ldots, N$ **do**
    Fine-tune $\mathcal{M}$ with $\mathcal{L}$.
    Evaluate $\mathcal{M}$ on $\mathcal{V}$ and log results.
    Move back any pseudosamples from $\mathcal{L}$ into $\mathcal{U}$.
    Move K samples from $\mathcal{U}$ into $\mathcal{L}$ based on a query strategy.
    Move pseudosamples with uncertainty below δ from $\mathcal{U}$ into $\mathcal{L}$.
    Update δ according to Equation 7.1.
end for

For diversity sampling, K-means clustering was used to sample diverse data points, deviating from uncertainty sampling where entropy was based on probabilities of the different classes. K-means was performed on the [CLS] token, which is a special classification token corresponding to the last hidden state in the DistilBERT model. K data points were then chosen to be sent to an oracle for labeling; for diversity sampling with K-means, the K data points were based on the smallest distance to each centroid, and for uncertainty sampling, the K most uncertain data points according to the classification by the currently fine-tuned model were chosen.

The CEAL approach required optimal values for the initial threshold δ_0 and the decay rate dr to be set in order to be implemented. The value δ sets the limit for the number of samples that are transferred to the labeled training dataset, and dr determines the rate of decay of δ over the number of epochs, as described by Equation 7.1. The decay rate dr was chosen to be 0.0033, as stated as the most optimal value according to the literature [37]. An initial threshold δ_0 of 0.35 was established through experimentation. The threshold allowed for the addition of pseudolabeled samples, that is, data points with entropy lower than the threshold are included in the training dataset. The addition of the pseudosamples had the potential of improving performance. Yet there was also a risk of decreased performance if incorrect labels were assigned.

The amount of initially labeled data and the acquisition size were varied to determine their impact on the model's performance. The amount of initially labeled data refers to the data used to train the model at the start of the experiment, and the acquisition size refers to the number of data points added each epoch. For the amount of initially labeled data, experiments were conducted with 0.05%, 0.1%, 0.5%, and 1% of the whole dataset, corresponding to 9, 19, 96, and 192 data points, respectively. The acquisition sizes tested in the experiments were 10, 25, and 50 data points. The validation set was set to be 4% of the whole dataset. To prevent misleading results due to lack of data in the validation set, a consistent amount of data was allocated for validation, regardless of the size of the training set. The conducted experiments were executed on a high-performance NVIDIA DGX A100 computing cluster consisting of eight NVIDIA A100 40 GB Tensor Core GPUs.

7.6 RESULTS

The four different query strategies are referred to as AL-entropy (uncertainty-based sampling), AL-random (random sampling used as baseline), AL-kmeans (diversity-based sampling), and CEAL-entropy (the CEAL strategy). The presented results referred to as CEAL-entropy are solely based on AL-entropy+CEAL-entropy. All combinations, that is, AL-entropy+CEAL-entropy, AL-random+CEAL-entropy, and AL-kmeans+CEAL-entropy, were tested, but due to their similar performance and space constraints, not all combinations are shown. As stated, the incorporation of pseudolabeled samples depended on the model's classification confidence to identify data points suitable for inclusion in the training dataset. Consequently, the results for CEAL-entropy are presented only for experiments in which the model displayed sufficient confidence in classifying data points. In the graphs presented in Figures 7.6–7.8, the x-axis denotes the number of labeled data points by the oracle, not the pseudolabeled samples.

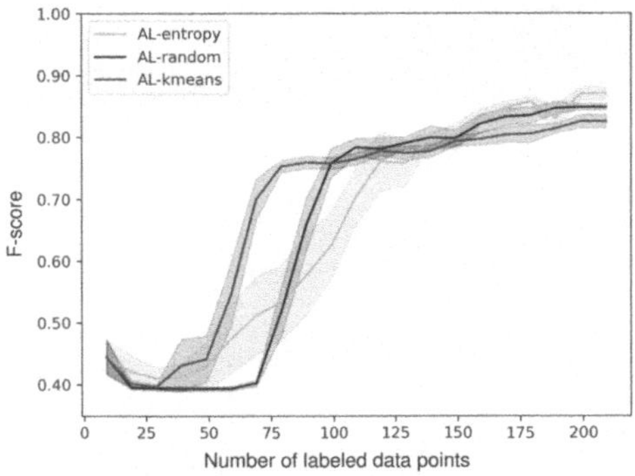

(a) Initially labeled data of 0.05%.

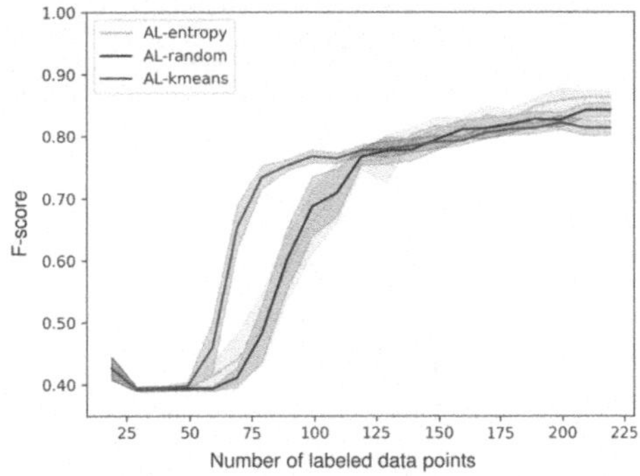

(b) Initially labeled data of 0.1%.

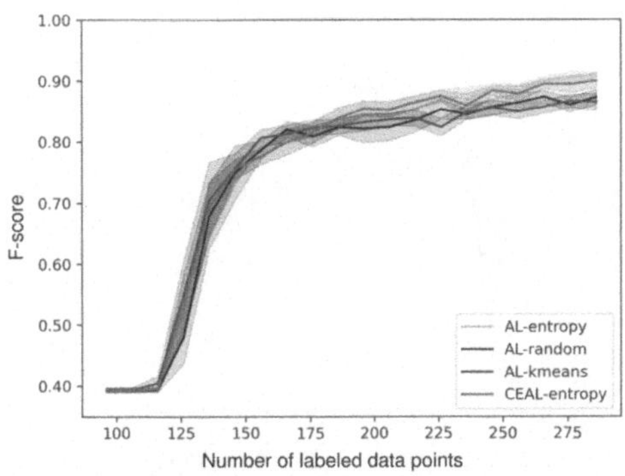

(c) Initially labeled data of 0.5%.

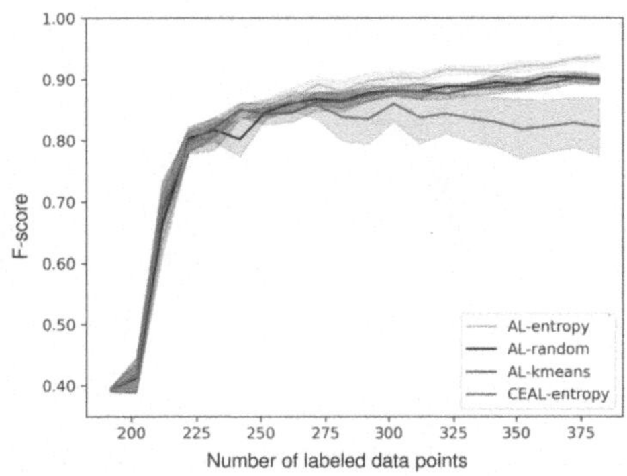

(d) Initially labeled data of 1%.

FIGURE 7.6 Average F-score of AL approaches and query strategies with different amounts of initially labeled data and acquisition size 10, averaged across 10 seeds and shown with 95% confidence intervals.

7.6.1 Acquisition Size 10

In Figure 7.6, four graphs are presented displaying the F-score of four scenarios with acquisition size 10 and different quantities of labeled data that the model had at its disposal for training. The experimental setup involved conducting experiments with an acquisition size of 10 over a span of 20 epochs, with varying amounts of initially labeled data. The total number of labeled data points was determined by adding the initially labeled data to the 200 data points (10 × 20) acquired from the pool of unlabeled data. This was the procedure for all query strategies, except when employing CEAL.

7.6.2 Acquisition Size 25

Figure 7.7 contains four graphs representing the F-score across four scenarios with acquisition size 25 and varying quantities of labeled data available for model training.

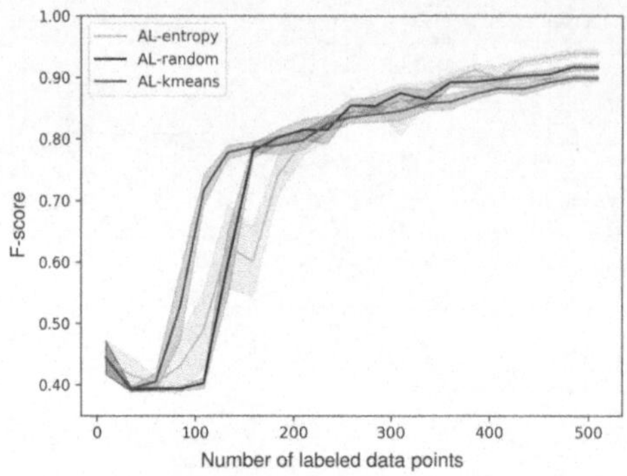

(a) Initially labeled data of 0.05%.

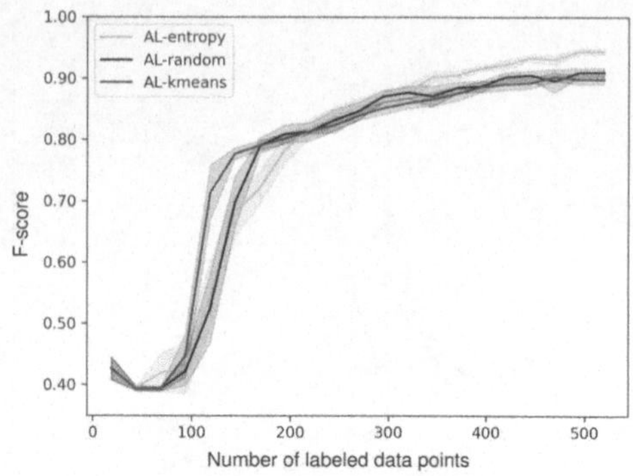

(b) Initially labeled data of 0.1%.

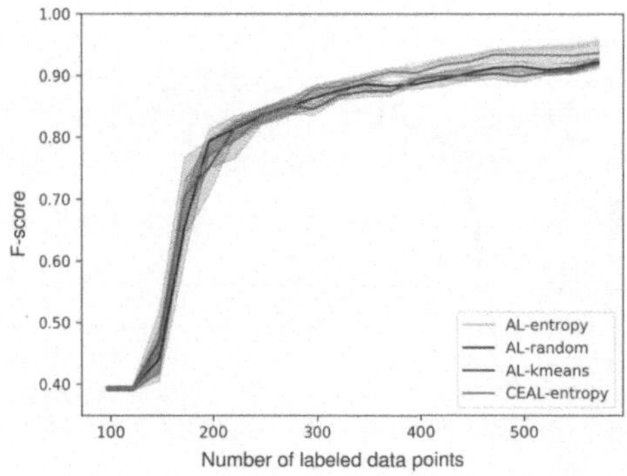

(c) Initially labeled data of 0.5%.

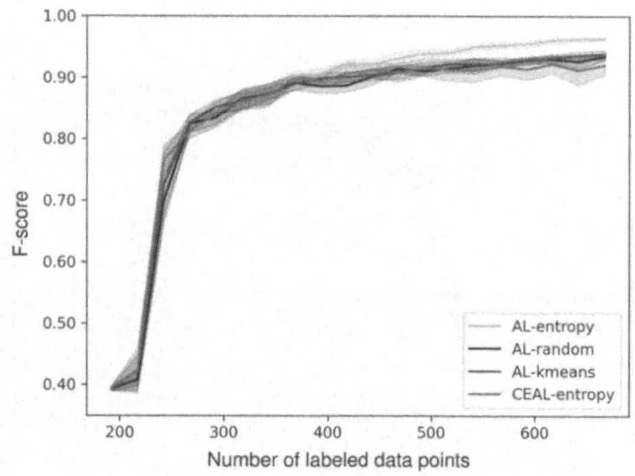

(d) Initially labeled data of 1%.

FIGURE 7.7 Average F-score of AL approaches and query strategies with different amounts of initially labeled data and acquisition size 25, averaged across 10 seeds and shown with 95% confidence intervals.

The experiments were conducted across 20 epochs, with varying amounts of initially labeled data. Similar to the previous section, the total number of labeled data points was determined by adding the initially labeled data to the 500 data points (25 × 20) acquired from the pool of unlabeled data.

7.6.3 Acquisition Size 50

In Figure 7.8, the F-scores for four separate scenarios with acquisition size 50 are displayed, characterized by the varying quantities of labeled data available to the model during the training process. As in the previous sections, the experiments were carried out over 20 epochs, with different amounts of initially labeled data. The total number of labeled data points was calculated by adding the initially labeled data to the 1,000 data points (50 × 20) acquired from the pool of unlabeled data, to provide a comprehensive understanding of the performance trends under this experimental condition.

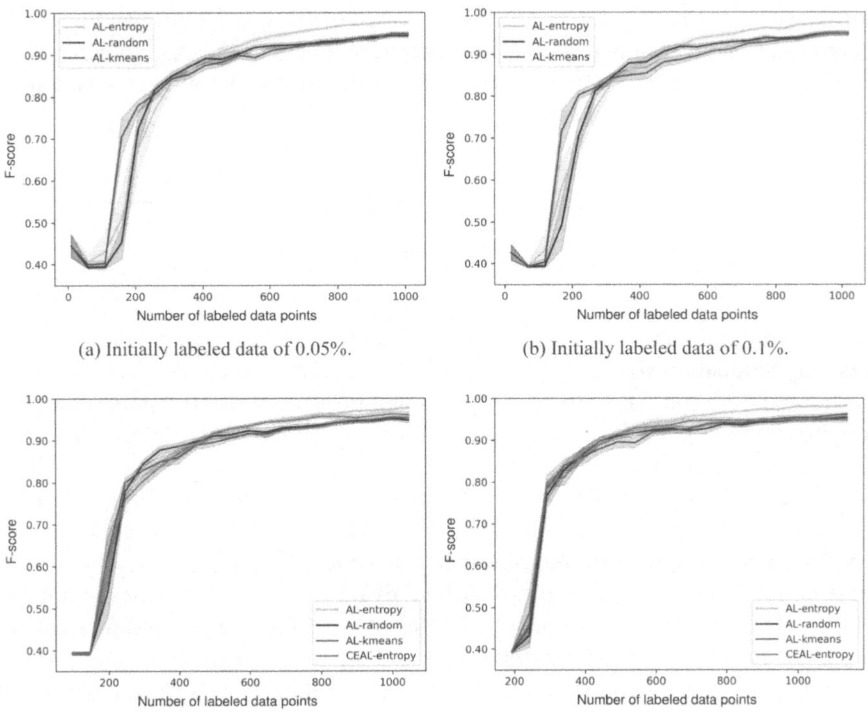

(a) Initially labeled data of 0.05%.

(b) Initially labeled data of 0.1%.

(c) Initially labeled data of 0.5%.

(d) Initially labeled data of 1%.

FIGURE 7.8 Average F-score of AL approaches and query strategies with different amounts of initially labeled data and acquisition size 50, averaged across 10 seeds and shown with 95% confidence intervals.

7.6.4 Comparison of Acquisition Sizes

Table 7.2 presents a comparison between acquisition sizes 10 and 50, using 0.5% of the initially labeled data (96 data points). It presents the effects of acquisition sizes on

TABLE 7.2 F-score for acquisition sizes 10 and 50, when trained on the same number of data points

DATA POINTS	*ACQ. SIZE*	*EPOCHS*	*AL-ENTROPY*	*AL-RANDOM*	*AL-KMEANS*	*CEAL-ENTROPY*
196	10	6	0.76	0.75	0.76	0.76
	50	2	0.39	0.39	0.39	0.39
246	10	11	0.86	0.82	0.83	0.85
	50	3	0.61	0.54	0.62	0.61
296	10	16	0.89	0.86	0.86	0.88
	50	4	0.76	0.78	0.80	0.76

performance, with the F-score for each of the tested query strategies given for a specific number of data points, thus resulting in different numbers of epochs. To compare how the F-score is affected by the same number of data points with varying acquisition sizes, the rows should be analyzed in pairs.

7.7 DISCUSSION

As can be seen in Figures 7.6–7.8, each of the tested query strategies yielded impressive results. For acquisition size 10, Figure 7.6(a) and (b) show that the AL-kmeans strategy outperformed both the AL-entropy and AL-random strategies until approximately 100 data points were labeled. After that point, all query strategies performed more or less equivalent to each other.

Similar to the results observed with an acquisition size of 10, Figure 7.7(a) and (b) display that even for acquisition size 25, the AL-kmeans demonstrates a tendency for improved performance in the initial stages of the training process. The availability of a larger initial dataset diminishes the advantage of the AL-kmeans approach, as evidenced in Figure 7.7(c) and (d). In the case of the highest number of initial data points, Figure 7.7(d) demonstrates that the AL-entropy query strategy marginally surpasses the AL-random and AL-kmeans strategies in performance, starting at approximately 400 labeled data points, while AL-kmeans and AL-random exhibit similar performance.

For acquisition size 50, the AL-kmeans strategy shows a performance slightly more advantageous than other query strategies up to 250 data points. However, as shown in Figure 7.8(a) and (b), the difference is subtle. Independent of the initial training data quantity, the AL-random and AL-kmeans strategies tend to converge toward each other, resulting in equivalent F-scores. Notably, similar to acquisition size 10, the AL-entropy strategy demonstrates a slightly higher F-score upon the model's training completion for all levels of initially labeled data.

This might suggest that employing a diversity-based method is most effective when only a limited amount of labeled data is available. Presumably, this approach succeeded in selecting data points that more accurately embodied the dataset compared to the uncertainty-based method. As the amount of training data increased, the importance of selecting diverse data points seemed to diminish. As a result, AL-random and AL-kmeans displayed similar behavior, whereas AL-entropy achieved a marginally higher F-score. Nonetheless, the difference can be considered minimal, and the similar results are likely influenced by the inherent simplicity of the problem, as all query strategies exhibit high performance.

Contrary to the initial assumption that CEAL would effectively increase the amount of training data and subsequently enhance the model's performance, this does not appear to be the case. In instances where 0.5% of all data were labeled before training, Figures 7.6(c), 7.7(c), and 7.8(c) display that the inclusion of pseudolabeled data points only had a marginal effect on the model's performance and achieved an F-score comparable to AL-entropy alone. This can be attributed to the fact that only a few pseudolabeled samples exhibited entropy below the threshold and were subsequently added to the training set. This is not surprising since the presented results for the CEAL approach combine AL-entropy+CEAL-entropy. Instead, when the model is provided with more initially labeled data, as can be seen in Figures 7.6(d), 7.7(d), and 7.8(d), it exhibits increased confidence in its predictions, leading to a greater number of data points falling below the threshold and being incorporated into the training set. This results in inferior performance compared to other strategies, as the model is not sufficiently confident in its predictions, causing data points to be assigned incorrect labels. In light of these findings, the optimality of setting an initial threshold and subsequently reducing it by a factor over the number of epochs according to Equation 7.1 can be questioned. However, the poor results from the CEAL approach might also be because of the small amount of data used for training. If there was more training data, the estimations might be better and more certain, resulting in a better output from CEAL. That is, CEAL might be a good choice if data is less scarce, but in this work the focus is on a scenario where data annotation can be expensive, and it is therefore of interest to limit the annotation cost. Based on the results, the threshold-setting method appears to be more sensitive than has been proposed in previous studies [37]. An alternative approach, in which the threshold is more flexible and adapted to the model's confidence, might have yielded a different outcome. Another possible way could involve training the model over a greater number of epochs, thereby increasing the likelihood of accurate label classification, while simultaneously allowing the threshold to be set at a lower value.

Upon examining Table 7.2 to analyze the impact of acquisition sizes, it becomes apparent that the choice of acquisition size can influence the performance of different query strategies. Uncertainty-based query strategies, such as AL-entropy and CEAL-entropy, achieved a higher F-score from a smaller acquisition size over a greater number of epochs. For example, after 296 labeled data points, AL-entropy achieved an F-score of 0.89 with an acquisition size of 10, and 0.76 with an acquisition size of 50. The smaller acquisition size also enhanced the performance of both AL-random and AL-kmeans strategies up to 246 labeled data points. When further increasing the amount of labeled data, the difference can be seen as negligible. Upon the model reaching a meaningful performance level, the impact of acquisition sizes on the convergence rate dropped to a barely noticeable level. However, uncertainty-based query strategies, such as AL-entropy and CEAL-entropy, seem to benefit from a smaller acquisition size over an extended number of epochs.

To address the research question, a trade-off between time and F-score must be made. AL-kmeans utilized the [CLS] token, which had a size equal to the number of data points $\times$ the number of hidden states, to select data points for labeling by the oracle. Consequently, AL-kmeans might not be an appropriate strategy when working with high-dimensional data, if time consumption is a performance requirement. In contrast, AL-entropy and AL-random selected their data points based on probabilities for each label and, therefore, did not necessitate selection of data points from this high-dimensional space.

7.7.1 Limitations

It can be argued that labeling data needs to include a nonbiased oracle. In this study, this concern has been mitigated, as the oracle is a computer software that simulates a human in providing correct labels. However, a broader perspective and a possible future scenario includes a human annotator as the oracle. In such scenarios, a malicious oracle may introduce bias, for example, by consistently mislabeling tweets referencing a particular threat as negative [44].

The overall performance of the various AL approaches and query strategies was notably high, with several strategies achieving an F-score exceeding 0.90. This raises the question of whether the classification task itself is relatively straightforward for a complex transformer such as DistilBERT. Moreover, this level of performance is expected, given the binary nature of the classification problem, compared to a multiclass problem. Another consideration is that the model potentially learned the precise names of the 70 distinct APTs, which might have limited its ability to generalize and maintain comparable performance if new data containing different APTs are introduced.

In this work, a relatively small amount of data is used to train an LLM model. The experiments are simulations using an already-annotated dataset, aiming to replicate a scenario with a human expert annotating the data samples gradually, as they are selected over time by the AL strategy. For AL in general, however, it is important to also take the annotation cost into account, and arguably even more so when expert knowledge is needed for the annotation process. With humans, the task becomes more complex, possibly introducing a varying labeling cost, different types of noise, disagreement about the labels, etc. [11, 12]. While simulations have the advantage of providing more control over the experiments, they also risk oversimplifying the real-world scenario that is intended to be replicated.

7.8 CONCLUSIONS

This work investigated the potential of AL and its effectiveness for continuous improvement of classification of APTs in tweets. The transformer model DistilBERT was employed to classify the tweets, and AL approaches were utilized to iteratively add new labeled data points to the training dataset. Different AL approaches, including uncertainty-based and diversity-based query strategies, were examined, with several strategies achieving high performance. The diversity-based query strategy K-means excelled in the early training stages with limited prelabeled data. However, as the volume of training data increased, the performance advantage diminished. Additionally, as the number of training epochs increased, the uncertainty-based strategy showed a marginally improved performance relative to the other strategies. Interestingly, the CEAL approach did not enhance the model's performance. The incorporation of data points with predicted labels often resulted in incorrect labels, thereby undermining the performance.

For future work, it would be interesting to explore the impact of combining the K-means strategy, which in this project demonstrated effectiveness when a minimal amount of labeled data was available, with uncertainty-based methods, such as entropy. This could be done by employing K-means to select *K* clusters and calculating entropy within each cluster, rather than on all data points in the unlabeled pool, which could potentially offer a more effective strategy for diverse sampling, while focusing on data points with higher model uncertainty. Additionally, assessing the generalizability of these findings across various datasets and distributions would be valuable. Moreover, experiments with human subjects in the annotation process would be of interest. This project focused on the performance of query strategies in a binary classification context, so extending the investigation to multiclass problems would be beneficial. Lastly, further examination of the potential of the CEAL approach is warranted, given its promising results in prior studies [37]. Exploring alternative methods for establishing the initial threshold, as well as for reducing the threshold, could prove beneficial.

ACKNOWLEDGMENTS

The authors would like to thank Katie Cohen and Leon Ericsson, Swedish Defence Research Agency, for comments on the manuscript.

NOTES

1. https://aclanthology.org/.
2. https://pypi.org/project/demoji/.

REFERENCES

1. U. Franke, A. Andreasson, H. Artman, J. Brynielsson, S. Varga, and N. Vilhelm, "Cyber situational awareness issues and challenges," in *Cybersecurity and Cognitive Science*, A. A. Moustafa, Ed. London, United Kingdom: Academic Press, 2022, ch. 10, pp. 235–265, doi: 10.1016/B978-0-323-90570-1.00015-2.
2. B. Settles, *Active learning* (Synthesis Lectures on Artificial Intelligence and Machine Learning #18). Cham, Switzerland: Springer, 2012, doi: 10.1007/978-3-031-01560-1.
3. P. Ren, Y. Xiao, X. Chang, P.-Y. Huang, Z. Li, B. B. Gupta, X. Chen, and X. Wang, "A survey of deep active learning," *ACM Computing Surveys*, vol. 54, no. 9, pp. 1–40, 2021, Art. no. 180, doi: 10.1145/3472291.
4. A. Carp, J. Brynielsson, and A. Tegen, "Active learning for improvement of classification of cyberthreat actors in text fragments," in *Proceedings of the 2023 22nd IEEE International Conference on Machine Learning and Applications (ICMLA 2023)*. Piscataway, NJ: IEEE, 2023, pp. 1279–1286, doi: 10.1109/ICMLA58977.2023.00193.

5. A. J. Joshi, F. Porikli, and N. Papanikolopoulos, "Multi-class active learning for image classification," in *Proceedings of the 2009 27th IEEE Conference on Computer Vision and Pattern Recognition (CVPR 2009)*. Piscataway, NJ: IEEE, 2009, pp. 2372–2379, doi: 10.1109/CVPR.2009.5206627.
6. D. Tuia, M. Volpi, L. Copa, M. Kanevski, and J. Muñoz-Marí, "A survey of active learning algorithms for supervised remote sensing image classification," *IEEE Journal of Selected Topics in Signal Processing*, vol. 5, no. 3, pp. 606–617, 2011, doi: 10.1109/JSTSP.2011.2139193.
7. H. M. S. Hossain, M. A. A. H. Khan, and N. Roy, "Active learning enabled activity recognition," *Pervasive and Mobile Computing*, vol. 38, part 2, pp. 312–330, 2017, doi: 10.1016/j.pmcj.2016.08.017.
8. A. Tegen, P. Davidsson, and J. A. Persson, "Activity recognition through interactive machine learning in a dynamic sensor setting," *Personal and Ubiquitous Computing*, vol. 28, no. 1, pp. 273–286, 2024, doi: 10.1007/s00779-020-01414-2.
9. F. Olsson, "A literature survey of active machine learning in the context of natural language processing," Swedish Institute of Computer Science, Kista, Sweden, SICS Tech. Rep. T2009:06, 2009. [Online]. Available: https://urn.kb.se/resolve?urn=urn:nbn:se:ri:diva-23510
10. Y. Fu, X. Zhu, and B. Li, "A survey on instance selection for active learning," *Knowledge and Information Systems*, vol. 35, no. 2, pp. 249–283, 2013, doi: 10.1007/s10115-012-0507-8.
11. A. Tegen, P. Davidsson, and J. A. Persson, "The effects of reluctant and fallible users in interactive online machine learning," in *Proceedings of the IAL@ECML PKDD 2020 Workshop on Interactive Adaptive Learning. CEUR Workshop Proceedings*, 2020, pp. 55–71. [Online]. Available: https://ceur-ws.org/Vol-2660/ialatecml_paper4.pdf
12. A. Tegen, P. Davidsson, and J. A. Persson, "Active learning and machine teaching for online learning: A study of attention and labelling cost," in *Proceedings of the 2021 20th IEEE International Conference on Machine Learning and Applications (ICMLA 2021)*. Piscataway, NJ: IEEE, 2021, pp. 1215–1220, doi: 10.1109/ICMLA52953.2021.00197.
13. A. Tegen, P. Davidsson, and J. A. Persson, "A taxonomy of interactive online machine learning strategies," in *Proceedings of the 2020 European Conference on Machine Learning and Knowledge Discovery in Databases (ECML PKDD 2020)*, vol. 2. Cham, Switzerland: Springer, 2020, pp. 137–153, doi: 10.1007/978-3-030-67661-2_9.
14. B. Settles, "From theories to queries: Active learning in practice," in *Proceedings of the AISTATS 2010 Active Learning and Experimental Design Workshop*. JMLR Workshop and Conference Proceedings, 2011, pp. 1–18. [Online]. Available: https://proceedings.mlr.press/v16/settles11a.html
15. W. Newhouse, S. Keith, B. Scribner, and G. Witte, "National initiative for cybersecurity education (NICE) cybersecurity workforce framework," National Institute of Standards and Technology, U.S. Department of Commerce, NIST Special Publication 800-181, 2017, doi: 10.6028/NIST.SP.800-181.
16. A. Sapienza, A. Bessi, S. Damodaran, P. Shakarian, K. Lerman, and E. Ferrara, "Early warnings of cyber threats in online discussions," in *Proceedings of the 2017 IEEE International Conference on Data Mining Workshops (ICDMW 2017)*. Piscataway, NJ: IEEE, 2017, pp. 667–674, doi: 10.1109/ICDMW.2017.94.
17. P. Chen, L. Desmet, and C. Huygens, "A study on advanced persistent threats," in *Proceedings of the 15th IFIP TC 6/TC 11 International Conference on Communications and Multimedia Security (CMS 2014)*. Berlin/Heidelberg, Germany: Springer, 2014, pp. 63–72.
18. Joint Task Force Transformation Initiative, "Managing information security risk: Organization, mission, and information system view," National Institute of Standards and Technology, U.S. Department of Commerce, NIST Special Publication 800-39, 2011, doi: 10.6028/NIST.SP.800-39.

19. A. Lemay, J. Calvet, F. Menet, and J. M. Fernandez, "Survey of publicly available reports on advanced persistent threat actors," *Computers & Security*, vol. 72, pp. 26–59, 2018, doi: 10.1016/j.cose.2017.08.005.
20. T. Mattern, J. Felker, R. Borum, and G. Bamford, "Operational levels of cyber intelligence," *International Journal of Intelligence and CounterIntelligence*, vol. 27, no. 4, pp. 702–719, 2014, doi: 10.1080/08850607.2014.924811.
21. B. Miller, F. Linder, and W. R. Mebane Jr., "Active learning approaches for labeling text: Review and assessment of the performance of active learning approaches," *Political Analysis*, vol. 28, no. 4, pp. 532–551, 2020, doi: 10.1017/pan.2020.4.
22. Z. J. Wang, D. Choi, S. Xu, and D. Yang, "Putting humans in the natural language processing loop: A survey," in *Proceedings of the First Workshop on Bridging Human–Computer Interaction and Natural Language Processing (HCINLP 2021).* Stroudsburg, PA: Association for Computational Linguistics, 2021, pp. 47–52. [Online]. Available: https://aclanthology.org/2021.hcinlp-1.8
23. Z. Zhang, E. Strubell, and E. Hovy, "A survey of active learning for natural language processing," in *Proceedings of the 2022 Conference on Empirical Methods in Natural Language Processing (EMNLP 2022).* Kerrville, TX: Association for Computational Linguistics, 2022, pp. 6166–6190, doi: 10.18653/v1/2022.emnlp-main.414.
24. N. Stiennon, L. Ouyang, J. Wu, D. M. Ziegler, R. Lowe, C. Voss, A. Radford, D. Amodei, and P. F. Christiano, "Learning to summarize from human feedback," in *Proceedings of the 34th Conference on Neural Information Processing Systems (NeurIPS 2020).* San Diego, CA: NeurIPS, 2020, pp. 3008–3021.
25. L. Zhang, J. Wu, D. Zhou, and G. Xu, "STAR: Constraint LoRA with dynamic active learning for data-efficient fine-tuning of large language models," 2024, arXiv: in *Findings of the Association for Computational Linguistics: ACL 2024.* Kerrville, TX: Association for Computational Linguistics, 2024, pp. 3519–3532, doi: 10.18653/v1/2024.findings-acl.209
26. Q. Hu, Y. Guo, X. Xie, M. Cordy, L. Ma, M. Papadakis, and Y. Le Traon, "Active code learning: Benchmarking sample-efficient training of code models," *IEEE Transactions on Software Engineering*, vol. 50, no. 5, pp. 1080–1095, 2024, doi: 10.1109/TSE.2024.3376964.
27. S. D. Bhattacharjee, A. Talukder, E. Al-Shaer, and P. Doshi, "Prioritized active learning for malicious URL detection using weighted text-based features," in *Proceedings of the 2017 IEEE International Conference on Intelligence and Security Informatics (ISI 2017).* Piscataway, NJ: IEEE, 2017, pp. 107–112, doi: 10.1109/ISI.2017.8004883.
28. J. Lin, R. Luley, and K. Xiong, "Active learning under malicious mislabeling and poisoning attacks," in *Proceedings of the 2021 IEEE Global Communications Conference (GLOBECOM 2021).* Piscataway, NJ: IEEE, 2021, pp. 1–6, doi: 10.1109/GLOBECOM46510.2021.9685101.
29. S. Moskal, and S. J. Yang, "Translating intrusion alerts to cyberattack stages using pseudo-active transfer learning (PATRL)," in *Proceedings of the 2021 IEEE Conference on Communications and Network Security (CNS 2021).* Piscataway, NJ: IEEE, 2021, pp. 110–118, doi: 10.1109/CNS53000.2021.9705037.
30. S. Pal, Y. Gupta, A. Shukla, A. Kanade, S. Shevade, and V. Ganapathy, "ActiveThief: Model extraction using active learning and unannotated public data," in *Proceedings of the 34th AAAI Conference on Artificial Intelligence (AAAI 2020)*, vol. 34, no. 1. Palo Alto, CA: AAAI Press, 2020, pp. 865–872, doi: 10.1609/aaai.v34i01.5432.
31. T. Li, Y. Hu, A. Ju, and Z. Hu, "Adversarial active learning for named entity recognition in cybersecurity," *Computers, Materials & Continua*, vol. 66, no. 1, pp. 407–420, 2021, doi: 10.32604/cmc.2020.012023.
32. S. Srivastava, D. Gupta, B. Paul, and S. Sahoo, "A reinforced active learning sampling for cybersecurity NER data annotation," in *Proceedings of the 2022 OITS International Conference on Information Technology (OCIT 2022).* Piscataway, NJ: IEEE, 2022, pp. 312–317, doi: 10.1109/OCIT56763.2022.00066.

33. B. Xie, G. Shen, C. Guo, and Y. Cui, "The named entity recognition of Chinese cybersecurity using an active learning strategy," *Wireless Communications and Mobile Computing*, vol. 2021, no. 1, pp. 1–11, 2021, Art. no. 6629591, doi: 10.1155/2021/6629591.
34. C. E. Shannon, "A mathematical theory of communication," *The Bell System Technical Journal*, vol. 27, no. 3, 4, pp. 379–423, 623–656, 1948, doi: 10.1002/j.1538-7305.1948.tb01338.x, 10.1002/j.1538-7305.1948.tb00917.x.
35. T. He, S. Zhang, J. Xin, P. Zhao, J. Wu, X. Xian, C. Li, and Z. Cui, "An active learning approach with uncertainty, representativeness, and diversity," *The Scientific World Journal*, vol. 2014, pp. 1–6, 2014, Art. no. 827586, doi: 10.1155/2014/827586.
36. D. Arthur, and S. Vassilvitskii, "k-means++: The advantages of careful seeding," in *Proceedings of the 18th Annual ACM-SIAM Symposium on Discrete Algorithms (SODA 2007)*. Philadelphia, PA: Society for Industrial and Applied Mathematics, 2007, pp. 1027–1035.
37. K. Wang, D. Zhang, Y. Li, R. Zhang, and L. Lin, "Cost-effective active learning for deep image classification," *IEEE Transactions on Circuits and Systems for Video Technology*, vol. 27, no. 12, pp. 2591–2600, 2017, doi: 10.1109/TCSVT.2016.2589879.
38. V. Sanh, L. Debut, J. Chaumond, and T. Wolf, "DistilBERT, a distilled version of BERT: Smaller, faster, cheaper and lighter," 2019, arXiv: 1910.01108.
39. T. Wolf, L. Debut, V. Sanh, J. Chaumond, C. Delangue, A. Moi, P. Cistac, T. Rault, R. Louf, M. Funtowicz, J. Davison, S. Shleifer, P. von Platen, C. Ma, Y. Jernite, J. Plu, C. Xu, T. Le Scao, S. Gugger, M. Drame, Q. Lhoest, and A. M. Rush, "Transformers: State-of-the-art natural language processing," in *Proceedings of the 2020 Conference on Empirical Methods in Natural Language Processing: System Demonstrations (EMNLP 2020)*. Stroudsburg, PA: Association for Computational Linguistics, 2020, pp. 38–45, doi: 10.18653/v1/2020.emnlp-demos.6
40. D. P. Kingma, and J. L. Ba, "Adam: A method for stochastic optimization," in *Proceedings of the 2015 Third International Conference on Learning Representations (ICLR 2015)*, 2015, arXiv: 1412.6980.
41. J. Devlin, M.-W. Chang, K. Lee, and K. Toutanova, "BERT: Pre-training of deep bidirectional transformers for language understanding," in *Proceedings of the 2019 Conference of the North American Chapter of the Association for Computational Linguistics: Human Language Technologies (NAACL-HLT 2019)*, vol. 1. Stroudsburg, PA: Association for Computational Linguistics, 2019, pp. 4171–4186, doi: 10.18653/v1/N19-1423.
42. H. Lilja, and L. Lundmark, "Tracking cyber threat actors in semi-automatic OSINT analysis," in *Proceedings of the IST-190 Symposium on Artificial Intelligence, Machine Learning and Big Data for Hybrid Military Operations (AI4HMO)*. NATO Science and Technology Organization, 2021, pp. 1–12, Art. no. 31.
43. A. Joulin, E. Grave, P. Bojanowski, and T. Mikolov, "Bag of tricks for efficient text classification," in *Proceedings of the 15th Conference of the European Chapter of the Association for Computational Linguistics (EACL 2017)*, vol. 2. Stroudsburg, PA: Association for Computational Linguistics, 2017, pp. 427–431. [Online]. Available: https://aclanthology.org/E17-2068/
44. B. Miller, A. Kantchelian, S. Afroz, R. Bachwani, E. G. Dauber, L. Huang, M. C. Tschantz, A. D. Joseph, and J. D. Tygar, "Adversarial active learning," in *Proceedings of the 2014 ACM Workshop on Artificial Intelligent and Security Workshop (AISec 2014)*. New York, NY: ACM, 2014, pp. 3–14, doi: 10.1145/2666652.2666656.

8 Enhanced Health Information Retrieval with Explainable Biomedical Inconsistency Detection Using Large Language Models

Prajwol Lamichhane, Indika Kahanda, Xudong Liu, Karthikeyan Umapathy, Sandeep Reddivari, and Andrea Arikawa

8.1 INTRODUCTION

The biomedical literature is expanding at an unprecedented rate, resulting in clinical inquiries that can return inconsistent information from several sources. This poses a

DOI: 10.1201/9781003570882-10

substantial challenge to practitioners of evidence-based medicine (EBM) (Yazi et al., 2021a) and professionals engaged in comparative effectiveness research and precision medicine (Tawfik & Spruit, 2018). The appearance of contradicting assertions in medical literature hinders healthcare professionals' decision-making processes, especially when the causes for these discrepancies are unclear. Furthermore, these inconsistent assertions impede a complete comprehension of the current state of medical knowledge, and practitioners in these fields should avoid making decisions based on out-of-date research (Sackett, 1997).

This chapter describes a research project that proposes novel techniques for predicting the occurrence of inconsistencies in the biomedical domain. Furthermore, it develops and tests a proof-of-concept prototype system that leverages information retrieval (IR) and machine learning (ML) to make these predictions. The IR system locates relevant research articles and extracts essential research statements based on a user's query. The ML model then predicts if each pair of statements on the same topic is contradictory to one another. At the end, explainable artificial intelligence (XAI), which is integrated with the ML classifier, provides deeper understanding of why specific research statements contradict each other by offering a thorough context of where these statements come from and how they appear in a text.

By leveraging the features of these advanced approaches, this chapter aims to lay the groundwork for the long-term goal of automatically discovering and classifying contradictions within biological texts, thereby addressing the issues raised by the growing volume of published research. The ultimate goal is to give health and medical researchers and practitioners a valuable tool for discovering and helping to understand inconsistencies, encouraging evidence-based decision-making, and broadening medical knowledge comprehension.

8.2 BACKGROUND

Biomedical research is being published at an unprecedented rate (Flier, 2023). For example, PubMed, which is a platform that provides access to MEDLINE (a very large database of biomedical publications), currently has 1.5 million items added each year (Novoa et al., 2023). Furthermore, the rate of publications is exponentially growing. While this is healthy for the progress of the research, this also means that biomedical researchers often find inconsistencies and contradictions reported across different publications (Carpenter et al., 2016). For example, one research study (Hotu et al., 2010) advocates for a novel care model's superiority in reducing systolic blood pressure and preventing cardiac and renal damage in high-risk patients, while a previous study (Matsui et al., 2009) has introduced a contradiction by suggesting potential adverse effects of a commonly used medication, doxazosin, on left ventricular structure and the risk of congestive heart failure. A significant number of contradictions is present in biomedical research, which poses serious challenges for researchers who want to stay up to date with biomedical knowledge to make informed decisions.

However, due to the rate of publication, it is no longer possible to manually read and discover these inconsistencies. This poses a significant challenge to scientific trustworthiness and medical decision-making (Irving, 1993). Contradiction/inconsistency detection, an important application of natural language processing, solves this problem by systematically identifying and resolving disputes in biomedical literature. The pipeline proposed in this chapter develops new techniques to systematically discover conflicts, ensuring biomedical knowledge's accuracy and reliability.

At the heart of the contradiction detection workflow is IR, which involves locating relevant biological literature from a large collection of research articles, clinical trials, and scientific databases (Schütze et al., 2008). The approach used in IR combines syntactic search, which accurately finds articles containing the exact keywords or phrases supplied in the query, with semantic search, which goes beyond keywords to understand the underlying meaning and context of biomedical concepts. This synergistic strategy guarantees that articles are not only relevant but also conceptually linked, giving the pipeline access to the most appropriate knowledge for contradiction identification.

ML is critical in the contradiction detection pipeline, notably with the use of bidirectional encoder representations from transformers (BERT), a sophisticated language model that excels at collecting contextual information from text (Vaswani et al., 2017). Specifically, four types of BERT models are explored in this study. These models are *pretrained* on large corpora of literature, including biomedical text, allowing them to accurately digest the intricacies of scientific language, expert terminology, and complex biomedical concepts. A secondary step is to *fine-tune* these models with a smaller domain-specific dataset targeting a specific application. This study fine-tunes BERT models for classifying a pair of sentences as contradictory.

XAI is a growing field that seeks to bring transparency and interpretability to ML models, allowing users to understand the reasoning behind the model's judgments (Arrieta et al., 2020). In the case of biological contradiction detection, XAI is especially important since it helps users comprehend how the system detects inconsistencies, increasing trust and confidence in the system's output. The pipeline suggested in this chapter uses XAI approaches to give users extensive explanations of the system's conclusions, including the exact contextual data and logic that resulted in the identification of conflicts.

8.3 LITERATURE REVIEW

Alamri and Stevenson (Alamri & Stevenson, 2015) conducted a series of research studies on identifying contradicting statements in the medical literature. Alamri and Stevenson in another research (Alamri & Stevenson, 2016) focused on creating corpora to identify contradicting statements in Medline abstracts related to cardiovascular disease, leveraging annotators' development of PICO (patient/population, intervention, comparison, and outcomes) questions and subsequent claim assessment.

Through their work (Alamri, 2016), two corpuses of claims were created: ManConCorpus, which was manually built, and AutoConCorpus, which was generated automatically from SemMedDB. In 2020, Halil Kilicoglu et al. (Kilicoglu et al., 2020) showed that SemMedDB uses a software called SemRep that can extract semantic relations from biological texts. SemMedDB was used to extract biological semantic connections for 20 illnesses (Rosemblat et al., 2019). Similar work by Lamichhane et al. (Lamichhane et al., 2023) classified conflicts between numerous biological claims using Subject, Predicate, and Object relations.

In 2018, a two-part contradiction model was developed by Tawfik and Spruit composed of (1) identifying abstract claims and (2) using a linear support vector machine to anticipate assertions (Tawfik & Spruit, 2018). Researchers retrieved many semantic characteristics from the Manual Contradiction Corpus (Alamri & Stevenson, 2016). In 2021, Yazi et al. (Yazi et al., 2021b) used ManConCorpus to identify inconsistencies by integrating deep neural network methods into their models. The authors concluded that, while the results of their ManConCorpus model testing are encouraging, the models' performance may be impacted by the corpus size.

Yazi et al. (2021a) carried out experiments to assess the impact of corpus size on the performance of a contradiction detection model. Surprisingly, the results indicated that larger training dataset sizes did not always translate into better performance; the model performed similarly on the medium-sized EBMSum corpus, as it did on the smaller ManConCorpus (Alamri, 2016).

The technique to contradiction detection developed by Kharrat, Hlaoua, and Romdhane (Kharrat et al., 2022) takes into account both semantic features and ambiguity in phrases, classifying contradictions as conformity, opposition, inconsistency, and conflict. In order to detect inconsistencies in sentence pairings, Wu, Niu, and Rahman (Wu et al., 2022) provide a framework that incorporates topological data analysis (TDA) representations into deep learning models. Their tests show encouraging outcomes in precisely identifying different kinds of inconsistencies in a variety of text genres. Building on the Stanford RTE system, Marneffe et al. (De Marneffe et al., 2008) provide a technique for identifying inconsistencies in the text.

Sosa et al. (2022) used pretrained BERT models that had been refined on a variety of datasets, including a new COVID-19 NLI dataset to detect conflicting claims in the COVID-19 medication efficacy literature. Their model proved to be effective in resolving conflicts in the difficult terrain of contradicting biomedical literature.

For the Spanish language, Sepulveda-Torres et al. (Sepulveda-Torres et al., 2021) created a contradiction data set that distinguishes between several kinds of contradictions. They found that although the model did not work well with other types of data, it did an effective job of capturing the inconsistencies in their data. Pielka et al.'s (Pielka et al., 2021) study on German texts and Rahimi et al.'s (Rahimi & ShamsFard, 2021) study on Persian texts were similar.

Yang et al. (Yang et al., 2023), provide a thorough examination of XAI, highlighting its critical function in building trust by facilitating a deeper comprehension of AI models. XAI is thoroughly explored by Das et al. (Das & Rad, 2020), with an emphasis on its importance in mission-critical applications where the black-box nature of deep neural networks presents ethical and trust-related difficulties. Their assessment stresses the dynamic character of the area and the importance of giving careful thought to the creation and selection of XAI methods, in addition to offering a thorough overview of currently available XAI approaches and highlighting new trends and concerns.

8.4 METHODOLOGY

8.4.1 Data

Data serve as a critical component within any ML pipeline. In this research, two distinct datasets are employed. The first dataset, known as ManConCorpus (Manually Extracted Contradiction Corpus), is readily accessible online and contains information pertaining to contradictions in cardiovascular diseases (Alamri, 2016). To expedite the contradiction detection pipeline development process, the ManConCorpus dataset is utilized for tuning and evaluating the individual components of the pipeline.

A second dataset is developed in-house by a team of experts focusing on common health concerns. The primary objective of this dataset is to collect a smaller set of health claims that is used for the overall (i.e., blackbox) evaluation of the pipeline. This includes an exploration of the explainable outputs displayed to the user about the contradictions within the realm of common health topics. This strategic approach is adopted to ensure the robustness of the pipeline by ensuring that there is no overlap between the dataset used for tuning the models and the evaluation of the pipeline. This eliminates any bias due to information leakage.

8.4.1.1 ManConCorpus dataset

The Manual Contradiction Corpus, or ManConCorpus for short, is a corpus created especially for research on contradiction identification in biomedical literature. Alamri and Stevenson (Alamri & Stevenson, 2016) created the ManConCorpus dataset in 2016, which is made up of contradicting research claims taken from 24 systematic reviews pertaining to cardiovascular issues. The dataset consists of 259 claims, each labeled with further details, such as the PMID (a unique identifier) of the related abstracts, assertion values, and clinical questions, based on the PICO framework. The assertion value indicates whether a claim affirms ("Yes") or rejects ("No") its related question, with 180 asserting positively and 79 asserting negatively. Each systematic review's PICO questions were formulated, and research claims were used as the responses to generate the corpus. Contradictions are assumed when claims answering the same inquiry have different assertion values. For example, Figure 8.1 shows several contradictory claims for the question "In women with preeclampsia, does treatment with L-arginine, compared to placebo, reduce blood pressure or preeclampsia?"

Each claim in the ManConCorpus dataset is associated with a PICO question, allowing for precise categorization and analysis of biomedical research. Additionally, each claim includes a PubMed ID (PMID), enabling researchers to access the original

<CORPUS>
<REVIEW REVIEW_PMID="23435582" REVIEW_TITLE="hyper_Arginine supplementation for improving maternal and neonatal outcomes in hypertensive disorder of pregnancy_ A systematic review">
<CLAIM ASSERTION="YS" PMID="21596735" QUESTION="In women with pre-eclampsia, does treatment with L Arginine, compared to placebo, reduce blood pressure or pre-eclampsia" TYPE="CAUS">Supplementation during 1pregnancy with a medical food containing L-arginine and antioxidant vitamins reduced the incidence of pre-eclampsia in a population at high risk of the condition.</CLAIM>
<CLAIM ASSERTION="NO" PMID="14678093" QUESTION="In women with pre-eclampsia, does treatment with L Arginine, compared to placebo, reduce blood pressure or pre-eclampsia" TYPE="CAUS">Oral L-arginine supplementation did not reduce mean diastolic blood pressure after 2 days of treatment compared with placebo in pre-eclamptic patients with gestational length varying from 28 to 36 weeks.</CLAIM>
</REVIEW>
</CORPUS>

FIGURE 8.1 A portion of ManConCorpus dataset.

study for further examination and verification. Each assertion in the ManConCorpus dataset provides a binary answer (Yes or No) to the PICO question, indicating whether the claim extracted from the corresponding PMID supports or refutes the hypothesis posed by the question.

8.4.1.2 In-house gold standard dataset

A gold standard dataset is an extremely dependable and carefully selected set of data that is used as the standard to assess how well models or algorithms perform in a given field. Due to time-consuming nature, ethical issues, patient privacy concerns, and the complicated nature of medical data, which frequently incorporates complex and dynamic variables, it can be difficult to curate a gold standard dataset in the scientific literature. Standardized, complete datasets are therefore difficult to create.

This work utilizes the approach discussed in Figure 8.2. for the development of the Gold Standard Dataset designed specially to identify inconsistencies in biomedical literature.

There are two phases involved in creating the dataset, the first of which uses cutting-edge AI technology. To get started, as shown in Figure 8.2, the popular OpenAI GPT4 model was utilized to compile a corpus of 200 topics that illustrate inconsistencies in biomedical fields. Some of the topics known to have inconsistencies in biomedical fields and are considered in the dataset are:

i. Benefits of intermittent fasting
ii. Low-carb diets and health
iii. Dairy products and bone health
iv. Benefits of a sugar detox

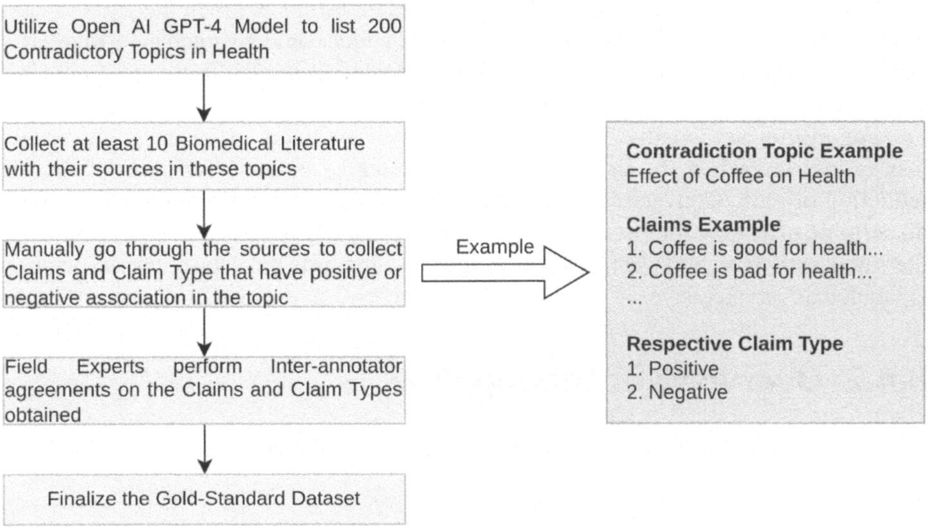

FIGURE 8.2 Pipeline for gold standard dataset creation.

Afterwards, in the first phase, a compilation of 10 biomedical research articles with their corresponding authors and sources was collected for a selected subset of studies. In the second phase, a thorough manual review was performed by three domain experts to extract phrases relevant to the designated subjects. For example, on the subject of "Benefits of Intermittent Fasting," two opposing assertions that were curated are given as shown in Table 8.1. To create a dataset that includes both positive and negative research claims on each selected topic, this manual approach involves a comprehensive assessment of at least 10 research articles per topic.

The final curated dataset comprises three main research topics: Artificial sweeteners and weight loss, red meat consumption and cardiovascular disease, and red meat and cancer risk. Besides a single article from 1997, all the others are published after the year 2000 as they reflect current practices, advancements in methodology and temporal trends in the respective fields.

TABLE 8.1 Two example claims on the topic "Benefits of intermittent fasting"

RESEARCH CLAIMS	*ASSERTION*
This overview suggests that intermittent fasting regimens may be a promising approach to losing weight and improving metabolic health for people who can safely tolerate intervals of not eating, or eating very little, for certain hours of the day, night, or days of the week (Patterson et al., 2015).	Positive
In this RCT, a prescription of time-restricted eating (TRE) group did not result in weight loss when compared with a control prescription of three meals per day (Lowe et al., 2020).	Negative

Seven studies revealed negative or indirect relationships between the pairs of entities (i.e., red meat and cancer risk). Conversely, 13 studies demonstrated a positive association. The remaining three studies yielded inconclusive results, failing to establish a clear relationship between the entity pairs examined.

For instance, a positive relationship between artificial sweeteners and weight loss is inferred when the use of artificial sweeteners either contributes to weight reduction or gain. Conversely, if the data indicate that artificial sweeteners have no discernible effect on weight loss, it is classified as a negative claim. In cases where the impact of artificial sweeteners on weight loss remains uncertain, the association is labeled as inconclusive.

8.4.2 Overview of the pipeline

The complete pipeline involves following three main components:

1. Information Retrieval
2. Machine Learning Classifier
3. Explainable AI (XAI)

The pipeline initiates with the information retrieval (IR) system made especially for biomedical data. The proposed architecture consists of three key components: a *datastore*, *syntactic* analysis, and *semantic* processing. These components are seamlessly integrated, allowing the system to address user questions about biomedical research with efficiency. By asking queries, users start the process, which sets off a pipeline in which the syntactic component finds relevant documents in the datastore. The semantic component then fine-tunes the analysis by identifying the most pertinent research claims related to the query. With its customizable characteristics, this flexible system may be tailored to satisfy a wide range of needs in the biomedical research area, resulting in optimal performance when collecting research claims and related documentation.

The second component of the pipeline is the ML classifier, which requires model training and dataset preparation. A methodical approach to the creation of datasets includes the deliberate identification, collection, and organization of relevant biomedical information from many sources. The data is subsequently cleaned, standardized, and organized using preprocessing techniques to prepare it for model training. Tokenization, text normalization, missing value handling, and noise reduction are all included in this. Preprocessed data is fed into LLMs like BERT during training so that the model can pick up on correlations, patterns, and contextual signals. The learned information from the training data enables the trained model to perform a range of tasks, including the classification of text for biomedical research.

The last component of the pipeline is XAI, which is a key component that improves the system's transparency and reliability. Deeply examining the choices made by the ML pipeline, XAI processes offer full explanations for their predictions. This method uses information from claim sources with full context to explain the output of ML classifiers. This focus on explainability not only makes the system more comprehensible but also gives users trust in the way its decisions are made.

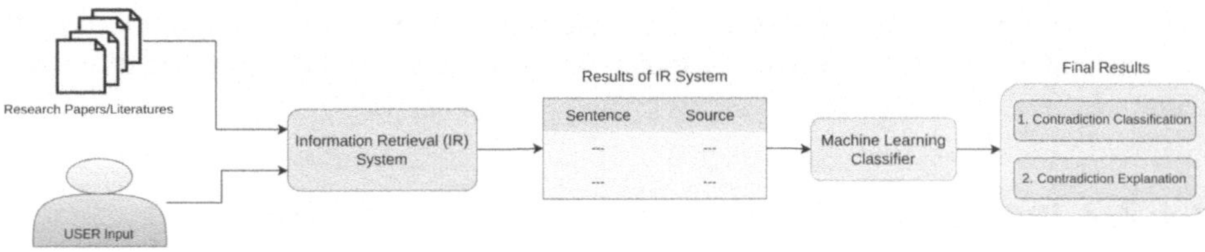

FIGURE 8.3 Full-phase contradiction detection pipeline.

The full-phase contradiction detection pipeline can be summed up by Figure 8.3. At first, relevant claims/contexts are extracted on the basis of user's input, using ML classifiers they are predicted as contradictory or not with respect to user's input. Finally, the results are further explained using XAI. In the following sections, we describe each component in detail.

8.4.2.1 Information Retrieval System

An IR system extracts meaningful information from vast volumes of data, such as research articles, based on user input or queries. For our IR system, the steps involved are data collection, cleaning, preprocessing, pipeline construction, and pipeline evaluation. As shown in Figure 8.4, the fundamental components of the information retrieval pipeline are created with Haystack (Haystack, 2018), an open-source Python framework based on LLMs that allows creation of unique natural language processing (NLP)-driven applications. The development of the initial working prototype of the IR pipeline is reported elsewhere (Lamichhane & Kahanda, 2023).

Furthermore, a number of measures were also utilized to further evaluate the system that was created. Basically, (I) IR development pipeline and (II) evaluation pipeline

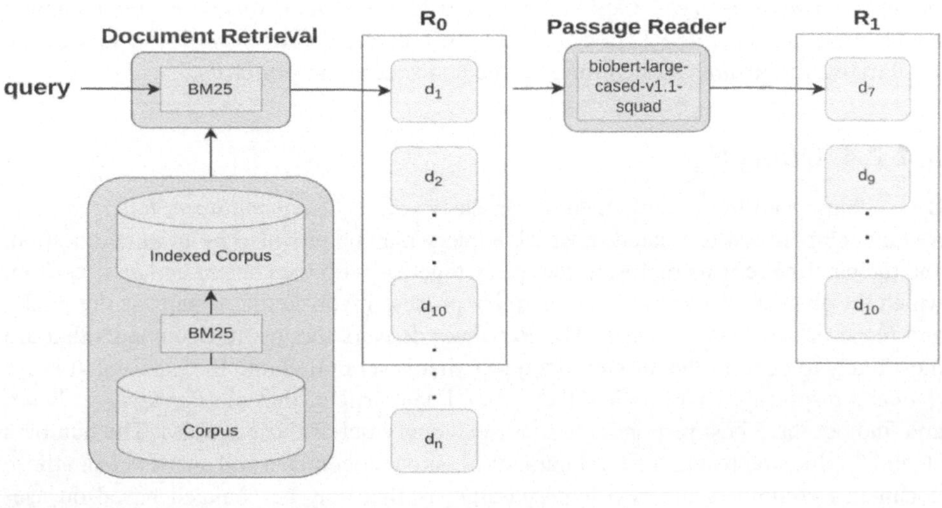

FIGURE 8.4 Multistage information retrieval pipeline.

make up the core components of the full-phase information retrieval system. The IR development architecture involves various processes and components that are discussed below.

8.4.2.2 Indexing documents in a DocumentStore

The first step was to create a *DocumentStore* for storing the biomedical articles that would be used to respond to user queries. In contrast to traditional methods, which usually pre-process and divide documents into segments before indexing, the approach utilized here in the pipeline handles individual articles as a single piece of text. This is because when a user queries into the system, it is required that only a single claim be recommended from a single article. On the contrary, if a document is segmented, each part can be thought of as a separate source of information and claims could be recommended from those parts coming from the same document or same research article. The goal of this choice is to make sure that when a user queries a document, the IR pipeline returns a variety of replies obtained from different articles, not from different segments of a single document.

8.4.2.3 Retriever

A *Retriever* was used to find relevant documents for a given query by going through all documents that are accessible and selecting just those that are pertinent to the query. Because it considers both the frequency of terms in documents and their length, *BM25Retriever* (Wang et al., 2021) was utilized in the retriever pipeline for measuring syntactic similarity in information retrieval tasks.

Following the *Retriever*'s selection of pertinent articles for the user query, a *Reader* was employed to read the article contents and identify the most promising text segments for potential responses. The *Reader* utilizes a BERT model to understand the semantics of the user's query for the recommendation. For the question-answering, a model known as *dmis-lab/biobert-large-cased-v1.1-squad*, a *FARMReader,* was used in conjunction with a large-cased *BioBERT* (Lee et al., 2020). Whereas the *Retriever* looks at syntactic similarities, the *Reader* concentrates on the text's semantic similarity.

8.4.2.4 Querying

By combining all three components: the elasticsearch *DocumentStore*, *Retriever*, and *Reader*, a pipeline was created in which a query may be provided as an input question, and the pipeline returns replies to the query together with the context and articles from which the answer was derived. The inquiry process involved many steps as the reader and retriever are separate parts. The *Retriever* delivers the top 10 documents that are most likely to contain the answer to a query that is submitted into the pipeline. It is the *Reader*'s responsibility to review the 10 candidate articles that the *Retriever* gathered and find out the 5 best responses to the user's query out of those articles. The numbers 10 and 5, which represent the total number of output documents and answers from those documents combined, are two hyperparameters that may be changed based on user requirements.

8.4.2.5 Evaluation

To determine the information retrieval system's efficacy and pinpoint areas for development, evaluation is required. It enables assessing the prediction quality of the system, which is crucial for determining its success. As mentioned before, the ManConCorpus dataset was used to test the components of the information retrieval system. The reader and the retriever function as separate sequential pipelines, necessitating different evaluation procedures.

8.4.2.5.1 Retriever Evaluation

At first, the *Retriever* used in the IR system was evaluated. A *Retriever*'s primary task is to recommend relevant documents based on user input. The pipeline is queried using PICO-structured questions extracted from the ManConCorpus to evaluate its performance. Subsequently, a comparison was made between the articles collected by the pipeline and those already present in the ManConCorpus. For clarity, Figure 8.5 illustrates the *Retriever* pipeline's functionality.

The ManConCorpus dataset, which consists of 24 PICO questions linked to several documents and related claims, was used to assess the *Retriever*'s efficacy. To get the documents that are most relevant to the user's query, the *Retriever* recommends documents depending on the query. Each PICO question in the ManConCorpus was fed to the *Retriever* in the experimental configuration, and it then recommended a group of documents that are most relevant to the question at hand. Then, each PICO question's recommended documents were compared to those listed in the ManConCorpus to assess how accurate the *Retriever*'s recommendations were.

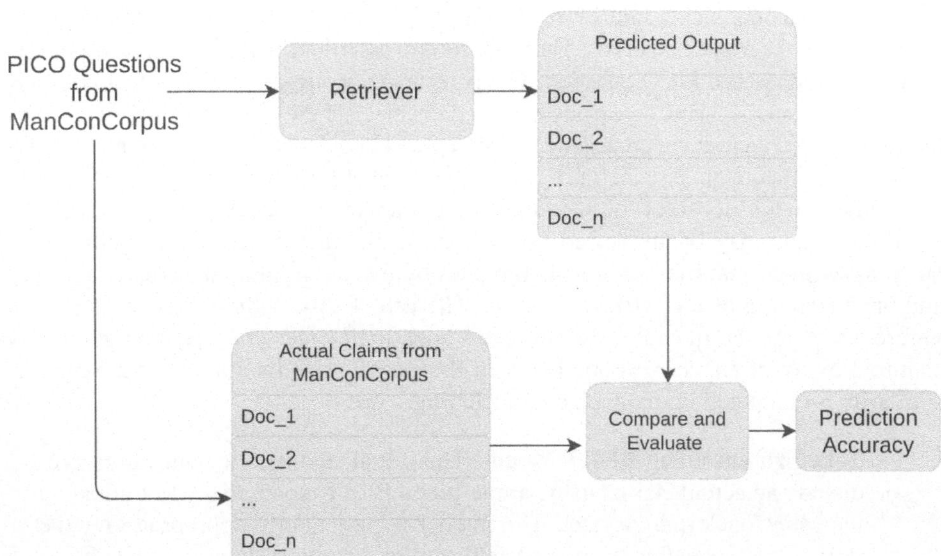

FIGURE 8.5 *Retriever* evaluation pipeline.

It is interesting to observe that not every PICO question in the ManConCorpus has the same number of documents associated with it; some questions include more or less documents than the *Retriever*'s default recommendation of 10. The assessment process was modified to take into consideration the varied document counts per question to resolve this disparity. In particular, the alignment between the suggested and actual documents out of all the accessible articles was evaluated for PICO queries that had less than 10 related documents.

A total of 78.37% of the 24 assessed PICO questions were answered correctly, according to the study. Interestingly, the *Retriever*'s suggestions precisely matched the gold-standard documents linked to the related PICO queries in 10 cases. Nonetheless, it is important to recognize some of the difficulties that arose throughout the assessment process.

Because the PICO questions on cardiovascular diseases had a similar theme, even when the keywords differed slightly, there were several instances of strong syntactic similarity among the questions. Sometimes this syntactic overlap resulted in subpar retrieval ability, especially when attempting to distinguish between items that were closely related. Future improvements to the *Retriever*'s suggestion system are thus necessary to solve these issues and further improve performance. These improvements should include a refinement of syntactic similarity measurements.

8.4.2.5.2 Reader Evaluation

The *Reader* concentrates on the semantic similarity between texts, namely the user's input and the documents. As the reader returned five research claims as output to user's query, the five predicted research claims were compared with five curated research claims from ManConCorpus on corresponding documents as a part of the evaluation process. Semantic similarity and exact matching were the two main metrics used in the information retrieval system's reader pipeline performance evaluation.

Exact matching was tested by comparing the sentences suggested by the reader to those found in the ManConCorpus. There are several assertions in the ManConCorpus that explicitly address each PICO question. The percentage of suggested claims that precisely matched those in the ManConCorpus was used to calculate the exact matching score. The findings showed an average precise matching score of about 48%, despite the inherent difficulties in predicting precise claims from a wide range of potential sentences.

Although the accuracy of the reader pipeline in reproducing claims from the ManConCorpus may be measured directly through exact matching, semantic similarity assessment plays an additional function by evaluating the contextual relevance and understanding of the retrieved claims. The use of this methodology, as shown in Figure 8.6, is crucial in capturing intricate semantic linkages that may not be entirely captured by exact matching alone between the expected and actual claims. Semantic similarity is computed in many phases, including:

1. **Tokenization using BERT Model:** The initial step is to tokenize both predicted and actual claims using a fine-tuned BERT model that was trained on the ManConCorpus dataset. This model is particularly appropriate for the task of claim prediction in the healthcare and medical domains since it has been directed to comprehend and handle biomedical text input.

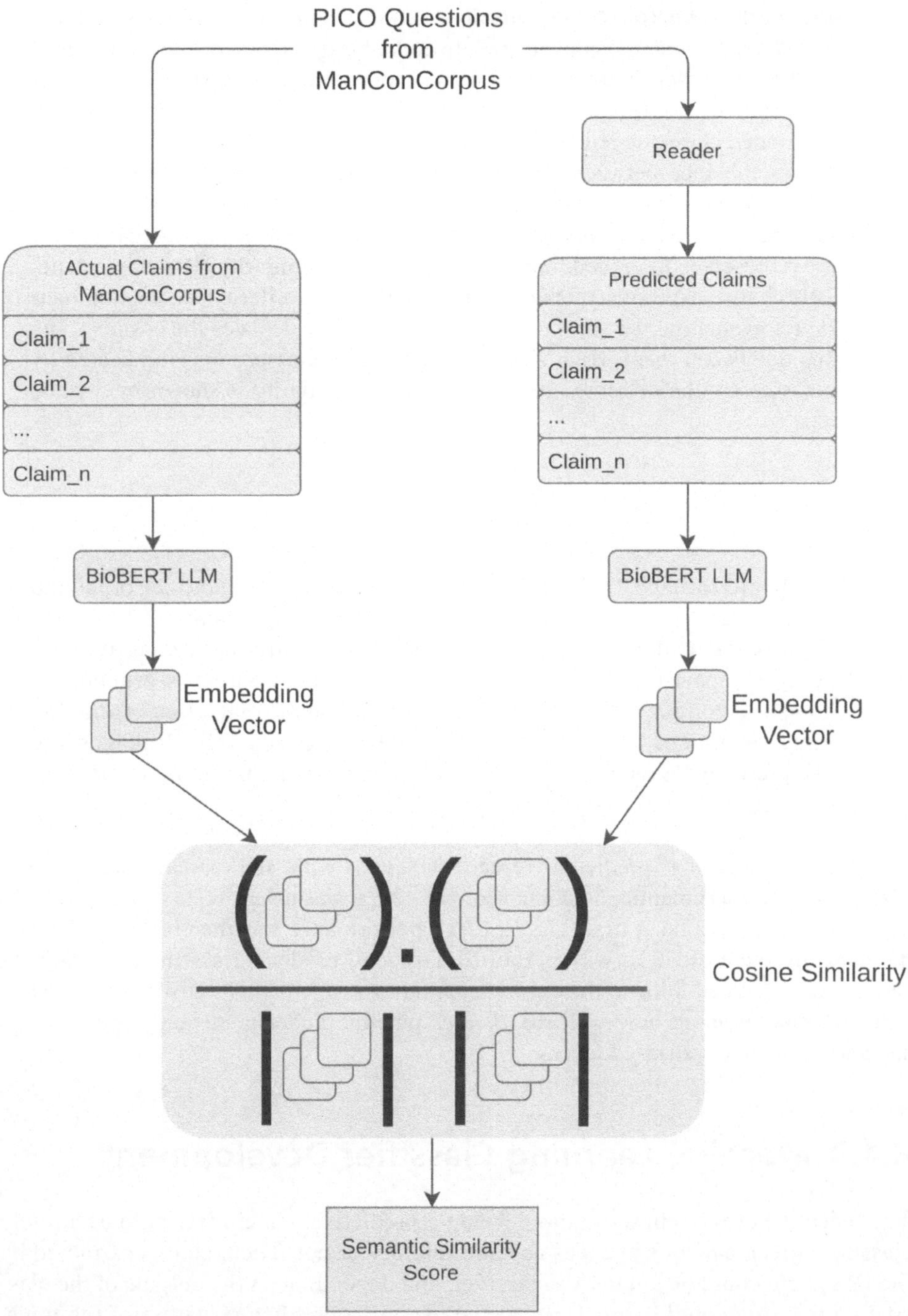

FIGURE 8.6 Reader evaluation pipeline using semantic similarity.

2. **Embedding Vector Representation:** After tokenization, the predicted and actual claims undergo conversion into embedding vectors utilizing the identical BERT model. Embedding vectors serve to represent textual data within a continuous vector space, wherein words possessing similar meanings are positioned closely together. This process facilitates the assessment of semantic resemblance between the predicted and actual claims, relying on their respective vector representations.
3. **Cosine Similarity Calculation:** Now that the predicted and actual claims are represented as embedding vectors, the cosine similarity (Rahutomo et al., 2012) between these vectors is calculated. Cosine similarity is a popular metric for evaluating the similarity of two vectors. It calculates the cosine of the angle between them. High similarity between the assertions is indicated by a cosine similarity score around 1, while dissimilarity is shown by a value near 0.

$$Cosine\ Similarity(a,b) = \frac{a \cdot b}{\|a\|\,\|b\|}$$

In this formula, "a" represents the actual claim from ManConCorpus and "b" represents the predicted claim based on the user question.
4. **Contextual Coherence Preservation:** Making sure that semantic similarity comparisons are carried out in the proper context is crucial. As a result, comparisons are limited to statements that come from the same document. This method guarantees that semantic linkages are assessed within the appropriate article's context and maintain contextual coherence.

While comparing predicted research claims with the actual claims from ManConCorpus, a semantic matching score of 42.3% was noted while using RoBERTa in the *Reader*'s pipeline at first. The *Reader*'s performance was then further enhanced by changing out RoBERTa with a BioBERT model, producing a semantic matching score of 63.2%. In addition to the exact matching assessment, this statistic offers important information about how well the *Reader* pipeline understands and replicates the semantic content of retrieved texts.

8.4.3 Machine Learning Classifier Development

The objective of this phase was to develop a predictive model that could accurately classify a given pair of sentences as contradictory or not. Then, claims extracted by the IR system could be fed to this classifier, and depending on the outputs of the classifier, a decision could be made about the overall contradictory nature of the topic. If the classifier classified all of the sentence pairs as noncontradictory, the topic was considered noncontradictory (or that there was not enough evidence to claim otherwise).

8.4.3.1 Data preparation for classifier training

For developing the ML classifier, we again used a ManConCorpus dataset. However, the ManConCorpus dataset was systematically arranged into three key columns: the *Contradiction Topic*, the *Claims*, and the *Claim Type*. In this case, the ManConCorpus Assertion Value was represented by the Claim Type. But the objective was to develop a classifier that could detect a contradiction between two given claims. Therefore, we prepared our own dataset by pairing up contradictions. In order to effectively identify contradictions, pairs of Claims were created only within every Contradiction Topic.

For example, suppose there were two Contradiction Topics, A and B, each of which had three research claims (A1, A2, A3 and B1, B2 and B3). Pairings were formed inside each topic, resulting in pairs A1|A2, A1|A3, A2|A3, B1|B2, B2|B3, and B1|B3. The "Contradiction" label for each pair was selected according to the type of assertion used in the claims. Pairs of sentences that had the same Claim Type were regarded as noncontradictory and cohesive. On the other hand, different Claim Types pointed to inconsistencies and differences in thought.

After this new dataset was prepared, a *preprocessing* step was designed to enhance the textual data in preparation for the next stage of model training. Using a special function applied to the Claim1 and Claim2 columns, nonalphanumeric and nonwhitespace characters were systematically removed. Next, Claim1 and Claim2 were concatenated and separated by the proper separator tokens ([CLS] and [SEP]) to create a new column called "Combined." Then, to create a structured training dataset, relevant columns were chosen, particularly "Combined" and "Assertion." The dataset rows were shuffled to add unpredictability and reduce potential biases during model training. Table 8.2 shows a pair of this prepared dataset.

After comparing the claims within the same topic in the ManConCorpus dataset, which originally consisted of 259 research claims from 24 systematic reviews, the previously described approach yielded a dataset of 1,775 samples. One-thousand and forty-seven samples in this dataset were noncontradictory and 728 samples were contradictory.

TABLE 8.2 A portion of the prepared dataset for classifier training

COMBINED	*ASSERTION*
[CLS] Supplementation during pregnancy with a medical food containing L-arginine and antioxidant vitamins reduced the incidence of preeclampsia in a population at high risk of the condition [SEP]. Oral L-arginine supplementation did not reduce mean diastolic blood pressure after 2 days of treatment compared with placebo in preeclamptic patients with gestational length varying from 28 to 36 weeks.	1
[CLS] Supplementation during pregnancy with a medical food containing L-arginine and antioxidant vitamins reduced the incidence of preeclampsia in a population at high risk of the condition [SEP]. We conclude that in women with preeclampsia, prolonged dietary supplementation with L-arginine significantly decreased blood pressure through increased endothelial synthesis and/or bioavailability of NO.	0

8.4.3.2 Development of predictive model

During the model training and development process, four distinct BERT models—Distil-BERT (Sanh et al., 2019), BioBERT(Lee et al., 2020), PubMedBERT (Gu et al., 2021), and Bioformer (Fang et al., 2023)—were explored, last three of which were biomedical BERT models. BioBERT is pretrained on a large-scale biomedical text corpus that includes PubMed abstracts and publications to capture domain-specific vocabulary used in the biomedical area. As the name implies, PubMedBERT receives its pretraining just from PubMed articles—a sizable collection of biological literature. In a similar vein, PubMed abstracts and full-text publications from PubMed Central are used to pretrain Bioformer.

These three models were especially useful for this current study since they have a solid basis for comprehending the nuances of medical and health-related language because they have been pretrained on large biomedical datasets. Choosing these models allowed us to use their prior understanding of biomedical language and complexities, improving their ability to identify inconsistencies in the context of general health, which was the main objective. Distil-BERT, which is a general-purpose LLM, was also an effective choice for the study because of its strong language representation, computational efficiency, and ability to efficiently manage large datasets. These features all enhanced the efficacy of the process of fine-tuning for the identification of contradictions in the field of common health.

As demonstrated in Figure 8.7, the initial phase in this method divided the preprocessed dataset into training and testing sets using the 80–20 train-test split. The next step was tokenization, where the textual data was transformed into tokenized input that was prepared for model ingestion by leveraging the advanced features of the model tokenizer. Then, each model was fine-tuned with the train data and evaluated with the test data.

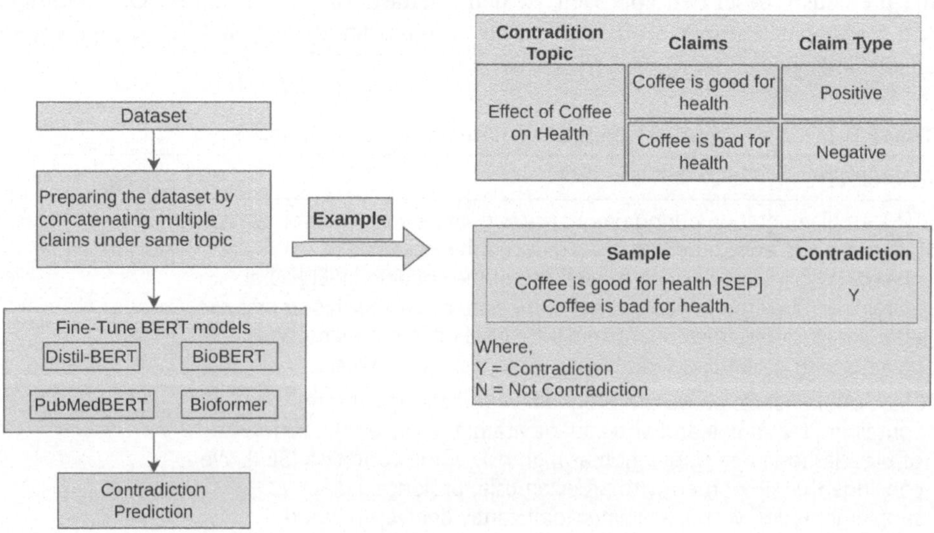

FIGURE 8.7 A simple example of predicting contradictions using fine-tuned BERT.

Using these as a foundation, a *TensorFlow* data pipeline was built, generating essential dataset objects from tokenized encodings. The critical hyperparameters were carefully assessed before a *TFTrainer* was constructed using TensorFlow's high-level API and the *Hugging Face* Transformers (Hugging Face) package. The number of iterations over the training dataset depends on the number of training epochs (num_train_epochs=2). To balance computational efficiency and model performance, batch sizes were carefully defined (per_device_train_batch_size=8 and per_device_eval_batch_size=8) for both training and evaluation.

Moreover, the addition of warm-up steps (warmup_steps = 500) and weight decay strength (weight_decay = 0.01) to the learning rate scheduler complicated the optimization procedure and improved the model's flexibility. Furthermore, evaluation steps (eval_steps = 10) were carefully set up to assess the model's performance on a frequent basis. Distil-BERT, BioBERT, PubMedBERT, and Bioformer were the four models that undergo fine-tuning in the same hyperparameter environment, guaranteeing a consistent and similar framework for the fine-tuning process.

8.4.3.3 Machine learning model results and discussion

The performance evaluation of the four BERT model utilized the ManConCorpus. The BERT model was trained and fine-tuned in two separate processes after the dataset had been preprocessed and made ready for training. First, the pretrained model was evaluated instantly after being loaded from the *Hugging Face* (Hugging Face, 2016) platform. The model was then loaded, and fine-tuned on the ManConCorpus dataset, and its performance was assessed in the second stage. By utilizing domain-specific data—in this example, the ManConCorpus dataset—to fine-tune the preexisting model, BERT may better identify inconsistencies in the unique context of cardiovascular disease.

Table 8.3 presents the classification results for pretrained BERT models tested on the ManConCorpus. With an accuracy of 0.41 and a flawless recall of 1.0, Distil-BERT performed exceptionally well, yielding an F1 score of 0.58. With an F1 score of 0.58, a precision of 0.43, and a high recall of 0.89, PubMedBERT likewise performed admirably. On the other hand, BioBERT and Bioformer demonstrated much lower F1 scores of 0.52 and 0.38, respectively, suggesting difficulties with precision and recall.

The classification results following BERT fine-tuning using the ManConCorpus are shown in Table 8.4., which shows significant performance gains for all models. With an F1 score of 0.97, which was attained by excellent precision (0.97) and recall (0.97), PubMedBERT was particularly noteworthy. BioBERT also showed a significant

TABLE 8.3 Classification results on pretrained BERT models tested on ManConCorpus

MODEL	*PRECISION*	*RECALL*	*F1 SCORE*
Distil-BERT	0.41	1	0.58
BioBERT	0.41	0.72	0.52
PubMedBERT	0.43	0.89	**0.58**
Bioformer	0.67	0.47	0.38

TABLE 8.4 Classification results on fine-tuned BERT models tested on ManConCorpus

MODEL	*PRECISION*	*RECALL*	*F1 SCORE*
Distil-BERT	0.87	0.97	0.92
BioBERT	0.92	0.97	0.95
PubMedBERT	0.97	0.97	**0.97**
Bioformer	0.85	0.97	0.91

improvement, achieving an F1 score of 0.95 and a high precision score of 0.92. With F1 values of 0.92 and 0.91, respectively, Distil-BERT and Bioformer had strong performance, demonstrating the effectiveness of fine-tuning in improving the models' classification abilities in the particular area of cardiovascular disease.

As mentioned before, the fine-tuning dataset used for these results consisted of 1,775 samples, 1,047 of which were noncontradictory and 728 of which were contradictory. Recall and F1-score were the main focus for the biomedical prediction job, where the dataset was unbalanced with a greater proportion of noncontradictory samples. Finding contradicting situations was essential, and a high recall rate should be the top concern. The F1-score was examined prior to recall because of the dataset's imbalance. The best model to predict contradiction in the sample dataset, according to these results, was PubMedBERT. This model was thus utilized in the proof-of-concept biomedical contradiction detection pipeline.

8.4.3.4 Explainability in AI

Once model predictions were obtained, the goal was to provide further context for the outcomes that the BERT models anticipated. *Explainability* increases dependability by helping to understand how BERT models identify contradictions in biomedical literature (Rasheed et al., 2022). Furthermore, it is essential in providing users with a thorough comprehension of the reasons behind the identification of particular statements as contradictory in relation to a particular query. Clarifying the fundamental reasoning behind the model's decision-making process was intended to improve transparency and allow for the educated interpretation of contradictions that were found.

To improve the explainability of the contradiction detection procedure inside the pipeline, a systematic technique was used. The IR system retrieved a collection of five relevant study claims in response to a user inquiry. To maintain uniformity, these claims were preprocessed using methods that aligned with the dataset training approach. Then, the BERT-based contradiction detection model was applied to the claims. The question was classified as a contradictory subject if the model found at least one contradiction between any two claims.

In the output, three sets of contradicting assertions are shown to the user in these situations to be examined. Additionally, contextual information was extracted from the original research materials to provide users with a better understanding of the contradictions that had been found. Specifically, the sentences that immediately appeared before and followed after the contradicting assertions were extracted. By placing the conflicting assertions into the larger framework of the initial research findings, this contextual information improves the transparency and interpretability of the contradiction detection process.

For each contradictory topic detected by the system, the output presented to the user is composed of three pairs of statements (labelled as Claim_1 and Claim_2), each opposing the other, from several research articles grouped as Contradiction #1, Contradiction #2, and Contradiction #3. In addition, Claim_1_Context and Claim_2_Context are provided as output to help researchers comprehend the context of these contradicting statements. These provide a good illustration of the relative positions of Claim_1 and Claim_2 inside the original text. In addition, the corresponding PMID hyperlinked to their original source articles are included. This allows the user an opportunity for further exploration if they think the provided contexts are not adequate. An actual output produced by the system is depicted in Figure 8.8.

USER QUERY: Could you please provide information on whether there is a causal relationship between red meat consumption and cancer development?

CONTRADICTIONS DETECTED

BioFactCheck has detected contradictions in the context of the user query. In other words, it appears that there are varying claims supported by experimental or clinical trial evidence that oppose each other. Following are some of the contradictory claims reported in literature.

DETECTED CONTRADICTION #1

CLAIM_1 (PMID: 21110906): The relationship between meat consumption and breast cancer has been the focus of several epidemiological investigations, yet there has been no clear scientific consensus as to whether red or processed meat intake increases the risk of breast cancer.

CLAIM_1_CONTEXT: The relationship between meat consumption and breast cancer has been the focus of several epidemiological investigations, yet there has been no clear scientific consensus as to whether red or processed meat intake increases the risk of breast cancer. We conducted a comprehensive meta-analysis incorporating data from several recently published prospective studies of red or processed meat intake and... -

CLAIM_2(PMID: 22412075): The corresponding HRs (95% CIs) were 1.18 (1.13-1.23) and 1.21 (1.13-1.31) for CVD mortality and 1.10 (1.06-1.14) and 1.16 (1.09-1.23) for cancer mortality.

CLAIM_2_CONTEXT: 5% CI) of total mortality for a 1-serving-per-day increase was 1.13 (1.07-1.20) for unprocessed red meat and 1.20 (1.15-1.24) for processed red meat. The corresponding HRs (95% CIs) were 1.18 (1.13-1.23) and 1.21 (1.13-1.31) for CVD mortality and 1.10 (1.06-1.14) and 1.16 (1.09-1.23) for cancer mortality. We estimated that substitutions of 1 serving per day of other foods (including fish, poultry, nuts, legumes, low-fat dairy, and whole grains) for 1 s...

DETECTED CONTRADICTION #2

FIGURE 8.8 Output for a query from in-house developed gold-standard dataset. (Continued)

CLAIM_1 (PMID: 21110906): The relationship between meat consumption and breast cancer has been the focus of several epidemiological investigations, yet there has been no clear scientific consensus as to whether red or processed meat intake increases the risk of breast cancer.

CLAIM_1_CONTEXT: The relationship between meat consumption and breast cancer has been the focus of several epidemiological investigations, yet there has been no clear scientific consensus as to whether red or processed meat intake increases the risk of breast cancer. We conducted a comprehensive meta-analysis incorporating data from several recently published prospective studies of red or processed meat intake and...

CLAIM_2 (PMID: 21674008): Here we update the evidence from prospective studies and explore whether there is a non-linear association of red and processed meats with colorectal cancer risk.

CLAIM_2_CONTEXT: ...in the 2007 World Cancer Research Fund/American Institute of Cancer Research report. Since then, ten prospective studies have published new results. Here we update the evidence from prospective studies and explore whether there is a non-linear association of red and processed meats with colorectal cancer risk. Relevant prospective studies were identified in PubMed until March 2011. For each study, relative risks and 95% confidence intervals (CI) were extrac...

===

DETECTED CONTRADICTION #3

===

CLAIM_1 (PMID: 21110906): The relationship between meat consumption and breast cancer has been the focus of several epidemiological investigations, yet there has been no clear scientific consensus as to whether red or processed meat intake increases the risk of breast cancer.

CLAIM_1_CONTEXT: The relationship between meat consumption and breast cancer has been the focus of several epidemiological investigations, yet there has been no clear scientific consensus as to whether red or processed meat intake increases the risk of breast cancer. We conducted a comprehensive meta-analysis incorporating data from several recently published prospective studies of red or processed meat intake and...

CLAIM_2 (PMID: 25941850): The potential relationship between red meat consumption and colorectal cancer (CRC) has been the subject of scientific debate.

CLAIM_2_CONTEXT: The potential relationship between red meat consumption and colorectal cancer (CRC) has been the subject of scientific debate. Given the high degree of resulting uncertainty, our objective was to update the state of the science by conducting a systematic quantitative assessme...

FIGURE 8.8 (Continued)

8.5 OVERALL RESULTS AND DISCUSSION

The fundamental goal of this study was to find conflicts within the domain of common health, which is defined as areas of general health concerns. Since a contradictory corpus in this field does not exist, the first efforts were concentrated on building a gold standard dataset to make machine learning model training easier. The process of developing a dataset is known to be tedious and time-consuming, especially when careful verification is needed, and several annotators are needed to establish an agreement. We developed an in-house, smaller, but high-quality dataset—consisting of 23 research articles addressing three different but contradictory topics (refer to the section "In-house Gold-Standard Dataset" for more details about this dataset). This dataset is regarded as the gold standard for evaluating the overall pipeline performance.

Using the ManConCorpus Dataset, the performance evaluation of each individual component of the pipeline was described in depth in the corresponding sections. The developed pipeline showed the potential to detect contradictions in any text comparison and retrieve documents based on queries, even if the primary focus was on common health. In order to verify the proof-of-concept pipeline's adaptability, it was tested on the in-house gold-standard dataset—a dataset previously unseen by the system. To do this, the IR pipeline's document store was loaded with the 23 research articles that annotators evaluated and verified. A user query was then created to evaluate the pipeline's capacity to extract relevant claims and identify inconsistencies. The output for one of the queries is displayed in Figure 8.8.

As can be seen in Figure 8.8, the system has detected contradictions related to the user query regarding the causal relationship between red meat consumption and cancer development. Three distinct contradictions have been presented to the user. The first pair is contradictory as Claim_1 questions the consensus on whether red or processed meat intake increases breast cancer risk, while Claim_2 provides data indicating an association between red meat consumption and higher mortality rates for cardiovascular disease and cancer.

In the second contradiction, while Claim_1 focuses on the relationship between meat consumption and breast cancer, specifically mentioning the lack of consensus on whether red or processed meat intake increases the risk, Claim_2 addresses the association of red and processed meats with colorectal cancer risk, indicating a different focus and outcome measure.

Lastly, in the third contradiction, according to Claim_1 there is a lack of clear evidence regarding the relationship between red meat consumption and cancer risk. But Claim_2 is an example of a misclassification by the system since the sentence does not support any conclusion (i.e. it is merely an objective statement).

8.6 CONCLUSIONS AND OPEN PROBLEMS

In this research, a complete explainable biomedical contradiction detection pipeline was proposed, prototyped, and tested, currently comprising a concatenation of three different pipelines. The IR system employed a retriever and a reader to grasp the syntactic and semantic relationship between the user's query and various biomedical literature, recommending the most appropriate biomedical claims to answer the query. These claims were then passed through the ML system, utilizing a fine-tuned BERT model to evaluate whether contradictions existed. If contradictions were predicted, further explanation was provided using a simple implementation of XAI, offering contextual references of the sources of the biomedical claims with contradictions.

However, it should be noted that, while the pipeline's performance was promising, it was trained on a rather limited dataset. In order to increase its robustness, using a larger dataset could produce more trustworthy results, considering the possible impact of the ManConCorpus's constrained size. Comprehensive hyperparameter tuning might help the ML contradiction classifier perform even better in future efforts. Investigating fine-tuning on more potent models, such as advanced GPT model, may also be beneficial. Furthermore, rating the outputs to a larger set of queries using an expanded gold-standard dataset should further demonstrate the utility of this approach. Prefiltering for entities or postprocessing of outputs to improve specificity would likely improve the overall performance and usefulness to the user.

Overall, this work signals a substantial advancement in the field of biomedical text contradiction detection, and additional study and improvement of the IR system should provide further enhancements. Enhancing AI's explainability and IR claims suggestion is a critical next step. With this method, end users should be better equipped to identify inconsistencies and understand the underlying causes of predictions. In addition, there is potential for effectively detecting discrepancies in real time by incorporating the established contradiction detection technique into online IR or online publication monitoring.

ACKNOWLEDGMENT

We are grateful for Dr. Jenifer Ross and Dr. Zhiping Yu from the Department of Nutrition and Dietetics at University of North Florida for their assistance in curating the in-house developed gold-standard dataset.

REFERENCES

Alamri, A. (2016). *The detection of contradictory claims in biomedical abstracts*. University of Sheffield.

Alamri, A., & Stevenson, M. (2015). Automatic identification of potentially contradictory claims to support systematic reviews. *2015 IEEE International Conference on Bioinformatics and Biomedicine (BIBM)*, 930–937.

Alamri, A., & Stevenson, M. (2016). A corpus of potentially contradictory research claims from cardiovascular research abstracts. *Journal of Biomedical Semantics*, *7*(1), 1–9.

Arrieta, A. B., Díaz-Rodríguez, N., Del Ser, J., Bennetot, A., Tabik, S., Barbado, A., & Herrera, F. (2020). Explainable artificial intelligence (XAI): Concepts, taxonomies, opportunities and challenges toward responsible AI. *Information Fusion*, *58*, 82–115.

Carpenter, D. M., Geryk, L. L., Chen, A. T., Nagler, R. H., Dieckmann, N. F., & Han, P. K. (2016). Conflicting health information: A critical research need. *Health Expectations*, *19*(6), 1173–1182.

Das, A., & Rad, P. (2020). Opportunities and challenges in explainable artificial intelligence (xai): A survey. *ArXiv Preprint ArXiv:2006.11371.*

De Marneffe, M.-C., Rafferty, A. N., & Manning, C. D. (2008). Finding contradictions in text. *Proceedings of Acl-08: Hlt*, 1039–1047.

Fang, L., Chen, Q., Wei, C.-H., Lu, Z., & Wang, K. (2023). Bioformer: an efficient transformer language model for biomedical text mining. *ArXiv Preprint ArXiv:2302.01588.*

Flier, J. S. (2023). Publishing biomedical research: A rapidly evolving ecosystem. *Perspectives in Biology and Medicine*, *66*(3), 358–382. https://doi.org/10.1353/pbm.2023.a902032

Gu, Y., Tinn, R., Cheng, H., Lucas, M., Usuyama, N., Liu, X., Naumann, T., Gao, J., & Poon, H. (2021). Domain-specific language model pretraining for biomedical natural language processing. *ACM Transactions on Computing for Healthcare (HEALTH)*, *3*(1), 1–23.

Haystack. (2018). An open source LLM framework. Haystack. https://haystack.deepset.ai/

Hotu, C., Bagg, W., Collins, J., Harwood, L., Whalley, G., Doughty, R., Gamble, G., Braatvedt, G., & DEFEND investigators. (2010). A community-based model of care improves blood pressure control and delays progression of proteinuria, left ventricular hypertrophy and diastolic dysfunction in Māori and Pacific patients with type 2 diabetes and chronic kidney disease: a randomized controlled trial. *Nephrology Dialysis Transplantation*, *25*(10), 3260–3266.

Hugging Face. (2016) The AI community building the future. Hugging Face. https://huggingface.co/

Irving, D. N. (1993). The impact of "scientific misinformation"; On other fields: Philosophy, theology, biomedical ethics, public policy. *Accountability in Research*, *2*(4), 243–272.

Kharrat, A. E., Hlaoua, L., & Ben Romdhane, L. (2022). Contradiction detection approach based on semantic relations and evidence of uncertainty. *International Conference on Computational Collective Intelligence*, 232–245.

Kilicoglu, H., Rosemblat, G., Fiszman, M., & Shin, D. (2020). Broad-coverage biomedical relation extraction with SemRep. *BMC Bioinformatics*, *21*(1), 28.

Lamichhane, P., & Kahanda, I. (2023). Enhancing health information retrieval with large language models: A study on MedQuAD dataset. *2023 International Conference on Machine Learning and Applications (ICMLA)*, 2147–2152. https://doi.org/10.1109/ICMLA58977.2023.00324

Lamichhane, P., Kahanda, I., Liu, X., Umapathy, K., Reddivari, S., Christie, C., Arikawa, A., & Ross, J. (2023). Poster: BioFactCheck: Exploring the feasibility of explainable automated inconsistency detection in biomedical and health literature. *2023 IEEE/ACM Conference on Connected Health: Applications, Systems and Engineering Technologies (CHASE)*, 196–197.

Lee, J., Yoon, W., Kim, S., Kim, D., Kim, S., So, C. H., & Kang, J. (2020). BioBERT: A pre-trained biomedical language representation model for biomedical text mining. *Bioinformatics*, *36*(4), 1234–1240.

Lowe, D. A., Wu, N., Rohdin-Bibby, L., Moore, A. H., Kelly, N., Liu, Y. E., Philip, E., Vittinghoff, E., Heymsfield, S. B., Olgin, J. E., Shepherd, J. A., & Weiss, E. J. (2020). Effects of time-restricted eating on weight loss and other metabolic parameters in women and men with overweight and obesity: The TREAT randomized clinical trial. *JAMA Internal Medicine*, *180*(11), 1491–1499.

Matsui, Y., Eguchi, K., Shimada, K., & Kario, K. (2009). Doxazosin and heart failure: to be or not to be. *Journal of Hypertension*, *27*(2), 434–435.

Novoa, J., Chagoyen, M., Benito, C., Moreno, F. J., & Pazos, F. (2023). PMIDigest: Interactive review of large collections of PubMed entries to distill relevant information. *Genes*, *14*(4), 942.

Patterson, R. E., Laughlin, G. A., Sears, D. D., LaCroix, A. Z., Marinac, C., Gallo, L. C., Hartman, S. J., Natarajan, L., Senger, C. M., Martinez, M. E., Villaseñor, A. (2015). Intermittent fasting and human metabolic health. *Journal of the Academy of Nutrition and Dietetics*, *115*(8), 1203.

Pielka, M., Sifa, R., Hillebrand, L. P., Biesner, D., Ramamurthy, R., Ladi, A., & Bauckhage, C. (2021). Tackling contradiction detection in German using machine translation and end-to-end recurrent neural networks. *2020 25th International Conference on Pattern Recognition (ICPR)*, 6696–6701.

Rahimi, Z., & ShamsFard, M. (2021). Contradiction detection in persian text. *ArXiv Preprint ArXiv:2107.01987.*

Rahutomo, F., Kitasuka, T., Aritsugi, M. (2012). Semantic cosine similarity. *The 7th International Student Conference on Advanced Science and Technology ICAST*, *4*(1), 1.

Rasheed, K., Qayyum, A., Ghaly, M., Al-Fuqaha, A., Razi, A., & Qadir, J. (2022). Explainable, trustworthy, and ethical machine learning for healthcare: A survey. *Computers in Biology and Medicine*, *149*, 106043.

Rosemblat, G., Fiszman, M., Shin, D., & Kilicoglu, H. (2019). Towards a characterization of apparent contradictions in the biomedical literature using context analysis. *Journal of Biomedical Informatics*, *98*, 103275.

Sackett, D. L. (1997). Evidence-based medicine. *Seminars in Perinatology*, *21*(1), 3–5. https://doi.org/10.1016/S0146-0005(97)80013-4

Sanh, V., Debut, L., Chaumond, J., & Wolf, T. (2019). DistilBERT, a distilled version of BERT: smaller, faster, cheaper and lighter. *ArXiv Preprint ArXiv:1910.01108.*

Schütze, H., Manning, C. D., & Raghavan, P. (2008). *Introduction to information retrieval* (Vol. *39*, pp. 234–265). Cambridge: Cambridge University Press.

Sepulveda-Torres, R., Bonet-Jover, A., & Saquete, E. (2021). "Here are the rules: Ignore all rules": Automatic contradiction detection in Spanish. *Applied Sciences*, *11*(7), 3060.

Sosa, D. N., Suresh, M., Potts, C., & Altman, R. B. (2022). Detecting contradictory COVID-19 drug efficacy claims from biomedical literature. ArXiv preprint ArXiv:2212.09867.

Tawfik, N. S., & Spruit, M. R. (2018). Automated contradiction detection in biomedical literature. *International Conference on Machine Learning and Data Mining in Pattern Recognition*, 138–148.

Vaswani, A., Shazeer, N., Parmar, N., Uszkoreit, J., Jones, L., Gomez, A. N., & Polosukhin, I. (2017). Attention is all you need. Advances in neural information processing systems, 30.

Wang, S., Zhuang, S., & Zuccon, G. (2021). Bert-based dense retrievers require interpolation with bm25 for effective passage retrieval. *Proceedings of the 2021 ACM SIGIR International Conference on Theory of Information Retrieval*, 317–324.

Wu, X., Niu, X., & Rahman, R. (2022). Topological analysis of contradictions in text. *Proceedings of the 45th International ACM SIGIR Conference on Research and Development in Information Retrieval*, 2478–2483.

Yang, W., Wei, Y., Wei, H., Chen, Y., Huang, G., Li, X., Li, R., Yao, N., Wang, X., Gu, X., & Amin, M. B. (2023). Survey on explainable AI: From approaches, limitations and applications aspects. *Human-Centric Intelligent Systems*, *3*(3), 161–188.

Yazi, F. S., Vong, W.-T., Raman, V., Then, P. H. H., & Lunia, M. J. (2021a). An experimental evaluation of deep neural network model performance for the recognition of contradictory medical research claims using small and medium-sized corpora. *Malaysian Journal of Computer Science*, 68–77.

Yazi, F. S., Vong, W.-T., Raman, V., Then, P. H. H., & Lunia, M. J. (2021b). Towards automated detection of contradictory research claims in medical literature using deep learning approach. *2021 Fifth International Conference on Information Retrieval and Knowledge Management (CAMP)*, 116–121.

Human-Like e-Learning Mediation Agents

9

Chukwuka Victor Obionwu,
Diptesh Mukherjee, Andreas Nurnberger,
Aarathi Vijayachandran Bhagavathi,
Aishwarya Suresh, Eathorne Choongo,
Bhavya Baburaj Chovatta Valappil,
Amit Kumar, and Gunter Saake

9.1 INTRODUCTION

Instructional feedback plays a crucial role in the process of learning by encouraging self-regulation, inspiring learners, and improving their overall learning experience [1].

Effective instructional feedback should be specific, timely, and constructive in order to guide students toward improvement and mastery of the material. While the primary objective of instructional feedback is to furnish learners with comprehensive information on their knowledge or performance, enabling them to make relevant enhancements and adjustments, it is important for educators to provide feedback that is actionable and encourages reflection on their own learning process. Instructional feedback can take numerous forms, such as written remarks, conversational talks, quizzes, or evaluations [2]. These forms of instructional feedback can either be mediated by a human instructor or automated through technology.

Table 9.1 shows some of the tasks available to students that enroll in our courses and the respective mediation strategies we adopt. For each of these tasks, we have developed and designated respective agents that are specifically suited to provide respective

DOI: 10.1201/9781003570882-11

TABLE 9.1 Possible tasks and mediation strategies in a learning environment

TASKS	*MEDIATION STRATEGY*
Knowledge Assessment Tasks	Chatting and Knowledge Reinforcement Tests
Decision support Tasks	Chatting and Deposition Surveys
Student Modeling Tasks	Personality Surveys
Study Evaluation Tasks	Weekly Test and Exams

support to students. Developing such a system in the past would have been very difficult, as training these agents would require lots of datasets and computer power, but with advancements in machine learning and natural language processing, we are now able to create more sophisticated and efficient systems. More specifically, our current conversational agent uses the generative pretrained transformer (GPT 3.5) turbo [3] and semantic search algorithm [4] to identify the intent for each student inquiry and provide human-like responses. This has allowed us to provide contextually relevant, human-like responses to students inquiries.

In the previous research effort, we described an instructional feedback strategy [5, 6]. Here, we introduce the task generation, evaluation, and assessment platform (TGEAS), Figure 9.1, in which the previous instructional feedback system has been extended to a knowledge agent, Figure 9.2. In the new strategy, students' engagement with Learning Artifacts will be supported by two AI systems: a Knowledge Agent System for interactive chat-based learning and a Task Generation and Evaluation Agent System for creating and assessing tasks. An instructor will supervise the process and intervene when needed, ensuring effective learning through a blend of technology and human guidance. The goal of the system is the provision of human-like automatic instructional feedback. To achieve this goal, we have revised our previous strategy and architecture to provide human-like and contextually relevant instructional feedback. Also in this paper, we will describe our strategy for task generation and assessment, which is also being extended to include generative artificial intelligence task engagement mediation. The rest of the paper will be devoted to the evaluation metrics we adopt to assess the viability of our strategy.

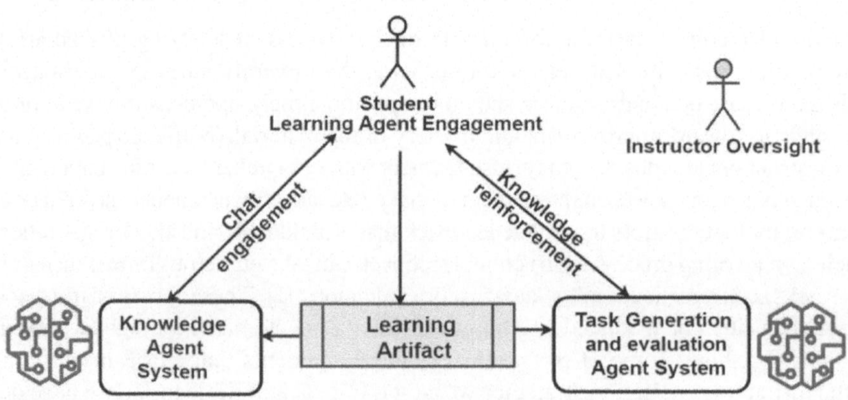

FIGURE 9.1 The task generation, evaluation, and assessment system.

The subsequent sections of the study have been delineated as follows: Section 9.2 provides a brief description of the dialogue system and Section 9.3 describes our retrieval augmented knowledge agent. Section 9.3 describes the task generation and assessment strategy. In Section 9.6, we describe the workings of our task generation and evaluation system (TGES), and in Section 9.5, the evaluation metrics we used for the TGES are described. The evaluation strategy for the TGES is described in Section 9.6, and the respective results are discussed in Section 9.8. We discuss the related works in Section 9.9 and in Section 9.10, we summarize our contributions and indicate directions for future efforts.

9.2 DIALOGUE SYSTEM

Dialogue systems are applications designed to facilitate communication between humans and machines through the use of natural language. They can be used in a variety of use cases, such as customer service, education, and entertainment. The authors Shukla et al. [7] proposed a specification of a task-oriented dialog system. Their system includes components such as natural language understanding, dialog management, and natural language generation to enable effective communication between the agent and the user.

A subset of dialogue systems is the conversational agent. Table 9.2 summarizes some of the main differences between conversational agents and other dialogue systems. Conversational agents are typically designed to mimic human conversation and

TABLE 9.2 Difference between conversational agent and dialogue system

	CONVERSATIONAL AGENT	*DIALOGUE SYSTEMS*
Focus	Primarily emphasize the interaction between a user and a computer system through natural language.	Aim to maintain coherence, context, and continuity across multiple exchanges.
Scope	Designed to simulate human-like conversations, often serving specific purposes such as answering queries, providing information, or executing tasks.	Encompass a broader framework designed to manage and structure extended conversations or dialogues.
Functionality	Uses NLP and AI to comprehend user inputs and generate appropriate responses.	Uses more sophisticated architectures and methodologies to understand user input, generate responses, and maintain the conversation thread, ensuring context and purpose are maintained throughout.
Purpose	Focus on individual interactions, addressing immediate user queries or commands without considering broader contextual continuity.	Create more natural, purposeful, and meaningful conversations, considering the broader context of the interaction beyond isolated queries or commands.

provide more engaging interactions. These applications can engage in both formal and informal discussions and further be tailored to deliver users with precise information about domain specific tasks. In such a use case, the agent is referred to as a task-oriented agent. Task-oriented agents, as the name suggests, are tailored to assist users in completing specific tasks [8]. They aim to deliver precise and relevant responses to specific inquiries, which they typically require. These agents employ state-of-the art models in natural language understanding (NLU), natural language generation (NLG) [9], and machine learning (ML) to determine the intent of a given user query and appropriate responses based on the provided context and the contextual limit of the agents knowledge base. In the next subsection, we describe the retrieval augmented conversational agent (RACG).

9.3 THE RETRIEVAL AUGMENTED KNOWLEDGE AGENT

Context determines meaning and is therefore important for effectively describing a subject or topic. It provides the background information and surrounding details that allow learners to grasp the meaning and significance of a subject [10]. It further helps bridge the gap between an instructor's knowledge and the student's understanding.

Thus, centering instructional feedback around the context of the topic can enhance learning outcomes. To this end, we used embedding models that have broad training approaches like retrieval, clustering, classification, and semantic textual similarity [11] as opposed to embedding models based on BERT, which might focus heavily on masked language modeling [12]. Our system employs the OpenAI Ada embedding model [13]. It has a context window size of 16,385 tokens, allowing for more comprehensive understanding of the input text. The model has been trained on a diverse range of text sources to improve its language capabilities. Figure 9.2 shows the retrieval augmented conversational agent.

The first task in this pipeline is splitting the input file into chunks. Several important factors must be considered for chunk size selection in a retrieval augmented generation (RAG) process. First is the balance between efficiency and accuracy when considering the size of the data chunks from the splitting task. Larger chunks can improve efficiency by reducing the number of retrieval steps needed. However, they might miss relevant information or context if the relevant passage falls across chunk boundaries. On the other hand, smaller chunks can capture finer details and context but can lead to more retrieval steps and an increase in the computational cost of the retrieval process [14]. Once the chunking process is done, the embedding model generates embeddings for each chunk of text. The model then calculates the similarity score between the user query embedding and the embedding of each potential reply chunk. This similarity score indicates how well the reply chunk matches the user's intent [6, 15, 16]. Thus, semantic and human-like feedback is returned to the user. To validate the retrieval augmented generation (RAG) system, we use a knowledge graph as ground truth.

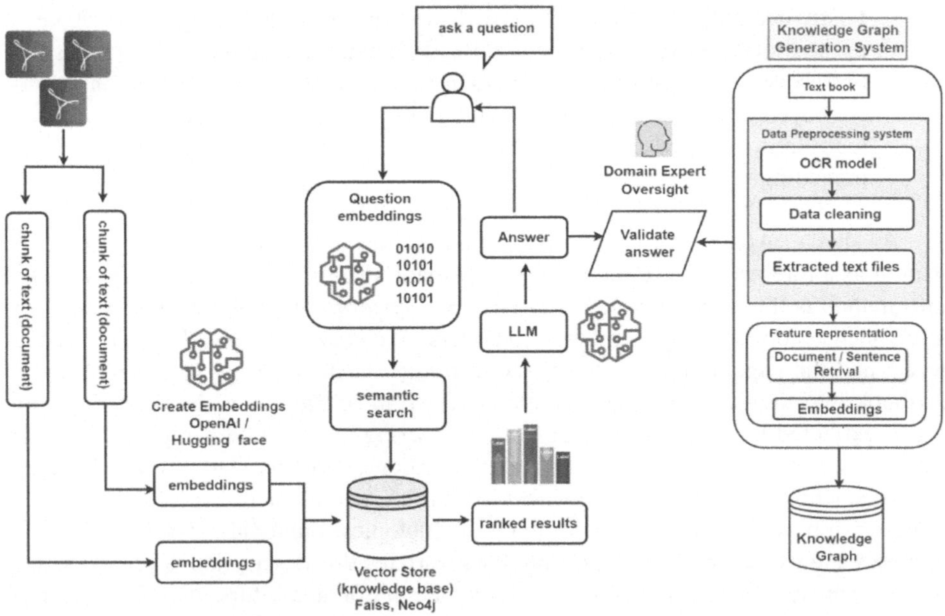

FIGURE 9.2 The knowledge agent.

The knowledge graphs generation is described in Section 9.3.1.

9.3.1 Knowledge Graph Generation System

Knowledge graphs (KGs) are used to encode factual information and relationships between entities. By validating the retrieved passages from the RAG system against a KG, one can ensure that a retrieved information align with established knowledge and minimize the risk of factual errors or hallucinations in the generated text. The first step in our KG generation pipeline is an optical character recognition (OCR) task. The various phases of our optical character recognition pipeline are as follows:

- **Preprocessing**: After images are acquired, several preprocessing procedures such as skew reduction, thinning, and noise removal are utilized are employed too.
- **Segmentation**: Here the characters are separated to make it more readable.
- **Feature Extraction**: Features from the segmented images are extracted, and these features aid in character recognition.
- **Classification**: Once the features are extracted, a classification algorithm is applied to identify and categorize the characters based on their unique characteristics. This step plays a crucial role in accurately recognizing and distinguishing different characters.

- **Postprocessing**: Extracted features from the segmented images contribute to the process of character recognition. Following the classification process, postprocessing techniques, such as error correction and verification, are employed to enhance the precision of character recognition. These strategies aid in reducing any misinterpretations or errors that may have arisen in the preceding steps.

As shown, **Algorithm 9.1: Data Preprocessing**, in the contents in *keyword.xlsx* is used as a reference to extract specific information from the textbook. The extracted information is then imported into a data frame *keyword mapping,* which contains columns such as topic, subtopic, and page numbers. Each row in *keyword mapping* is analyzed, and the page numbers are utilized to transform the appropriate pages from the textbook into images using the function *conv pdf to image.* After the conversion, the text is extracted using the *pytesseract.image to string* method and is repeated for each subtopic, culminating in comprehensive information extraction. The extracted content is subsequently compiled into *data dict* which is a nested dictionary. Top-level keys represent topics. Second-level keys represent subtopics. Subtopic entry includes page numbers and extracted texts. Algorithm 9.1 illustrates this description. The next step is feature representation, which aims to extract meaningful and informative features from the data, eliminating the need for manual feature engineering. By learning representations directly from the data, feature learning algorithms can adapt to different tasks and improve performance in various domains.

As elaborated in **Algorithm 9.2:Feature Representation**, *topic.txt* contains the text file that contains the description of the mentioned topic, which is used to create the knowledge graph. Figure 9.3 shows a section of the generated knowledge graph. It currently has 196 nodes, 30 labels, 166 explanations, 277 relationships, and 111 relationships.

Each line of the text file is used to create the corresponding line embedding using the function *create_embedding. dict_elem* dictionary contains the key and the values as the line and its corresponding embedding. The key is just the line number. The final dictionary *node data* contains the key as the topic name and the value as the *dict elem* dictionary.

Having described the knowledge agent system part of the task generation, evaluation, and assessment system, Figure 9.1, attention will be shifted to the task generation and evaluation subsystem in the next section.

Algorithm 9.1 Data Preprocessing

Require: textbook, keyword.xlsx

1: **Initialize dictionary** *data_dict* = {}
2: *keyword_mapping* ← *read_file* (*keyword.xlsx*)
3: **for** *row* **in** *keyword_mapping* **do**
4:     extract *topic, subtopic, page_numbers* **from** *row*
5:     Initialize string *extracted_text*
6:     **for** *page_num* **in** *page_numbers* **do**
7:         *page_img* ← *conv_pdf_to_image* (*page_num*)

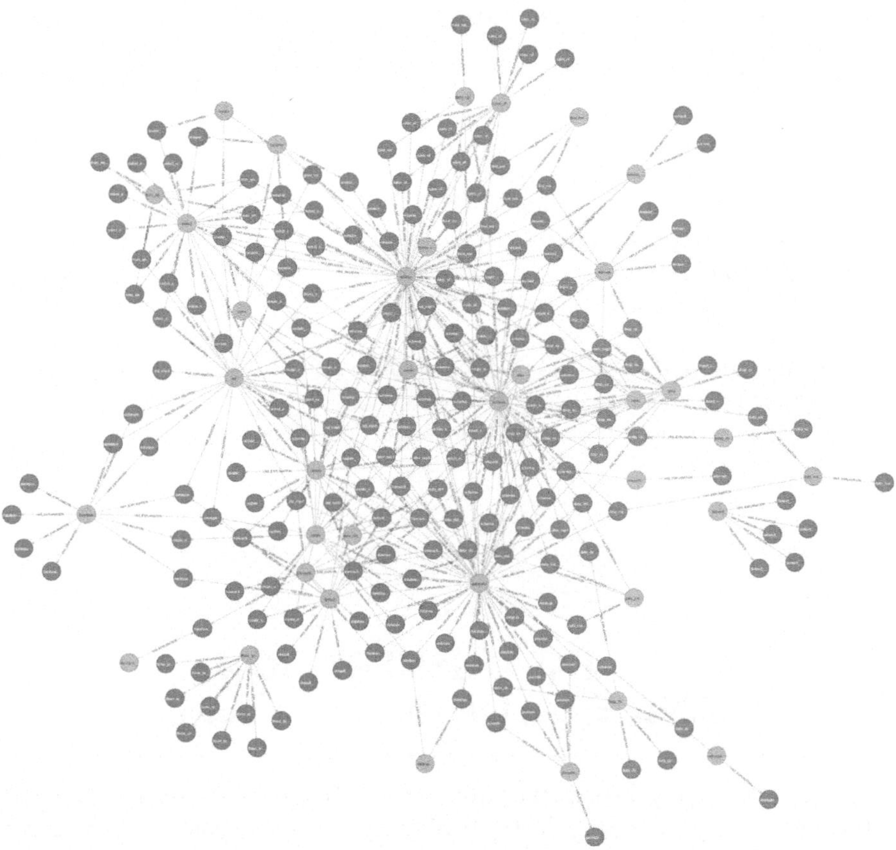

FIGURE 9.3 Knowledge graph of database concepts lecture.

text ← *pytesseract.image_to_string*
(*page_img*)
extracted_text append *text*
end for
data_dict append
{*topic*: [*subtopic, page_num, extracted_text*]}
end for
return *data_dict*

Algorithm 9.2 Feature Representation

Require: text file *topic.txt*
foreach *topic.txt* **do**
initialize dictionary *node_data*
node_data ← {'*label*': *topic*}

```
file_lines ← read_lines (topic.txt)
Initialize count ← 0
for lines in file_lines do
    count ← count + 1
    line_emb ← create_embedding (line)
    initialize dictionary dict_elem
    key ← generate_key (count)
    dict_elem append {key: {'disc':
    line,' emb': emb}}
    node_data append dict-elem
end for
create_node (NEO4J_CREDS, node_data)
initialize start_node ← < topic >
initialize end_node with keys in the node_data dictionary except topic
create_relationships (NEO4J_CREDS,
start_node, end_nodes)
end foreach
```

9.4 TASK GENERATION AND EVALUATION SYSTEM

The approach we employ involves the utilization of a Question-Answer Pair Generation pipeline, specifically employing the t5-small model, as well as an Answer Evaluation pipeline utilizing the SBERT model. The process is initiated by the admin user, who provides a PDF file and specific parameters that determine the generation of a question-answer pair. This procedure can be iterated indefinitely with any alternative PDF file of our preference, utilizing either identical or distinct parametric parameters.

9.4.1 Task Generation and Evaluation System Workflow

The TGES is depicted in Figure 9.4.

The task flow proceeds as follows: Once the student enters the lecture name, semester, exercise group, and username into the question generation form and submits it, a PDF and respective parameters are sent to the question-and-answer pair generation pipeline. The respective parameters are the number of pages from where the questions would be generated and also the number of generated questions that were given by the admin user or tutor beforehand and were saved in the system. This will trigger the generation of a question-answer pair for the student. Once the internal similarity check is completed, the resulting information is passed on to the pipeline responsible for generating question-answer pairs. The created list of questions is now transferred to the

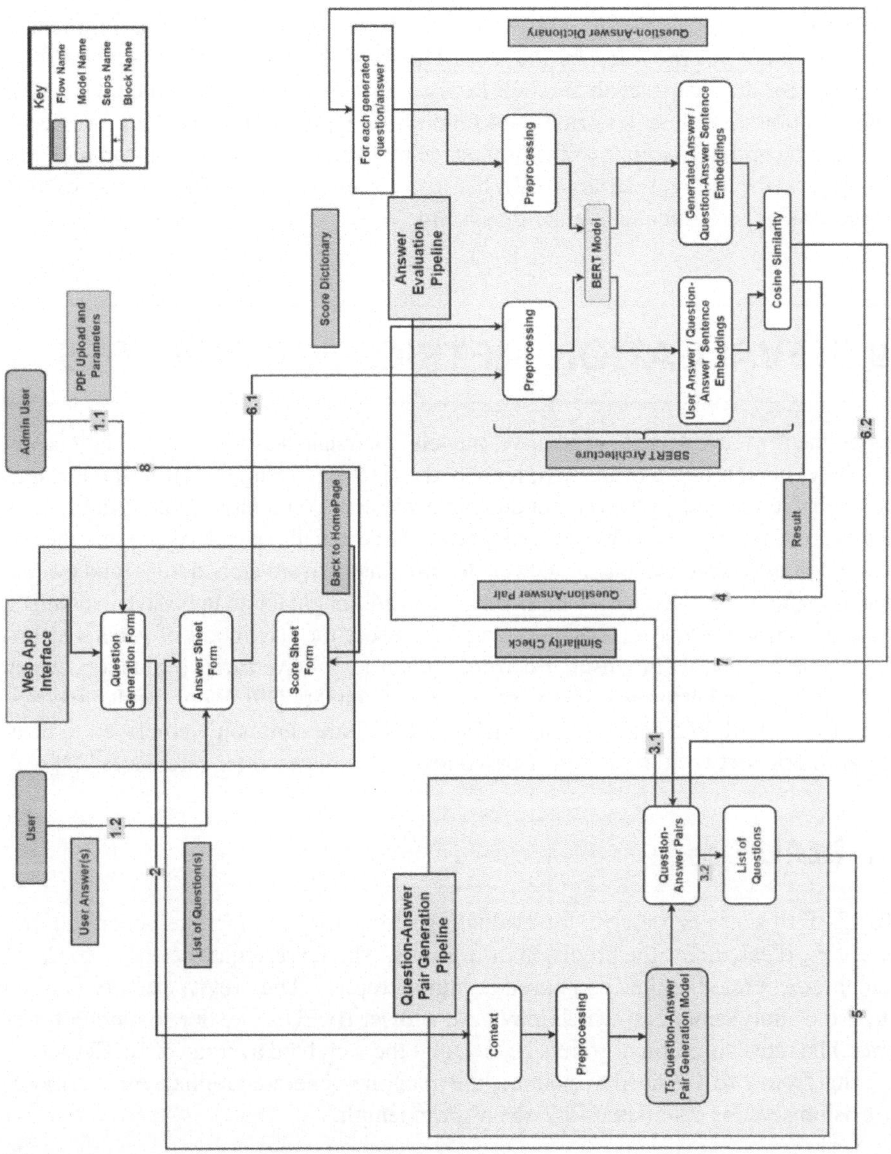

FIGURE 9.4 Task generation and evaluation system.

answer sheet form for the user to respond to. Once the user submits their answers, both the answers and the produced question-answer pair are transferred to the answer evaluation pipeline for additional processing. The assessment of the responses is conducted by comparing the student's answer to each generated question with a previously generated answer to the corresponding question. The comparison is conducted by utilizing semantic similarity [17] through the answer-evaluation model, resulting in the assignment of a score to each answer provided by the user. This score is delivered to the score sheet form in order for the user to evaluate their performance. The score sheet includes scores for each question as well as a normalized final score. Upon reviewing the score sheet, the user has the option to return to the homepage by utilizing the Home button included in the user interface to retake the test. In the next section, we describe metrics used to access the task generation and evaluation system.

9.5 EVALUATION METRICS FOR THE TGES

The assessment of text-to-text generative models encompasses a range of approaches. Among these, human-based evaluation, notably through the Turing test [18], stands out as the most reliable method. However, conducting evaluations on a large scale using human annotators proves costly and lacks scalability beyond a certain threshold. As a result, there is a necessity for automated evaluators such as discriminators, word embeddings, and evaluators that measure words overlaps. Automated metrics are sought for an unbiased assessment. Discriminative evaluation uses machine judges to assess the likelihood of generated text fooling a classifier and distinguishing it from human text. Adversarial evaluation, which maximizes adversarial error, is used to assess sentence quality [19]. Word overlap metrics, including BLEU [20], ROUGE [21], and METEOR [22], are commonly employed to measure the similarity between machine-generated text and human-written references [23].

9.5.1 BLEU Score

The BLEU [20] score is a metric for evaluating the quality of machine-generated dialogue replies. It calculates the precision of n-grams, which are sequences of n words, in the machine-generated replies compared to human replies. The brevity penalty is often used to avoid short sentences, which may have a lower BLEU score due to having fewer n-grams. The most common n-gram size is 4, and the weighted average of BLEU scores for n-grams from 1 to 4 is used to evaluate the machine-generated replies. First, a modified precision score is calculated for each n-gram length:

$$P_n(r,\hat{r}) = \frac{\sum_{ngr} \text{Count}_{\text{matched}(ngr)}}{\sum_{k} \text{Count}(ngr)} \tag{9.1}$$

where *ngr* represents all possible *n*-grams of length *n* in hypothesis sentences. Later, BLEU-4 is computed as:

$$\text{BLEU-4} = b(r,\hat{r})\exp\left(\sum_{n=1}^{4} 0.25\log P_n(r,\hat{r})\right) \tag{9.2}$$

[24] To prevent short sentences from being favored, the modified precision score is frequently multiplied by a brevity penalty to obtain the final score.

9.5.2 ROUGE Score

The ROUGE [21] score calculates the recall of n-grams in machine-generated dialogue replies. ROUGE-L is a more robust metric that is not as sensitive to the choice of n-grams. It is calculated based on the longest matching sequence (LCS), which is the longest substring that appears in both the machine-generated reply and the human reply.

$$R = \max_j \frac{\sum_l r, r^{\hat{ij}}}{\left|r^{\hat{ij}}\right|} \tag{9.3}$$

$$P = \max_j \frac{\sum_l r_i, r^{\hat{ij}}}{\left|r_{ij}\right|} \tag{9.4}$$

ROUGE is calculated as follows:

$$ROUGE = \frac{(1+\beta^2)RP}{R+\beta^2 P} \tag{9.5}$$

9.5.3 METEOR Score

The METEOR score, as proposed by Lavie and Agarwal [22], quantifies the degree of similarity between responses generated by humans and those generated by machines. This is achieved through the alignment of words present in both responses. The process of alignment can be performed on a word-by-word basis, enabling the consideration of exact matches, Porter stemming matches, or WordNet synonym matches. The METEOR score is computed by taking the parametric harmonic mean (Fmean) of the unigram recall and unigram precision values of the two responses.

$$P = \frac{m}{t} \tag{9.6}$$

$$R = \frac{m}{r} \tag{9.7}$$

The Fmean is calculated:

$$Fmean = \frac{P \cdot R}{\alpha P + (1-\alpha)R} \tag{9.8}$$

Penalty term (Pen) is computed as:

$$Pen = \gamma \left(\frac{ch}{m} \right)^{\theta} \tag{9.9}$$

Finally, METEOR is calculated as:

$$METEOR = (1 - Pen)Fmean \tag{9.10}$$

where t and r are the total numbers of unigrams in the translation and the reference and m represents the number of mapped unigrams between the reference and hypothesis sentences.

9.5.4 BERT Score

BERT Score [25] is a metric used to assess the quality of a candidate sentence (referred to as $\hat{x}$) in comparison to a reference sentence (referred to as x). It employs contextual embeddings to represent individual tokens in these sentences and computes their similarity using cosine similarity, optionally weighted by inverse document frequency scores. This approach allows for a more sophisticated evaluation that takes into account the contextual meaning of words.

BERT Score computes both recall and precision for each token in x and $\hat{x}$ to calculate an F1 measure. It employs a greedy matching strategy to maximize the matching similarity score, where each token is matched to the most similar token in the other sentence. The recall, precision, and F1 scores are computed as follows:

$$\text{Recall(RBERT)} = \frac{1}{|x|} \sum_{x_i \in x} \max_{\hat{x}_j \in \hat{x}} x_i \hat{x}_j$$

$$\text{Precision(PBERT)} = \frac{1}{|\hat{x}|} \sum_{\hat{x}_j \in \hat{x}} \max_{x_i \in x} x_i \hat{x}_j$$

$$\text{F1 Score (FBERT)} = \frac{2 \cdot \text{PBERT} \cdot \text{RBERT}}{\text{PBERT} + \text{RBERT}}$$

BERT Score also allows for the incorporation of importance weighting, which considers that rare words can be more indicative of sentence similarity than common words. Importance weighting can be based on inverse document frequency (idf) scores computed from the test corpus. Given a set of M reference sentences $\{x(i)\}^{M}{}_{i=1}$, the idf score of a word-piece token is:

$$idf(w) = -\log\left(\frac{1}{M}\sum_{i=1}^{M} I\left[w \in x(i)\right]\right) \tag{9.11}$$

So, recall with idf, weighting is:

$$RBERT = \frac{\sum_{xi \in x} idf(xi)\max_{\hat{x}_j \in \hat{x}} xi\hat{x}_j}{\sum_{xi \in x} idf(xi)} \tag{9.12}$$

To ensure that BERT Score values fall within a readable numerical range, a baseline rescaling is applied. This involves rescaling BERT Score linearly with respect to its empirical lower bound (baseline, denoted as b). The rescaled scores are typically between 0 and 1, improving score readability without affecting the ranking ability and human correlation of BERT Score.

Our project involves fine-tuning T5-small and T5-large models using the RACE dataset, along with utilizing the non-fine-tuned GPT model through OpenAI's Application Programming Interface (API) as a performance benchmark. In this work, the use of GPT-3.5 Turbo was limited due to the cost restrictions associated with accessing OpenAI's API, resulting in GPT-3.5 Turbo not being fine-tuned, whereas the other models were fine-tuned. The primary objective of our model is to assist students in addressing contextually relevant questions in a distributed data management course. To achieve this, we opted to re-fine-tune the T5-small (2) model on a dataset we developed ourselves. Given the absence of a suitable reading comprehension from examinations (RACE)-like dataset within a DBMS domain, we took the initiative to curate our own dataset. This bespoke dataset consists of approximately 180 rows with 3 columns each. It draws its content from research papers, encompassing contextual input as well as question-and-answer pairs for the desired output. Our paper selection process was tailored to cater to student needs, with a focus on DBMS-related research. Notably, the parameters and tokenizer from the preceding model were integrated for further training on our tailored dataset. Given the scope of our project and the volume of data needed, extending the custom-made dataset would have required an impractical amount of human resources and time, so the dataset was kept small.

So our evaluation consists of three pillars: human evaluation, evaluation using a word overlap evaluator, and lastly evaluation using dynamic word embedding. We are not using the discriminator here because it has been seen from the literature survey that they are not well correlated with human judgment as adversarial accuracy is not the correct measure when realism is the main concern [23].

9.6 TGES EVALUATION STRATEGY

Before implementing the evaluation pipeline Figure 9.5, our initial step involves the training of both T5-small and T5-large models on the RACE-M dataset. Our choice of the RACE dataset over the Squad dataset stems from our focus on assessing models for their ability to generate abstractive questions. With its diverse range of topics and genres, the RACE dataset holds a prominent position as a widely recognized benchmark in the domains of natural language processing (NLP) and machine learning. This dataset is partitioned into two main segments: RACE-M (tailored for middle school levels) and RACE-H (geared toward high school levels). The questions within the dataset are strategically designed to assess various dimensions of reading comprehension, encompassing comprehension of main ideas, inferential abilities, detail identification, and drawing conclusions from provided text passages. Due to the significant computational demands associated with larger datasets like RACE All (97.7K) and RACE High (69.4K), RACE M (28.3K) was selected to optimize computational efficiency and reduce training time. The smaller dataset size of RACE M allows for more rapid experimentation and model development without compromising essential data diversity for the specific research objectives.

For evaluation, the RACE dataset was divided into three parts: G-train (70%), G-valid (10%), and G-test (20%)[23]. T5-large and T5-small were trained on G-train and subsequently validated using G-valid and tested on the testing set. Then T5-small was again trained on the custom-made dataset and all the model outputs were added to a new G-test which consists of both the questions and context from the custom dataset and the questions generated by the model and its corresponding context from where it is generated.

- **Human Evaluator:** In the first strategy, **Algorithm 9.3**, a simple methodology was employed to assess the performance of various models. This method involved creating a mixture comprising a collection of unclassified questions. This mixture was then distributed among a group of 10 skilled annotators, with each annotator receiving four distinct sets of 20 unclassified questions (i.e., 10 real and 10 fake) with the context from which they were generated, each set corresponding to one of the models under consideration.

  The primary task assigned to these annotators was to meticulously evaluate and rate the responses generated by the different models. To facilitate this evaluation, a modified Likert scale, as in Table 9.3, was designed to capture the

TABLE 9.3 Likert scale for human annotators

SCORE	*ESTABLISHED MEANING*	*MODIFIED MEANING*
1	Very Unsatisfied	Easily Distinguishable
2	Unsatisfied	Distinguishable
3	Neutral	Neutral
4	Satisfied	Near Undistinguishable
5	Very Satisfied	Perfectly Undistinguishable

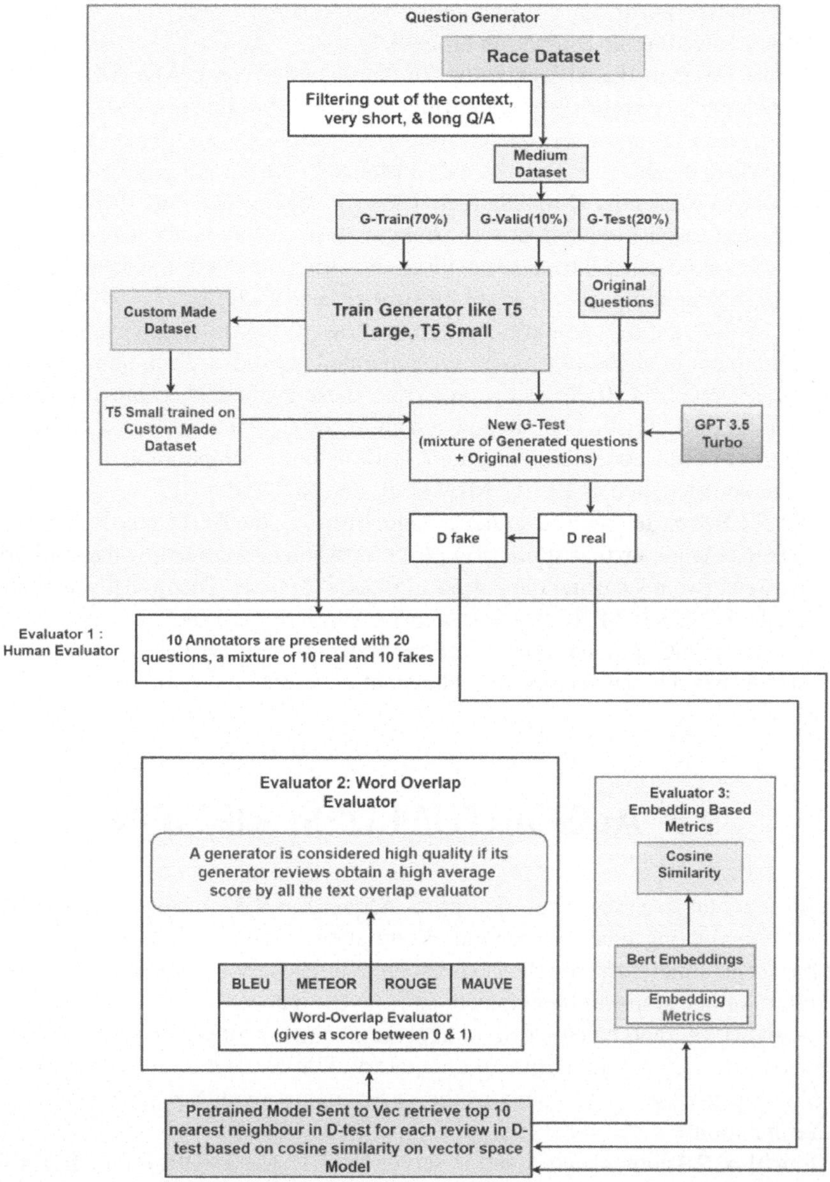

FIGURE 9.5 TGES evaluation strategy.

nuances of model performance. This scale allowed the annotators to express their judgments in a more nuanced manner than a simple binary choice.

Upon receiving the responses from the various annotators, the next step involved aggregating their ratings. This aggregation process entailed calculating the average rating assigned by all 10 annotators for each model's set of questions. This calculated average rating provided a consolidated measure of the model's performance, accounting for the perspectives of all the annotators involved in the evaluation process.

- **Word Overlap Based Evaluator:** In the second strategy, **Algorithm 9.4**, the evaluation process involves employing word overlap evaluators. To facilitate this assessment, a pretrained model known as "sent2vec" is utilized. This model calculates the distance between each instance in the D-fake dataset and the 180 instances within the D-real dataset. The resulting distances are then organized in an increasing order, effectively arranging the most similar sentences from the D-real dataset. These 10 positions, denoting the relative 10 nearest neighbors, are meticulously recorded for further analysis [23].

  Following this preparatory step, a series of word overlap evaluators are used to comprehensively assess the generated questions. Each question in the dataset receives 10 distinct scores from these evaluators, capturing various aspects of linguistic overlap and semantic correspondence. These individual scores are subsequently aggregated, leading to the computation of key evaluation metrics such as BLEU, METEOR, and ROUGE.
- **BERT Score:** In the third strategy, **Algorithm 9.5**, the BERT model carries out STS (semantic textual similarity) [26] task wherein it compares the similarity between two texts using the cosine similarity measure. The model was trained on the STSB and SICK datasets. A cosine similarity of 1 indicates that the vectors are pointing in the same direction (perfect similarity), while a similarity of 0 indicates that the vectors are orthogonal (completely dissimilar).

9.7 ALGORITHM DESCRIPTION

The human evaluator assessment algorithm, **Algorithm 9.3**, outlines a comprehensive process for evaluating machine-generated output from different models using human annotators. The annotators rate the Likert scores for each model's output in comparison to a mixture of questions and real questions from the dataset.

The algorithm then calculates and aggregates these scores to generate an average Likert score for each model across all annotators. This average score serves as a quantitative metric to assess and compare the performance of different models based on human evaluations.

Algorithm 9.4 computes evaluation scores (BLEU, METEOR, ROUGE) for Dfake sentences using a pretrained sentence embeddings model (sent2vec) and word overlap evaluators with respect to similar sentences from the D-real dataset. It leverages sentence embeddings and cosine distances to identify semantically similar D-real sentences. Then, for each D-fake sentence, it calculates word overlap and subsequently the desired evaluation scores, providing a comprehensive assessment of the quality of generated sentences based on their similarity and overlap with real sentences.

Algorithm 9.5 calculates BERT scores for generated sentences by computing cosine similarity of their BERT embeddings with the BERT embeddings of reference sentences. The reference sentences are coming in the same way via nearest neighbor method and sent to vec model from D-real and the target is coming from the D-fake. BERT embeddings capture contextual information and semantic meaning, allowing the algorithm to assess the similarity between D-fake and D-real sentences in a dynamic manner.

Algorithm 9.3: Human Evaluator Assessment:

Require: 10 human annotators, a set of 20 unclassified questions per model, and the same 20 unclassified questions per annotator

Ensure: Modified Likert scale table

```
Calculate the Average Likert Score for Each Model and Each Annotator:
for each annotator i do
    for each model m do
        Sum_likert_score ← Sum of Likert scores for machine-generated outputs
        by annotator i for model m
        Num_original_questions ← Number of machine generated questions from
        model m provided to annotator i
        Average_likert_score_model_m_annotator_i ← Sum_likert_score/
        Num_original_questions
    end for
end for
for each model m do
    Sum_avg_likert_score ← Sum of the Average Likert scores for model m
    across all annotators
    Num_annotators ← Total number of annotators
    Average_likert_score_model_m ← Sum_avg_likert_score/Num_annotators
end for
Return [Avg_Likert_Score_1, .., Avg_Likert_Score_m]
```

9.8 EVALUATION AND RESULT

9.8.1 Human Evaluator Results

Among the diverse array of text generators, GPT-3.5 Turbo stands out as a frontrunner, consistently earning the highest average Likert score of 4.8. This supremacy is vividly corroborated by the superior quality of coherent questions generated by the GPT-3.5 Turbo model. Following GPT-3.5 Turbo, T5-Small(2) emerges as a noteworthy contender, having undergone fine-tuning on a customized dataset, and securing the second-highest average score of 4. The ranking hierarchy further extends to T5-Large, positioned as the third-ranking generator, and subsequently to T5-Small(1).

An observation surfaces Figure 9.6, when examining the role of model parameters in evaluation scores. Despite the T5-Large model boasting an expansive parameter count of 770 million, it obtains a lower rank than T5-Small(2), which is equipped with a mere 64 million parameters. This discrepancy in ranking suggests that factors beyond

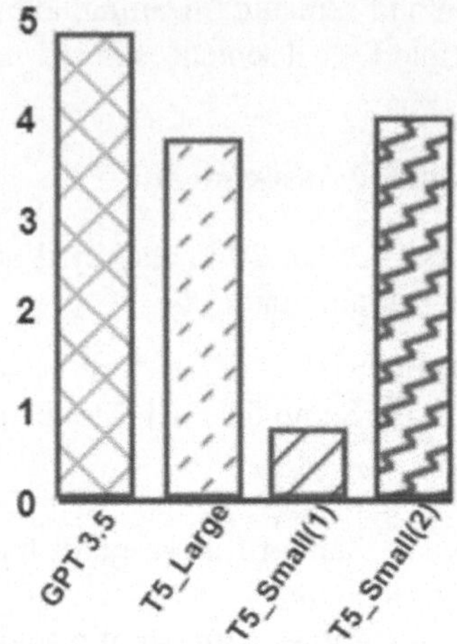

FIGURE 9.6 Ranking of models on Likert score.

sheer model complexity contribute to the perceived quality of text generation, this factor is attributed to training T5-Small(2) on domain-specific custom-made dataset. This also translates to our key findings that a customized large language model (LLM) with a focus on generating a particular type of task can be effectively utilized in educational settings. By tailoring the model specifically for educational purposes, such as generating quizzes, summarizing educational content, or providing targeted feedback, we can achieve highly relevant and efficient outputs. These customized LLMs do not need to be huge models that require expensive infrastructure to handle and maintain. Instead, smaller customized models can perform very efficiently, often matching or even surpassing the performance of larger non-fine-tuned general-purpose LLMs on specific educational tasks. This efficiency not only reduces the cost and complexity of deployment but also ensures that educational institutions can implement these solutions more easily and at a lower cost. User bias, a pervasive phenomenon in rating systems, introduces an additional layer of complexity to the evaluation process. While the aforementioned ranking reflects the collective assessment of users, it fails to account for the potential biases users may introduce. Cultural, emotional, and cognitive biases, among others, can significantly sway user judgments and distort the objectivity of ratings.

Algorithm 9.4: Word Overlap Embeddings Process with Pretrained Sentence Embeddings

Require: List of 10 D-fake sentences, list of D-real sentences, pretrained sentence embeddings model (i.e., sent2vec)

Ensure: Evaluation scores (BLEU, METEOR, ROUGE)

1: Initialize an empty list eval_scores.
2: **for** each D-fake sentence in List of D-fake sentences **do**
3: Initialize an empty list similar_real_positions.
4: Calculate sentence embeddings for D-fake sentence using the pretrained sent-2vec model.

5: **for** each D-real sentence in List of D-real sentences **do**
6: Calculate the cosine distance between the embeddings of D-fake and D-real sentences.
7: Append the distance and the position of D-real sentence to similar_real_positions.
8: **end for**
9: Sort similar_real_positions based on increasing cosine distances.
10: Select the positions of the top 10 D-real sentences.
11: Initialize a list of D-real sentences using the selected positions.
12: Initialize a list of D-rake words.
13: Tokenize and preprocess D-fake sentence, and add words to the list of D-fake words.
14: Initialize a list of evaluation scores for the current D-fake sentence.
15: **for** each selected D-real sentence **do**
16: Initialize a list of D-real words.
17: Tokenize and preprocess the current D-real sentence, and add words to the list of D-real words.
18: Calculate word overlap between D-fake words and D-real words.
19: Calculate BLEU, METEOR, and ROUGE scores using the word overlap.
20: Append the evaluation scores to the list of evaluation scores.
21: **end for**
22: Calculate the average of the evaluation scores for the current D-fake sentence.
23: Append the average evaluation score to eval_scores.
24: **end for**
25: **Return** eval_scores.

Algorithm 9.5: Dynamic Word Embedding Process Using BERT Score

Require: List of Reference Sentences (D-real), List of Generated Sentences (D-fake), Pretrained Sentence Embeddings Model (i.e., sent2vec), BERT Model

Ensure: BERT Scores for Each Generated Sentence
1: Initialize an empty list bertscores.
2: **for** each D-fake sentence in List of Generated Sentences **do**
3: Initialize an empty list bertsimilarities.
4: Calculate BERT embeddings for D-fake sentence using the BERT model.
5: **for** each D-real sentence obtained from sent2vec model in List of Reference Sentences **do**
6: Calculate BERT embeddings for D-real sentence using the BERT model.
7: Calculate cosine similarity between BERT embeddings of D-fake and D-real sentences.
8: Append the cosine similarity to bertsimilarities.
9: **end for**
10: Calculate the average cosine similarity from bertsimilarities.
11: Append the average cosine similarity (BERT score) to bertscores.
12: **end for**
13: **Return** bertscores.

9.8.2 Word Overlap Evaluator Results

The Bleu scores of models Fig 9.7 shows a lack of correlation with Likert scale rankings in Fig.6, exemplified by the unexpected higher ranking of T5-Small(1) over T5-Large. Bleu-4, which calculates a score for up to 4-grams using uniform weights, is used for evaluation. Shifting focus to the Rouge-1 scale, an evaluation metric that assesses models based on precision, recall, and F1 score, a contrasting pattern emerges. Here we are using Rouge-1 score which measures the overlap of unigram (single word) units between the generated text and the reference text Fig 9.8. T5-Large attains the highest precision, a result that contradicts expectations. However, the recall-based ranking aligns more closely with the anticipated model performance. Rouge-1 F1 score Fig.9 could capture the ranking as it is in Likert Scale. Conversely, Meteor, a word overlap evaluation metric, emerges as a more reliable indicator of model performance. This is attributed to its robust correlation with human judgment, signifying its ability to align with human perceptual assessment. Unlike Bleu score, which fails to capture the true discrepancy between T5-Large and T5-Small(1), Meteor successfully reflects the subtle differences that are evident from the Likert scale evaluations. Intriguingly, despite a marginal disparity between T5-Large and T5-Small(1) as compared to the Likert scale evaluations, it unveil a more pronounced distinction. This emphasizes the importance of incorporating diverse evaluation methodologies that encompass both quantitative and qualitative aspects, allowing for a more comprehensive understanding of model performance. The Meteor score exhibits a notably robust correlation with the assessment provided by human evaluators, a correlation that is reaffirmed by our own evaluation findings Fig 9.10. While the distinction between the performance of T5 small and T5 large models may not be substantial, this alignment becomes particularly pronounced when juxtaposed with the corresponding likert scores. A summary Table 9.4 is provided, considering the 95% confidence interval.

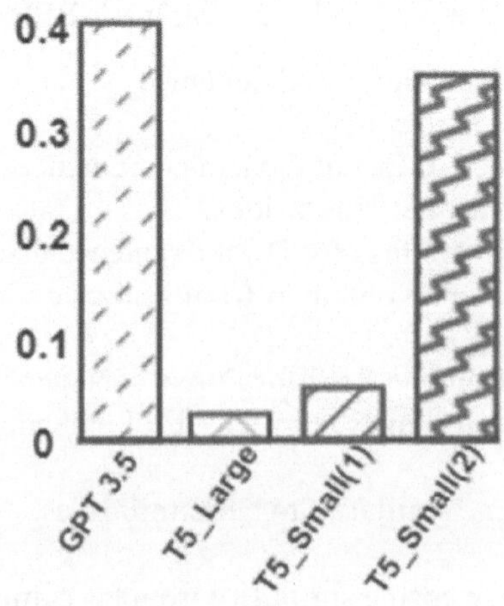

FIGURE 9.7 Ranking of models on Bleu score.

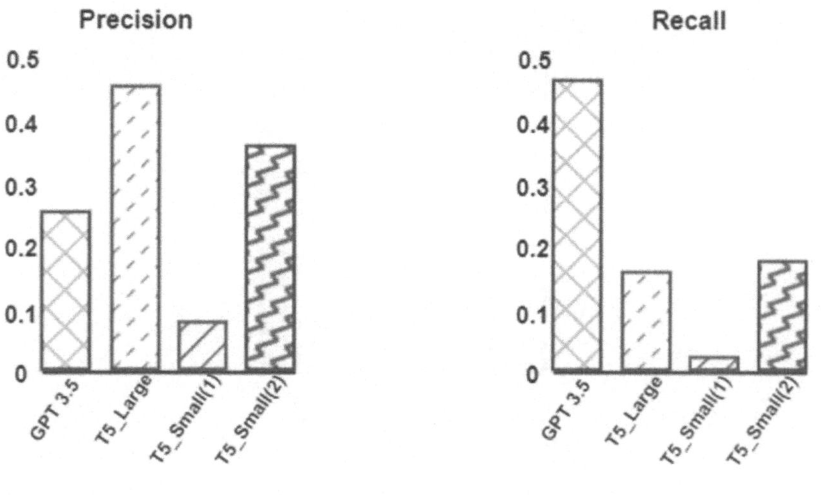

FIGURE 9.8 Ranking of models on Rouge-1 score.

In contrast, Rouge-1, which captures unigram overlaps, aligns its F1 score (Figure 9.9) well with Likert scale rankings compared to the Meteor score ranking (Figure 9.10). A summary Table 9.4 is provided, considering the 95% confidence interval.

9.8.3 Word Embeddings Evaluation

The convergence of the BERT Score Table 9.5 with the Likert scale and its notable proximity to human evaluator assessments emphasize its role as a bridge between objective automated evaluation and the inherently subjective nature of human evaluation. This convergence not only signifies the utility of the BERT Score as a robust evaluation metric but also sheds light on its potential to streamline and enhance the development and refinement of automated text evaluation algorithms.

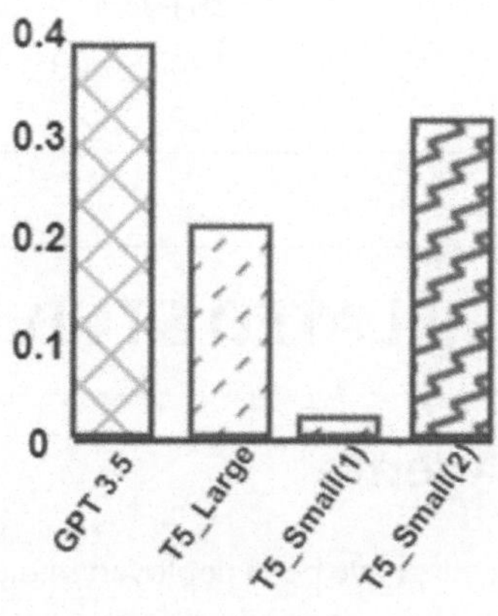

FIGURE 9.9 Ranking of models on Rouge-1 F1 score.

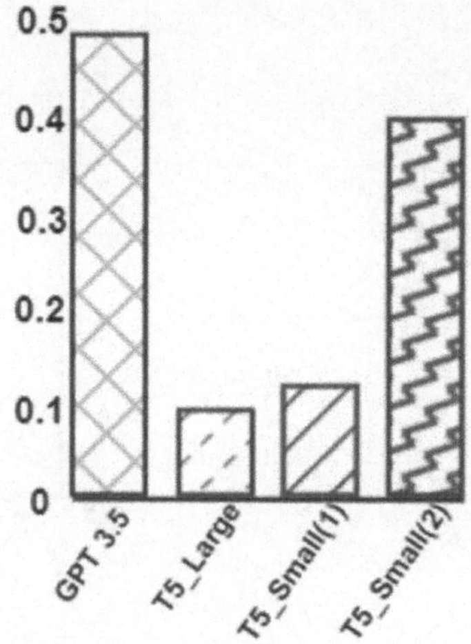

FIGURE 9.10 Ranking of models Meteor score.

TABLE 9.4 Word overlap metrics scores

MODEL NAME	*AVERAGE LIKERT*	*BLEU*	*R1 F1*	*R1 RECALL*	*R1 PRECISION*	*METEOR*
GPT-3.5	4.8±0.58	0.42±0.02	0.36±0.09	0.48±0.05	0.30±0.17	0.51±0.11
T5-Large	3.6±1.76	0.02±0.05	0.22±0.12	0.14±0.09	0.50±0.21	0.09±0.07
T5-Small(1)	0.8±0.73	0.07±0.03	0.01±0.03	0.01±0.04	0.09±0.11	0.12±0.07
T5-Small(2)	4±0.88	0.35±0.23	0.31±0.05	0.17±0.08	0.36±0.09	0.44±0.03

TABLE 9.5 BERT scores for different models

MODEL NAME	*BERT SCORE*
GPT-3.5	0.41±0.12
T5-Large	0.36±0.04
T5-Small(1)	0.1247±0.09
T5-Small(2)	0.480±0.08

9.9 RELATED STUDIES

9.9.1 Knowledge Agents

Several forms of knowledge agents have been deployed in the commercial and industrial sectors. While this is mostly not the case in education, the recent adoption of the blended instructional strategy has opened up avenues for the integration of conversational agents.

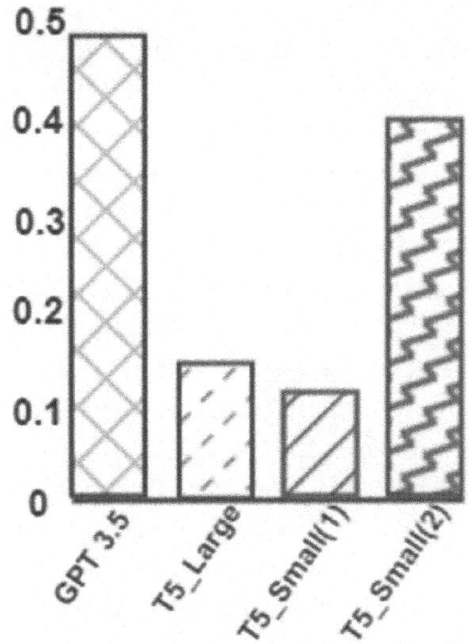

FIGURE 9.11 Ranking of models: BERT.

Wambsganss et al. [27] developed a digital assistant that can provide students with formative comments on their essays. To evaluate students' impressions of their conversational agent, a survey-based assessment was undertaken. Participants were asked to write an argumentative paragraph and use the conversational agent to receive customized feedback. A further effort by Tellols et al. [28] enhanced conversational agents with machine learning technology, thus giving them sentient skills. They showcased their methodology by integrating a virtual instructor into children's educational software. During their assessment, they conducted a comparison between two conversational agents. Based on their findings, the conversational agent, which was augmented with machine learning capabilities, exhibited superior performance and user satisfaction.

Compared to these agents, our knowledge agent utilizes large language models with specialized embedding models that are capable of providing human-like instructional feedback to students during their study engagement. Thus, our knowledge agent is able to analyze and understand students' language queries, allowing it to provide factual and contextually accurate responses. We further employ knowledge graphs to verify the responses generated by the knowledge agent to ensure that the information provided is accurate and reliable. This approach effectively solves the hallucination challenge that is found in systems that employ language models. This comprehensive approach enhances the overall learning experience for students by offering personalized and precise support throughout their educational journey.

9.9.2 Task Generation and Evaluation System

In the following, we review related works on question generation. Zhang et al. [25] introduced the BERTSCORE, a novel metric for evaluating generated text against gold-standard references. The study demonstrates that BERTSCORE addresses the limitations of common metrics, particularly in challenging adversarial examples. We explored various configuration choices for BERTSCORE, revealing its effectiveness for model selection. While no single configuration clearly outperforms others, the study provides insights into different trade-offs based on domains and languages. A further effort by Liu et al. [29], presents a systematic investigation into the correlation between automatic ROUGE scores and human evaluation for meeting summarization. According to experimental results, the correlation between ROUGE scores and human evaluation is generally low, with the ROUGE-SU4 score showing a better correlation than the ROUGE-1 score. Denkowski et al. [30] showed that Meteor1.3 has a high correlation with human judgments, though overfitting issues arise from skewed tuning data. For a phrase-based system with small Urdu-English data, the tuning version of Meteor outperforms BLEU in minimum error rate training, and we expect it to generalize well in other tuning scenarios. In conversation systems that are not task-focused, word overlap metrics often do not align well with human judgments during task testing. User studies should therefore complement these systems. On the other hand, [31] said that text overlap metrics can show how people make decisions in task-oriented dialogue settings, especially when datasets have a lot of ground truth references. The study challenges the strong correlation between BLEU scores and human judgment in translation quality, highlighting limitations in BLEU's model of translation variation. It suggests that while BLEU's advantages, like cost-effectiveness and feasibility, remain, its appropriate use should be carefully considered [32]. The author also suggested that an increase in BLEU score might not necessarily lead to significant system improvement, advocating for human evaluation instead. Currently, we have not found any research article on task generation and evaluation systems that are used in the educational sector.

9.10 SUMMARY AND FUTURE WORK

In this paper, we introduced the knowledge agent and task generation, evaluation, and assessment system (TGEAS). The knowledge agent facilitates the integration of a conversational infrastructure that allows us to provide human-like instructional feedback interaction to students during their study engagements, and the TGEAS creates and evaluates. The validation of the feedback from the knowledge agent against a knowledge graph (KG) removes the risk of hallucination. In the evaluation, we found that the Rouge-1 F1 score exhibited a robust correlation with human evaluations, emphasizing the importance of incorporating diverse evaluation methodologies. Also, the BERT Score showed notable proximity to human evaluator assessments, serving as a bridge between objective automated evaluation and subjective human evaluation. Our future effort is aimed at developing an intelligent agent mediated learning environment and personalization system. This system will leverage the insights gained from the evaluation process to tailor learning experiences to individual needs and preferences. By combining automated assessment tools with human feedback, we aim to create a more personalized and effective learning environment for users.

REFERENCES

1. E. D. Toit, "Constructive feedback as a learning tool to enhance students' self-regulation and performance in higher education," *Perspectives in Education*, vol. 30, no. 2, pp. 32–40, 2012.
2. N. Frey, and D. Fisher, *The formative assessment action plan: Practical steps to more successful teaching and learning.* ASCD, 2011.
3. J. Ye, X. Chen, N. Xu, C. Zu, Z. Shao, S. Liu, Y. Cui, Z. Zhou, C. Gong, and Y. Shen *et al.*, "A comprehensive capability analysis of gpt-3 and gpt-3.5 series models," *arXiv preprint arXiv:2303.10420*, 2023.
4. W. Wei, P. M. Barnaghi, and A. Bargiela, "Search with meanings: An overview of semantic search systems," *International Journal of Communications of SIWN*, vol. 3, pp. 76–82, 2008.
5. C. V. Obionwu, T. Tiwari, B. B. C. Valappil, N. Raikar, D. S. Walia, S. L. Abbas, C. Okafor, D. Broneske, and G. Saake, "A domain specific students' assistance system for the provision of instructional feedback," in *2023 International Conference on Machine Learning and Applications (ICMLA)*. 2023, pp. 2065–2070.
6. C. V. Obionwu, B. Valappil, M. Genty, M. Jomy, V. Padmanabhan, A. Suresh, S. Bedi, D. Broneske, and G. Saake., "Expert agent guided learning with transformers and knowledge graphs," in *Proceedings of the 13th International Conference on Data Science, Technology and Applications - DATA, INSTICC*. SciTePress, 2024, pp. 180–189.
7. S. Shukla, L. Liden, S. Shayandeh, E. Kamal, J. Li, M. Mazzola, T. Park, B. Peng, and J. Gao, "Conversation learner–a machine teaching tool for building dialog managers for task-oriented dialog systems," *arXiv preprint arXiv:2004.04305*, 2020.
8. N. Luz, N. Silva, and P. Novais, "A survey of task-oriented crowdsourcing," *Artificial Intelligence Review*, vol. 44, pp. 187–213, 2015.
9. E. Reiter, "Natural language generation," in *The handbook of computational linguistics and natural language processing*, Wiley Online Library, pp. 574–598, 2010.
10. E. B. Johnson, *Contextual teaching and learning: What it is and why it's here to stay.* Corwin Press, 2002.
11. S. S. Birunda, and R. K. Devi, "A review on word embedding techniques for text classification," in *Innovative Data Communication Technologies and Application: Proceedings of ICIDCA 2020*. 2021, pp. 267–281.
12. D. Nozza, F. Bianchi, and D. Hovy, "What the [mask]? making sense of language-specific Bert models," *arXiv preprint arXiv:2003.02912*, 2020.
13. D. Machlab, and R. Battle, "LLM in-context recall is prompt dependent," *arXiv preprint arXiv:2404.08865*, 2024.
14. Y. Gao, Y. Xiong, X. Gao, K. Jia, J. Pan, Y. Bi, Y. Dai, J. Sun, and H. Wang, "Retrieval-augmented generation for large language models: A survey," *arXiv preprint arXiv:2312.10997*, 2023.
15. P. Xia, L. Zhang, and F. Li, "Learning similarity with cosine similarity ensemble," *Information Sciences*, vol. 307, pp. 39–52, 2015.
16. L. Wu, C. Quan, C. Li, Q. Wang, B. Zheng, and X. Luo, "A context-aware user-item representation learning for item recommendation," *ACM Transactions on Information Systems (TOIS)*, vol. 37, no. 2, pp. 1–29, 2019.
17. C. V. Obionwu, R. Kumar, S. Shantharam, D. Broneske, and G. Saake, "Semantic relatedness: A strategy for plagiarism detection in SQL assignments," in *2023 6th World Conference on Computing and Communication Technologies (WCCCT)*. IEEE, 2023, pp. 158–165.

18. C. Machinery, "Computing machinery and intelligence—AM Turing," *Mind*, vol. 59, no. 236, p. 433, 1950.
19. S. Riezler, and Y. Goldberg, "Proceedings of the 20th SIGNLL conference on computational natural language learning," in *Proceedings of the 20th SIGNLL Conference on Computational Natural Language Learning*. 2016.
20. K. Papineni, S. Roukos, T. Ward, and W.-J. Zhu, "Bleu: A method for automatic evaluation of machine translation," in *40th Annual Meeting of the Association for Computational Linguistics (ACL)*. ACL, 2002, pp. 311–318.
21. C.-Y. Lin, "Rouge: A package for automatic evaluation of summaries," in *Text summarization branches out*, 2004, pp. 74–81.
22. A. Lavie, and A. Agarwal, "Meteor: An automatic metric for mt evaluation with high levels of correlation with human judgments," in *Proceedings of the Second Workshop on Statistical Machine Translation*. 2005.
23. C. Garbacea, S. Carton, S. Yan, and Q. Mei, "Judge the judges: A large-scale evaluation study of neural language models for online review generation," *arXiv preprint arXiv:1901.00398*, 2019.
24. E. Merdivan, D. Singh, S. Hanke, J. Kropf, A. Holzinger, and M. Geist, "Human annotated dialogues dataset for natural conversational agents," *Applied Sciences*, vol. 10, no. 3, p. 762, 2020.
25. T. Zhang, V. Kishore, F. Wu, K. Q. Weinberger, and Artzi, "Bertscore: Evaluating text generation with Bert," *arXiv preprint arXiv:1904.09675*, 2019.
26. J. Wang, and Y. Dong, "Measurement of text similarity: A survey," *Information*, vol. 11, no. 9, p. 421, 2020.
27. T. Wambsganss, S. Guggisberg, and M. Söllner, "Arguebot: A conversational agent for adaptive argumentation feedback," in *Innovation through information systems: Volume II: A collection of latest research on technology issues*. Springer, 2021, pp. 267–282.
28. D. Tellols, M. Lopez-Sanchez, I. Rodríguez, P. Almajano, and A. Puig, "Enhancing sentient embodied conversational agents with machine learning," *Pattern Recognition Letters*, vol. 129, pp. 317–323, 2020.
29. F. Liu, and Y. Liu, "Correlation between rouge and human evaluation of extractive meeting summaries," in *Proceedings of ACL-08: HLT, Short Papers*. 2008, pp. 201–204.
30. M. Denkowski, and A. Lavie, "Meteor 1.3: Automatic metric for reliable optimization and evaluation of machine translation systems," in *Proceedings of the Sixth Workshop on Statistical Machine Translation*. 2011, pp. 85–91.
31. S. Sharma, L. E. Asri, H. Schulz, and J. Zumer, "Relevance of unsupervised metrics in task-oriented dialogue for evaluating natural language generation," *arXiv preprint arXiv:1706.09799*, 2017.
32. C. Callison-Burch, M. Osborne, and P. Koehn, "Re-evaluating the role of bleu in machine translation research," in *11th Conference of the European Chapter of the Association for Computational Linguistics*. 2006, pp. 249–256.

PART THREE

Deep Learning for Real-World Predictive Modeling

10 Transformer Graph Neural Networks (T-GNNs) for Home Valuation

Faraz Moghimi, Reid Johnson, and Andy Krause

10.1 INTRODUCTION

The real estate market possesses distinct characteristics that set it apart as a unique asset class in the marketplace:

1. Uniqueness: Unlike other markets, there are no identical properties available simultaneously. Even if two houses or apartments appear very similar, they are never exact replicas at any given time.
2. Inefficiency in Transactions: Real estate transactions are not as streamlined as those in other asset classes of comparable value. Factors such as lower liquidity, higher transaction costs, and a decentralized marketplace highlight this inefficiency when compared to stocks and bonds.
3. Primary Investment Choice: Real estate remains the primary investment or purchase for American households, with 65% of the population owning a portion of the $43.4 trillion U.S. residential real estate market [1, 2].

DOI: 10.1201/9781003570882-13

Therefore, precise and reliable real estate pricing (appraisal) is crucial for establishing a fair and efficient market environment for all stakeholders in the housing sector. The goal of real estate pricing (appraisal) is to determine the genuine value of a house in the present market [3]. Although real estate pricing techniques have been developed over time [4], the evolution of machine learning and diverse estimation tools offer promising opportunities in this field [5].

Numerous studies in prior literature have explored the application of machine learning (ML) in real estate pricing [6–8]. Additionally, professionals from diverse sectors are keen on addressing this particular challenge. Delivering accurate estimates of home prices would bring value to a wide range of stakeholders in the real estate market, including bankers, brokers, buyers, and sellers. For example, the Zestimate home valuation model serves as a successful and widely adopted consumer-oriented implementation of real estate valuation. This model offers up-to-date valuation estimates for nearly all residential properties throughout the United States [9].

However, numerous significant challenges in the field complicate the development of effective machine learning solutions. The concept of an intrinsic value for a house is inherently ambiguous [10, 11]. For instance, the same property can be assigned vastly different values in different locations and time periods, with these valuations also subject to change over time. Additionally, while some price disparities may be rooted in legitimate economic factors (such as superior neighborhood amenities), others may arise solely from sometime "irrational" behavioral decisions among market participants. Finally, real estate prices are predominantly relative, influenced by both temporal and geographic variations (when and where).

Real estate transaction data exhibits high sparsity and irregular timing across different locations. For example, a house may change hands multiple times in a year, while another property may stay off the market for extended periods. Moreover, transaction density varies between urban and rural areas, each with unique pricing mechanisms. These factors pose challenges for using traditional time series models like Recurrent Neural Network (RNNs) and transformers to analyze temporal information in real estate pricing. Considering these challenges, the following properties are extremely desirable for any universal and scalable machine learning solution for real estate pricing:

- Dynamic Spatial Representation: Failure of an ML model to grasp this concept from the data and adapt over different location may lead to bias towards specific locations, limiting its universal applicability.
- Evolving Temporal Representation: Without a dynamic understanding of time, the model may struggle to extrapolate effectively from historical data to present or future time frames across diverse locations.
- Time-Dependent Performance Consistency: It is crucial for our model to maintain consistent performance over time without being overly swayed by specific time periods.
- Location-Dependent Performance Stability: A model with unbiased performance across locations is preferred, aiming for stable performance variance across different geographical areas.

- Context-Specific Interpretability: While ML interpretability often relies on concepts like feature importance (SHAP, LIME, etc.) [12], real estate interpretability typically involves generating a list of similar or "comparable" homes for each prediction [13]. Essentially, for users to trust the prediction, the model must demonstrate how nearby, recently transacted homes are valued.

We are motivated to leverage graph neural networks (GNNs) [14] for this task because of their inherent compatibility with these desirable characteristics; we will delve deeper into this alignment in the methodology section. It is worth mentioning that although convolution-based GNNs have been sporadically studied in previous research [15, 16], the concept is still relatively new, particularly in its applications to achieve the aforementioned desirable properties.

We summarize our contributions as follows: (1) Introducing a novel transformer graph neural network (T-GNN) for real estate pricing. (2) Designing an intuitive graph representation of real estate transactions to tackle sparsity and latency issues, enabling the T-GNN model to learn dynamic time and space representations. (3) Evaluating the framework on the King County, WA, dataset. (4) Demonstrating significantly improved performance (40%–50%) compared to tabular benchmark models (XGBoost, FCN). (5) Investigating the reasons for T-GNN's significantly better performance where we attribute it to learning a better dynamic representations of time and space. (6) Providing empirical and theoretical evidence supporting the effectiveness of transformer GNN blocks over convolutional-based GNNs for this problem. (7) Exploring how the T-GNN model can enhance missing feature imputation and overall performance in the presence of missing data.

10.2 METHODOLOGY

10.2.1 Real Estate Transaction Data

We leverage real estate transaction data in the development and validation of real estate pricing models. These datasets contain a variety of information, including property-specific details like square footage, number of bedrooms and bathrooms, as well as location identifiers such as latitude, longitude, city, and zip code. Temporal markers, such as transaction dates, and additional factors like previous tax valuations and sales history, further enrich the dataset. The ultimate target variable in this context is the actual recorded sale price of the properties.

We use a publicly available longitudinal dataset that captures residential real estate transactions in King County, Washington, for our experiments and prototyping [17]. This dataset spans over two decades and includes home sales data from multiple cities within the county. It is important to note that the dataset has sufficient variation across both time and space. For a comprehensive exploration model, we refer readers to the following blog post [18].

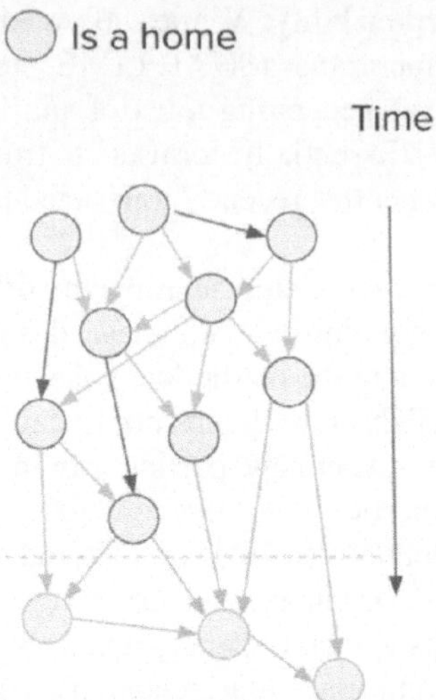

FIGURE 10.1 Representing house transactions as a graph. Each node (circle) represents a transaction, with links (arrows) representing connections from earlier to later transactions.

10.2.2 Representing Real Estate Transactions as a Graph

GNNs leverage message passing to combine information from interconnected entities in a graph structure. We follow a similar process to [19] for converting our transaction data into a graph as demonstrated in fig 10.1. Specifically, each home transaction is a node that would be connected to the closest neighboring home transactions that occurred in the past. For our final model, we use a K-nearest neighbor search (K = 200) for selecting the neighbors. This captures spatial and temporal dynamics affecting future house prices, accommodating data sparsity and irregular timing patterns.

10.2.3 Transformer Graph Neural Network

Our model leverages a transformer-based GNN operator, where multihead attention considers both edge attributes and node features, calculated at the edge level [20]. Specifically, the transformer GNN employs the following message passing mechanism for each target node i, which is connected to a set of neighboring source nodes j.

$$x_i' = W_1 x_i + \sum_{j \in \mathcal{N}(i)} \alpha_{i,j} \left(\mathbf{W}_2 \mathbf{x}_j + \mathbf{W}_3 \mathbf{e}_{ij} \right) \tag{10.1}$$

where x_i or x_j represent the node features for our target and source nodes, respectively; *e*i,j represents the edge attributes of the edges connecting each respective source node j to

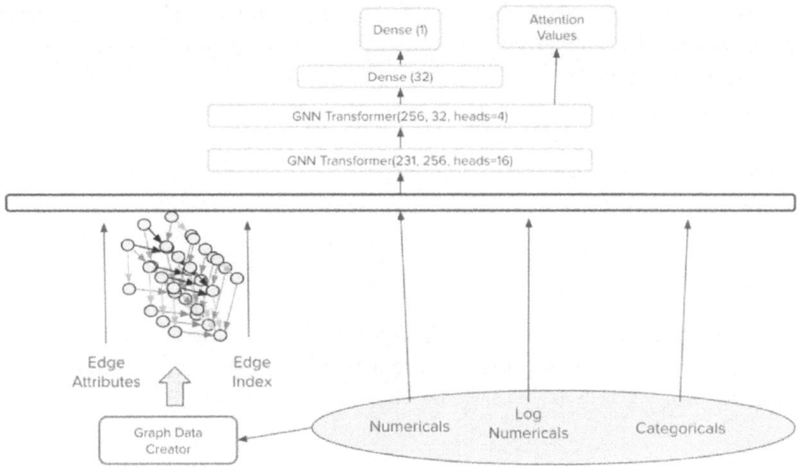

FIGURE 10.2 T-GNN architecture for house pricing.

the target node I; and $\alpha_{i,j}$ represents the multiheaded-attention value for each connection (edge). The attention values are derived as follows and aggregated via a mean operation:

$$\alpha_{i,j} = softmax\left(\frac{\left(W_{1x_i}\right)^{\top}\left(W_{2x_j} + W_3 e_{ij}\right)}{\sqrt{d}}\right) \tag{10.2}$$

Our decision to adopt a transformer-based GNN is driven by the limitations of convolution-based GNNs. Convolution-based GNNs typically assume a fixed and homogeneous underlying graph structure [21]. However, the meaning of connections between home transactions in different locations, such as New York City versus suburban Texas, are inherently different. Similarly, the meaning of connections change in different time periods. By employing a T-GNN operator, we aim to capture the intricate nuances and relationships within real estate transactions more effectively. Our results support this notion.

We combine our graph generation process with two T-GNN blocks followed by two dense layers, as illustrated in Figure 10.2.

10.3 EXPERIMENTS

10.3.1 Data Preprocessing

We perform the following preprocessing steps:

1. Log transformation: We apply a log transformation to skewed numerical distributions in our data, including the target price value.

2. Encoding sparse categorical variables: Sparse categorical variables, such as city, are encoded using the price statistics of each category from a time period preceding our training set.
3. Trigonometric transformation: We conduct a trigonometric transformation on transaction timestamps to capture seasonality effects during training, and we include the year as a separate feature.
4. Handling missing values: Missing values are replaced with the median of the training samples.
5. Normalization: We normalize our features to have zero mean and unit variance.
6. Edge features: For our primary T-GNN model, we incorporate geographic distance, time, and numerical feature differences between node pairs as edge features.

10.3.2 Model Experiments: T-GNN Versus Tabular Models

We benchmarked our T-GNN model with a four-layer Fully Connected Network (FCN) and XGBoost. The test data is from 2021 and the training data is from 2018 to 2020. Notably, the test set includes post-COVID-19 transactions, which saw a positive shock to home prices. Models were trained using Mean Squared Error (MSE), on log-transformed prices.

For accuracy, we primarily rely on mean absolute percentage error (MAPE) and median absolute percentage error (MdAPE) metrics. Additionally, median percentage error (MdPE) helps us assess the bias of the models, indicating if a model systematically under- or over-prices. We report the results in Table 10.1 and find that T-GNN outperforms the benchmark models significantly in all metrics, with MdAPE improving by 40% and the bias score by over 280%.

TABLE 10.1 Performance comparison between the T-GNN, XGBoost, and fully connected neural network

MODEL	*XGBOOST*	*FCN*	*T-GNN*	*IMPROVEMENT (%)*
Count	30,995	30,995	**30,995**	-
R2	0.59	0.704	**0.80**	11.97%
MAE	228646	208261	**162826**	-27.90%
MdAE	106399	108155	**78571**	-37.65%
MPE	-0.089	-0.079	**-0.01**	-1405.92%
MdPE	-0.1	-0.104	**-0.03**	-287.62%
MAPE	0.1759	0.17	**0.14**	-21.63%
MdAPE	0.139	0.136	**0.10**	-39.78%

10.3.3 Graph Constriction Schemes

A crucial decision when utilizing GNNs is constructing the underlying graph. We explore two approaches for constructing the graph:

1. **Fixed Radius:** Setting a fixed radius (e.g., 0.5 kilometers) where all neighboring transactions within the radius would connect to a home node.
2. **K-Nearest Neighbor (KNN):** Connecting each home to a fixed number of homes in the vicinity (K = 200).

In this section, we aim to compare the model performance when constructed using a radius of 0.5 km and the KNN approach with 200 closest neighbors. The results are presented in Table 10.2. We find that the KNN setup outperforms the radius model across all metrics. This outcome aligns intuitively with the idea that the KNN setup would offer more robust representations across diverse neighborhoods and housing densities in various locations. In contrast, the radius setup may tend to overemphasize dense neighborhoods while potentially overlooking sparser areas.

10.3.4 Why T-GNN Over Other GNN Architectures?

In the previous section, we delved into the theoretical rationale behind our decision to opt for a T-GNN instead of a convolution-based GNN. In this section, we empirically test this choice by performing a comparison between our T-GNN and two other GNN variants, namely graph convolution networks (GCN GNNs) and GraphSAGE GNN. We report the results in Table 10.3. Consistent with our theoretical notion, we find that the GCN architecture fails to converge entirely throughout the training process, and T-GNN significantly outperforms the GraphSAGE GNN.

TABLE 10.2 Effect of graph construction on model performance

TRAINING START (YEAR)	*MODEL*	*COUNT*	*MdPE*	*MdAPE*	*MAPE*
2018	XGBoost	30,995	-0.100	0.139	0.18
2016	**XGBoost**	**30,995**	**-0.127**	**0.150**	**0.18**
2018	FCN	30,995	-0.105	0.136	0.17
2016	**FCN**	**30,995**	-0.109	**0.156**	**0.192**
2018	T-GNN	30,995	-0.03	0.097	0.14
2016	**T-GNN**	**30,995**	**-0.015**	**0.0933**	**0.14**

TABLE 10.3 Performance of transformers, convolution, and graph SAGE GNNs

MODEL	*TRANSFORMER GNN*	*GCN*	*GRAPHSAGE GNN*
Count	30995	30995	30995
R2	0.76	Not Converging	0.72
MAE	171127.39	Not Coverging	188318.88
MdAE	81740.25	Not Coverging	94004.38
MPE	–0.01	Not Converging	0.02
MdPE	–0.03	Not Converging	–0.02
MAPE	0.15	Not Converging	0.17
MdAPE	0.10	Not Converging	0.11

10.3.5 Why T-GNN Performs Better Than Tabular Models: T-GNN Performance Across Time and Space

To investigate why the GNN model outperforms the benchmarks significantly, we hypothesize that learning a better representation of time and space is the key mechanism behind the better performance of the T-GNN model.

In this section, we delve deeper into this concept. Specifically, while aggregated metrics such as MAPE and MdAPE provide a good overview of a model's performance, the robustness of a model across both time and space are additional desirable properties in the home pricing models. Essentially, we aspire for models to demonstrate consistent performance across diverse locations. Furthermore, we seek a model that exhibits minimal volatility in various market conditions captured along the temporal axis. As such, if learning a better representation time and space is the driving factor behind T-GNN's better performance, we would expect the performance to be more consistent across time and space.

10.3.5.1 The locations axis

Here, we analyze and compare the performance of the T-GNN and XGBoost models based on the MAPE score, categorized by different cities within our test sample from the state of Washington. Our focus is on assessing the consistency of errors exhibited by each model across different cities. We report the results in Figure 10.3, with the blue bars indicating the T-GNN performance and the orange bars representing the XGBoost performance.

Overall, we note that the T-GNN model exhibits significantly less performance volatility across different locations. Interestingly, both models show similar performance in Seattle, which is the majority group in our dataset. However, the T-GNN consistently outperforms the XGBoost model in almost all other more sparse locations. This highlights that tabular models like XGBoost perform well on the dominant sample but

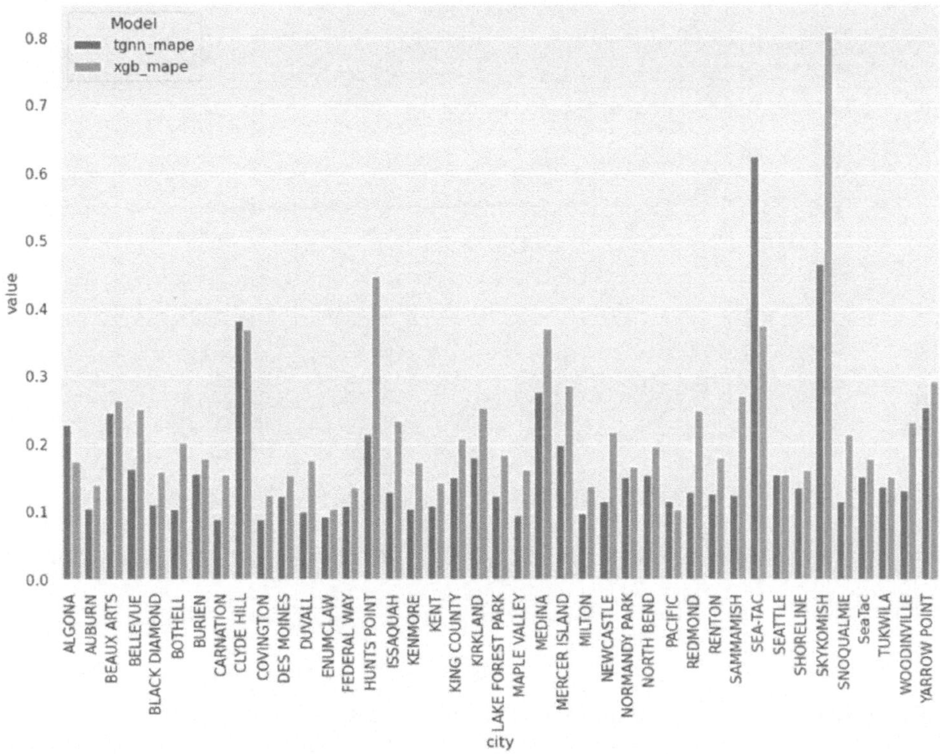

FIGURE 10.3 MAPE score of the T-GNN model and XGBoost model broken down by city regions.

may struggle to capture the pricing dynamics of minority cities. It reinforces the idea that the strength of the T-GNN lies in its superior understanding of location factors. Furthermore, it supports the notion that the T-GNN model presents a more robust solution from the perspective of location bias in the home pricing domain.

10.3.5.2 The time axis

In this domain, another crucial aspect to consider is the notion of time. Specifically, our modeling solution should demonstrate consistent performance and adaptability across different time periods. To evaluate this concept, we analyze the performance of both T-GNN and XGBoost based on the MAPE score broken down into weekly periods. It is important to acknowledge that the overall model performance is anticipated to decline over time, as we move further into the future beyond the data encompassed in the training set. However, a slower and less volatile decay is a desirable attribute of any modeling solution. The results of this analysis are presented in Figure 10.4. Consistent with the hypothesis that T-GNN more effectively grasps the concept of time, we observe that the decay in T-GNN performance and the subsequent volatility in performance over time are notably smaller compared to the XGBoost baseline.

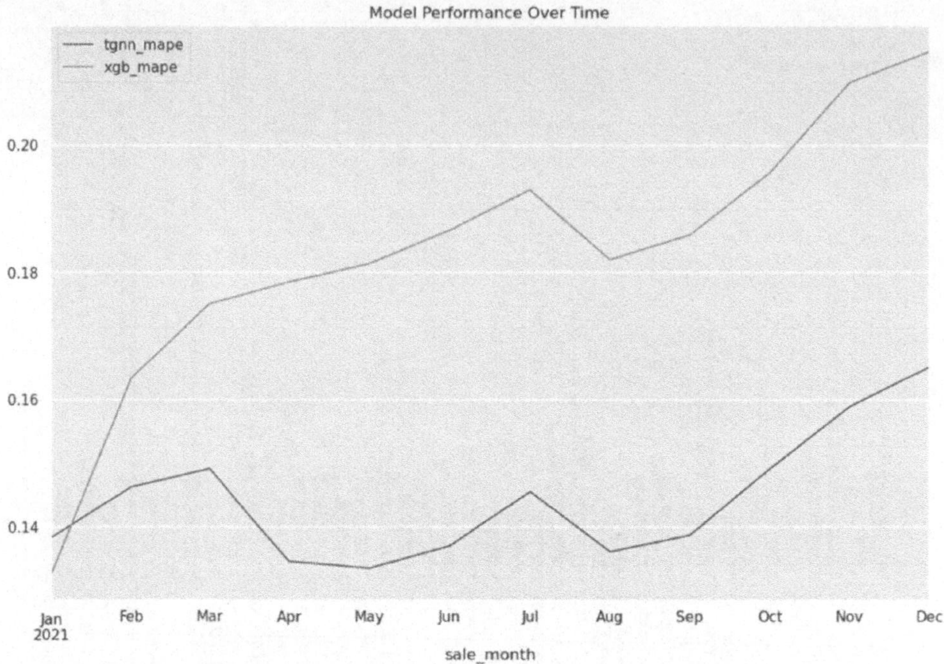

FIGURE 10.4 MAPE score of the T-GNN model versus XGBoost model across time.

10.3.6 Feature Imputation via T-GNN

Handling missing features poses a common challenge in machine learning applications. Although it is not the primary focus of our study, there have been intriguing studies in the literature introducing the concept of missing feature propagation based on a graph structures. The concept behind this approach is that utilizing the graph structure to impute missing features can be a more effective method for handling missing data, as the model can leverage information from neighboring nodes.

Specifically, we leverage the method proposed in [22] called feature propagation. Feature propagation is a general approach for handling missing features in graph machine learning applications based on minimizing the Dirichlet energy and leveraging diffusion-type differential equations on the graph to address the partial availability of node or edge features.

To test the effectiveness of the feature propagation method, we randomly remove 20% of all features from the training set. Subsequently, we establish a performance baseline for the model in two scenarios: 1. filling missing values with sample medians, and 2. employing the feature propagation method to impute missing values. The outcomes are presented in Table 10.4. We see that utilizing the graph-based feature propagation leads to slight performance improvement, suggesting potential additional advantages in applying the T-GNN model in the realm of home pricing.

TABLE 10.4 T-GNN model performance with artificial missing data filled with both sample median, and propagated via the graph structure

MODEL	*T-GNN*	*T-GNN*
FEATURE IMPUTATION	*MEDIAN*	*GRAPH FEATURE PROPAGATION*
Count	30995	30995
R2	0.720	0.777
MAE	188319	169226
MdAE	94004	85798
MPE	0.021	–0.028
MdPE	–0.023	–0.050
MAPE	0.173	0.147
MdAPE	0.111	0.106
%within 30%	0.861	0.912
%within 10%	0.457	0.474

10.4 CONCLUSION

In this chapter, we introduced a novel approach to address the real estate pricing problem using a T-GNN. Our method surpasses the benchmark models (XGBoost, FCN) with a notable 40% increase in accuracy and a 280% improvement in bias reduction. We attribute the success of the T-GNN to its ability to learn a more effective representation of time and space, thereby capturing nuanced impacts on home prices. Additionally, we provide both theoretical justification and empirical evidence supporting why transformer-based GNNs may outperform convolution-based GNNs in this context. Furthermore, we explore the potential of leveraging GNN structures for enhanced missing feature imputations in real estate pricing, uncovering promising insights. Our study demonstrates the efficacy of the T-GNN framework in improving real estate pricing accuracy and robustness, showcasing its potential for advancing real estate valuation methodologies.

ACKNOWLEDGMENT

This project was conducted with funding from Zillow Group.

REFERENCES

1. M. Taylor, Homeowner data and statistics 2023, [online] Available: https://www.bankrate.com/homeownership/home-ownership-statistics

2. T. Manhertz, U.S. housing market has doubled in value since the great recession Gaining\$6.9 Trillion in 2021, 2022, [online] Available: https://www.zillow.com/research/us-housing-market-total-value-2021–30615/.
3. E. Pagourtzi, V. Assimakopoulos, T. Hatzichristos and N. French, "Real estate appraisal: a review of valuation methods", Journal of Property Investment & Finance, 21(4), 383–401, 2003.
4. American Institute of Real Estate Appraisers. The Appraisal of Real Estate. The Institute, 1952.
5. A. Baldominos, I. Blanco, A. Moreno, R. Iturrarte, O. Bernardez and C. Afonso, "Identifying real estate opportunities using machine learning", Applied Sciences, 8(11), 2321, 2018.
6. B. Trawinski, Z. Telec, J. Krasnoborski, M. Piwowarczyk, M. Talaga, T. Lasota, et al., "Comparison of expert algorithms with machine learning models for real estate appraisal", IEEE International Conference on Innovations in Intelligent Systems and Applications (INISTA), 2017.
7. Y. Yu, J. Lu, D. Shen and B. Chen, "Research on real estate pricing methods based on data mining and machine learning", Neural Computing and Applications, 33, 3925–3937, 2021.
8. F. Lorenz, J. Willwersch, M. Cajias and F. Fuerst, "Interpretable machine learning for real estate market analysis", Real Estate Economics, 51(5), 1178–1208, 2023.
9. Zillow Group, "What is a Zestimate?", Zillow Group Available: https://www.zillow.com/z/zestimate/.
10. E. Hargrove, "Weak anthropocentric intrinsic value", The Monist, 75(2), 183–207, 1992.
11. M. Sachdeva, S. Fotheringham and Z. Li, "Do places have value?: Quantifying the intrinsic value of housing neighborhoods using MGWR", Journal of Housing Research, 31(1), 24–52, 2022.
12. P. Linardatos, V. Papastefanopoulos and S. Kotsiantis, "Explainable AI: A review of machine learning interpretability methods", Entropy, 23(1), 18, 2020.
13. J. Bradley, "Real estate comps: How to find them", bankrate, [online] Available: https://www.bankrate.com/real-estate/how-to-find-real-estate-comps/
14. F. Scarselli, M. Gori, A. Tsoi, M. Hagenbuchner and G. Monfardini, The graph neural network model, 2008.
15. W. Zhang, H. Liu, L. Zha, H. Zhu, J. Liu, D. Dou, et al., "MugRep: A multi-task hierarchical graph representation learning framework for real estate appraisal", Proceedings of the 27th ACM SIGKDD Conference on Knowledge Discovery & Data Mining, 2021.
16. C. Li, W. Wang, W. Du and W. Peng, "Look around! A neighbor relation graph learning framework for real estate appraisal", arXiv, 2022.
17. A. Krause, King County (WA) Home Sales, [online] Available: https://www.kaggle.com/datasets/andykrause/kingcountys
18. R. Johnson, House price EDA and modeling with python, [online] Available: https://www.kaggle.com/code/reidjohnson/house-price-edaand-modeling-with-python.
19. S. Yun, M. Jeong, R. Kim, J. Kang and H. Kim, "Graph transformer networks", Advances in neural information processing systems.
20. J. Zhou, G. Cui, S. Hu, Z. Zhang, C. Yang, Z. Liu, et al., "Graph neural networks: A review of methods and applications", AI Open, 1, 57–81, 2020.
21. Y. Shi, Z. Huang, S. Feng, H. Zhong, W. Wang and Y. Sun, Masked label prediction: Unified message passing model for semi-supervised classification, 2020.
22. F. Moghimi, R. A. Johnson, and A. Krause, "Rethinking real estate pricing with Transformer Graph Neural Networks (T-GNN)", 2023 International Conference on Machine Learning and Applications (ICMLA). IEEE, 2023.

11 Model Error Clustering Approach for HVAC and Water Heater in Residential Subpopulations

Viswadeep Lebakula, Eve Tsybina, Jeff Munk, and Justin Hill

11.1 INTRODUCTION

In the view of the shift in energy resources and the changing electricity market landscape, thermostatically controlled residential loads are increasingly considered grid support instruments and cost optimisation items. They are argued to help reduce peak charges, defer infrastructure investments, and provide local grid relief [1–3]. Along with the adoption of load control policies, utilities and researchers face the challenge of doing so in an economically efficient way. This, in turn, caused a significant volume of research to explore predictive control methods in managing fleets of thermostatically controlled devices, such as HVAC and water heaters [4–10]. But how do you provide the forecast for the controller?

DOI: 10.1201/9781003570882-14

Theoretical research suggests a few approaches, such as model-based reinforcement learning [11] or Q-learning [12, 13]. The available experimental studies adopt machine learning and rule-based approaches [14, 15]. The important takeaway from these studies is that model predictive controllers are sensitive to high-magnitude unsystematic errors and systematic errors, which are difficult to address in a fleet of devices. Unsystematic errors emerge when an autoregressive process encounters a random event, which disappears in the next period, but the model propagates it because it already noticed the deviation. Such events may occur due to outlier behaviour of residents or data or communication issues. Systematic errors occur because larger fleets are characterised by unobserved heterogeneity, which may be difficult or costly to control for. For instance, if certain subpopulations of homes are heavy users of HVAC or water heaters, the utility will not know this until several months later. And even then, it may be too expensive to generate a separate model to control for that specific group of customers as opposed to having a model for an average home.

It is difficult to develop a completely error-free model that can account for all the variables. Models can predict with acceptable accuracy (low error) most of the time but in some instances with lower accuracy (high error). These errors may be due to several factors, such as sensor malfunction, changes in occupant behaviour, and seasonality. Because of the presence of errors, error detection and correction are important tasks in designing predictive controllers. Put simply, isolating high error instances will allow researchers to investigate the cause, which may provide insights to better modelling approaches. Error detection and correction exist in time series research, for instance, in finance studies [16] or manufacturing research [17]. But this area is not well addressed in load control studies. In our earlier research, we attempted to produce a testable methodology for detecting and isolating whole groups of device instances into non-anomaly (high error) and anomaly (low error) groups [18, 19]. Here we expand the earlier approach to multiple subpopulations of devices. We applied the error detection methodology to isolate the undesired instances (high error) for the five new subpopulations based on location of the home (corner and interior) and water usage (high, low, and medium). This approach was tested on the observed behaviour of a residential neighbourhood in Atlanta, GA. We report the running time, number of clusters, and characteristics of error of the identified cluster centres. .

11.2 METHODOLOGY

11.2.1 Detection of Anomalies through Clusters of Devices

It was already argued in [18, 19] that normal operation of devices would be characterised by a small error between the forecast value and the observed value. If a device deviates from the forecast, the error will produce a large square distance from the anticipated value. Simultaneously, this error will not occur for all devices in a fleet. Rather, it will affect a subpopulation of devices, and the devices will be affected to a varying extent. Some of the devices may require immediate intervention while others may be monitored until the next hour. Detecting the affected subpopulations and classifying them according to the extent of anomalies can be done through several unsupervised learning algorithms. Here we used three common (previously used for anomaly detection) clustering algorithms: K-means clustering with Euclidean distance (KME), Ward algorithm, and K-means with dynamic time warping (KMDTW) [18–22].

11.2.2 Model and Experimental Setup

The proposed methodology is applied to the residential neighbourhood in Atlanta, GA. The neighbourhood consists of 46 townhomes with identical HVAC systems and water heaters. Each HVAC system is a 4.5-tonne multizoned two-stage heat pump with three thermostats, one for each floor. Detailed information about the neighbourhood can be found in [23].

The overall optimisation of HVAC systems is performed individually for each HVAC system according to Equation 11.1.

$$min\ W_p\left(\sum_{i=0}\rho_t^c\ P_t^{ac}+\sum_{i=0}\rho_t^c\ P_t^{hp}\right)+W_{temphi}\sum_{t=0}T_{tz}^{auxcoolmax}+W_{templo}\sum_{t=0}T_{tz}^{auxcoolmin}$$

$$+W_{temphi}\sum_{t=0}T_{tz}^{auxheatmax}+W_{templo}\sum_{t=0}T_{tz}^{auxheatmin} \tag{11.1}$$

The controller minimises the total cost of electricity calculated as the active power used for heating, P_t^{hp}, or cooling, P_t^{ac}, multiplied by the price signal, ρ_t^c, a synthetic value designed to enforce a desired shape of HVAC load for the neighbourhood. While each townhome receives the same ρ_t^c, floor temperatures and user preferences result in varying costs and varying operational patterns of individual HVAC systems. In order to enforce the temperature setpoints, the total cost also includes weighted variables for minimum and maximum setpoints for cooling and heating, denoted as $T_{tz}^{auxcoolmax}$, $T_{tz}^{auxcoolmin}$, $T_{tz}^{auxheatmax}$, and $T_{tz}^{auxheatmin}$. The optimisation is subject to constraints which pertain to user comfort and HVAC operation. Equations 11.2–11.5 show comfort constraints for the cooling mode; the variables for the heating mode are symmetrical.

$$T_{tz}^{auxcoolmax}\geq T_{tz}^{in}-T_{setp}^{cool} \tag{11.2}$$

$$T_{tz}^{auxcoolmax}\geq 0 \tag{11.3}$$

$$T_{tz}^{auxcoolmin}\geq T_{tz}^{in}-\left(T_{setp}^{cool}-T_{band}\right) \tag{11.4}$$

$$T_{tz}^{auxcoolmin}\geq 0 \tag{11.5}$$

Specifically, upper comfort band cannot be smaller than the difference between the current temperature, T_{tz}^{in}, and the cooling setpoint, T_{setp}^{cool}, and must be a number larger than zero. The owner comfort band cannot be smaller than zero or smaller than the difference between the current temperature, T_{tz}^{in}, and the cooling setpoint after correcting for the operating margin of the HVAC unit, T_{band}, set at 3F by the utility. The discussed Model Predictive Control (MPC) function is calculated every 10 minutes, resulting in control signals that are dispatched to each HVAC system based on individual optimisation results.

It can be observed that the controller has a time component, t, and a thermostat component, tz. As has been discussed above, there are three thermostats on three floors of the house, which cause the controller to include one variable per floor. The resulting model error observations include three absolute errors for each timestamp, as described in Equation 11.6.

$$AE_{f,t} = \left|Forecast_{f,t} - Observation_{f,t}\right| \tag{11.6}$$

The optimisation of water heaters is performed according to Equation 11.7. Similarly to Equation 11.1, the price signal, ρ_t^c, is same for the entire neighborhood, but each water heater responds to it individually and the optimisation is done separately for each water heater.

$$min\left(\sum_{t=0} \rho_t^c\ On_t\ heat_{wh} + \rho_t^c On_t^{element}\ heat_{element} + W_{mode} \sum_{t=0} Chg_t^{mode}\right) \tag{11.7}$$

The total cost of operating a water heater is calculated based on $heat_{wh}$, power consumed by the heat pump, and $heat_{element}$, power of the resistance heating element. Both variables are set in accordance with manufacturer specifications, but there is a significant difference between the two. Depending on the water heater model, $heat_{wh}$ can be on the order of 400–800 W and $heat_{element}$ can reach 4500 W. In the model, the manufacturer values are converted into loads using variable On_t, a binary variable that is 1 if the mode is "on" and 0 for the mode "off." To prevent devices from switching modes uncontrollably, W_{mode} is used to put a penalty on the changing mode; Chg_t^{mode} is a binary variable that takes the value of 1 when the water heater changes the mode.

Similar to the HVAC optimisation, the water heater is optimised subject to operating and temperature comfort constraints. Equations 11.8–11.12 provide details on temperature constraints.

$$T_t^{limit} = T_{set} - mode_t^{shed}\ DB_{shed} - mode_t^{normal}\ DB_{normal} - mode_t^{load}\ DB_{load} \tag{11.8}$$

$$T_t^{limit} - t_{t0} \leq M \times DBbin_t^{low} \tag{11.9}$$

$$t_{t1} - T_t^{limit} \leq M\ \left(1 - DBbin_t^{low}\right) \tag{11.10}$$

$$\left(T_{set} - DB_{elementOn}\right) - t_{t1} \leq M \times DBbin_t^{low,elementOn} \tag{11.11}$$

$$t_{t1} - \left(T_{set} - DB_{elementOn}\right) \leq M\ \left(1 - DBbin_t^{low,elementOn}\right) \tag{11.12}$$

The water heaters are designed to operate based on the temperature setpoint, T_{set}, and the deadband limit, DB, denoted as T_t^{limit}. The temperature is required to stay within the deadband limit, which is achieved through a very large constant weight, M.

The model temperature is calculated for six nodes assigned to a water heater tank. Having multiple nodes better accounts for the fact that water heater tanks are not heated homogenously. Higher temperatures are observed at the top of the tank compared to the bottom of the tank. As a result, the temperature for the bottom of the tank is calculated according to Equation 11.13.

$$T_{[t+1]j} = T_{tj} + \frac{ts}{60C_j}\left(heat_{wh}\ On_t - q_{tj} + UA_j\ \left(T_{amb} - T_{tj}\right) + draw_t * 8.35 * 6\left(T_{cold} - T_{tj}\right)\right) \quad (11.13)$$

Temperature for the top of the tank is calculated according to Equation 11.14.

$$T_{[t+1]j} = T_{tj} + \frac{ts}{60C_j}\left(heat_{wh}\ On_t + heat_{element}\ On_t + q_{tj-1} + UA_j\ \left(T_{amb} - T_{tj}\right)\right.$$
$$\left. + draw_t * 8.35 * 6\left(T_{cold} - T_{tj}\right)\right) \quad (11.14)$$

Water temperature projection for the next timestamp $T_{[t+1]j}$ is calculated based on a number of water tank constant characteristics, including the specific heat capacity of the water in tank C_j, surface area A_j, heat transfer coefficient U which is the same for all water heaters, q_{tj} indicating the movement of hot water upwards, and individual characteristics related to water usage. Specifically, it uses existing temperature T_{tj}, water usage $draw_t$, the difference between ambient temperature and water temperature $\left(T_{amb} - T_{tj}\right)$, and the difference between the inlet water temperature and the temperature of water in the heater $\left(T_{cold} - T_{tj}\right)$. Temperature forecast errors are thus calculated for bottom and top parts of the water heater according to Equations 11.15–11.16.

$$e_{t\_up} = t\_up_{obs} - t_{up\ forecast} \quad (11.15)$$

$$e_{t\_low} = t\_low_{obs} - t\_low_{forecast} \quad (11.16)$$

An additional forecast value is the operating status of a water heater, indicating if a water heater is operating at a given timestamp.

In order to calculate the deviation of observed temperature and operating status findings from the model predictions, we use the data from the vendor API. HVAC temperature data was collected in June–August 2022, from 10 a.m. to 10 p.m. on weekdays. The sampling was made according to the forecast resolution, at 10-minute intervals, which generated 72 observations for each day. The observations for water heaters were collected in October–December 2022, sampled at 10-minute intervals, from 5 a.m. to 10 p.m. Monday through Friday. We were able to derive the temperature from the anticipated length of discharge, according to Equations 11.17–11.18.

$$t\_low = \frac{storage_{total} - storage_{present}}{a} + b \quad (11.17)$$

$$t\_up = t\_low + c \quad (11.18)$$

Present energy storage capacity, $storage_{present}$, and total energy storage capacity, $storage_{total}$, were converted into top water temperature, t_up, and bottom water temperature, t_low, using empirical coefficients $a = 132$, $b = 21.5$, and $c = 5$.

11.3 RESULTS

11.3.1 Observed Error Distribution

We start with a discussion of the observed error distribution. Figure 11.1 shows the HVAC errors for the observed period for each floor. It is helpful to recall that the first floor has a garage and a bonus room, the second floor has a kitchen and a living room, and the third floor has bedrooms. The errors tend to be lower for the third floor, where little temperature disturbance exists. The majority of observations stay along the 45-degree line, indicating that the observed temperature was about the same as the predicted temperature, though there are still observed instances when the errors are moderate or high. Also, the error may be grouped into more than two categories: low error (LE), high error (HE), and extremely high error (EHE).

Since heat transfer is the main contributor to temperature changes inside the house, it is helpful to see if the distribution of errors would change for interior or corner units. Interior units have neighbouring townhome units on each side, while corner units have one additional wall exposed to ambient conditions. Under normal operation, a utility may not know if a given customer is highly exposed to temperature changes; for instance, as a result of insulation or location of the house. It is of interest if the error correction is capable of isolating and correcting errors in the event of unobserved heterogeneity. We use the information on the location of townhomes in the experimental neighbourhood and split the dataset into two groups: one for interior townhomes and one for corner townhomes. Figure 11.2 shows the two floors which are of main interest for the study, the second floor and the third floor, for corner and interior units. Corner

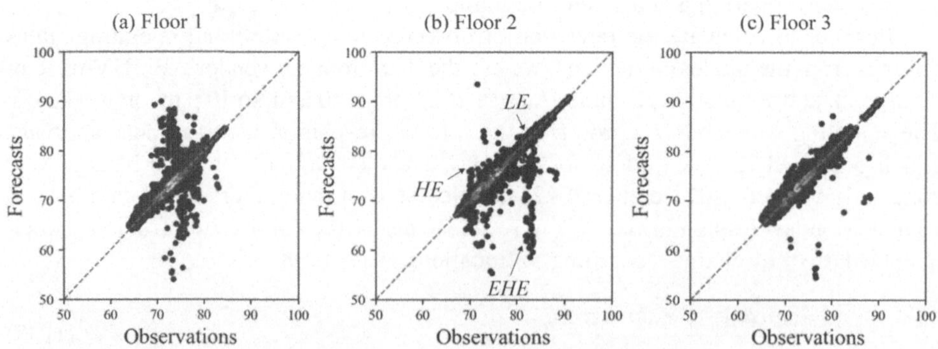

FIGURE 11.1 Scatter plot of forecast and observed indoor temperature, degrees F, by floor. The dotted line is a 45-degree line. This figure is from [19].

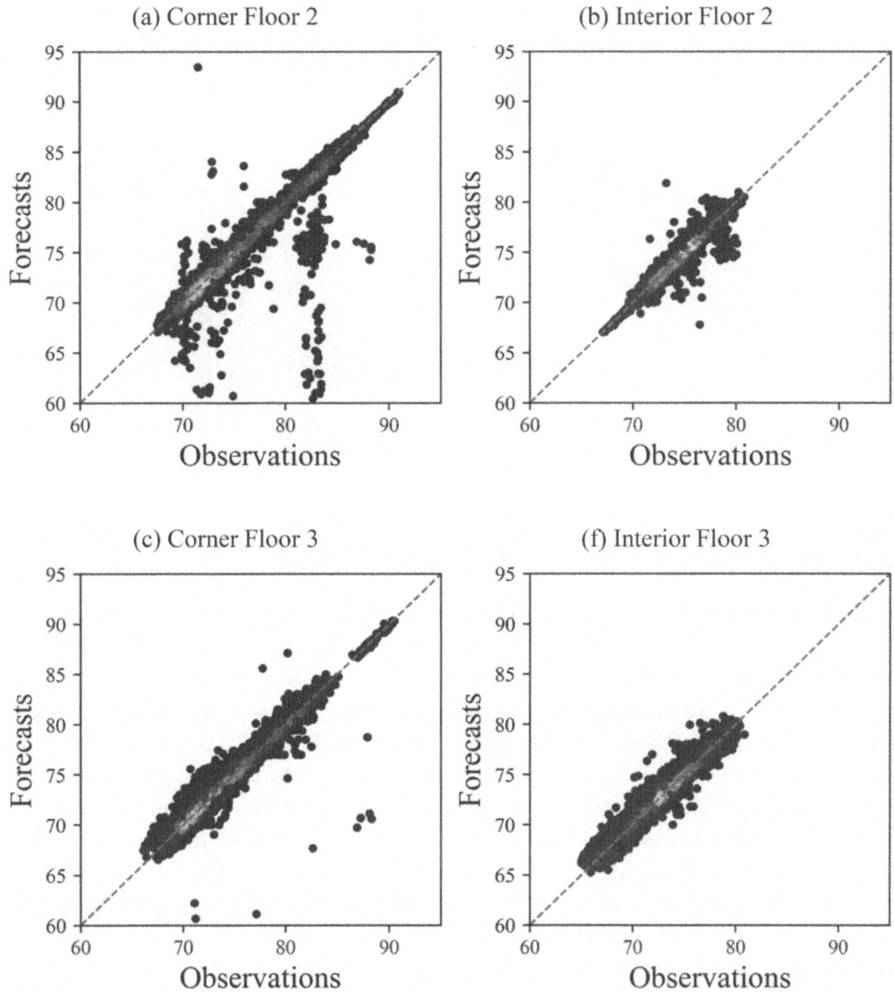

FIGURE 11.2 Scatter plot of forecasts and observations of corner ((a), (c)) and interior homes ((b), (d)) for floor 2 and floor 3.. The dotted line is a 45-degree line.

units do tend to have a larger number of outliers, although the number is not as high as the high-density blue and yellow areas are still concentrated along the 45-degree line. It is possible to say that the variance increases as a function of the corner unit, making the dataset heteroskedastic.

A similar look at the distribution for water heaters is provided in Figure 11.3. It also shows that most observations are concentrated along the 45-degree line. The majority of observations for the upper limit tend to be concentrated along the 45-degree line. Only a smaller part of it is scattered in an unsystematic way both above and below the line. The lower limit has deviations in the upper half of the graph. The blue and yellow incidence which runs parallel to the 45-degree line indicates that a large number of observations have a slightly positive error. It also has a larger scatter in the lower half of

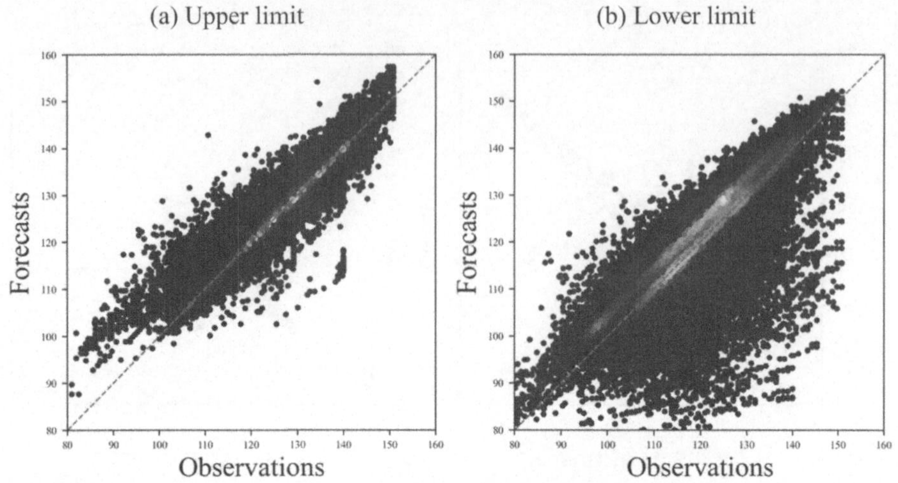

FIGURE 11.3 Distribution of temperature errors, (a) upper limit (t_up), (b) lower limit (t_low). The dotted line is 45-degree line. This figure is from [18].

the graph, which means that a small number of observations have a large negative error. This presents an interesting observation for error correction, as we will be clustering large unsystematic errors below and small to moderate errors above the 45-degree line.

The errors in water temperature are affected mostly by water usage. If a given townhome uses a lot of hot water, the temperature will both stay below that of an average townhome and become more unpredictable. The important part about water usage is that it is usually an unobserved variable that is not known at the beginning of the development of the control system. Where possible, utilities can second-guess water usage of a given house, but it is highly probable that a utility will not know for sure what kind of a household it is serving in terms of water usage. As a result, it is important to have a system that is capable of addressing forecast errors in situations where unobserved variables are present in the data.

We include the water usage to see if the situation is going to change for high water usage townhomes versus low water usage townhomes. Figure 11.4 shows that the distribution of errors indeed changes for higher usage townhomes and that the change is more pronounced for lower bound than for upper bound.

11.4 ERROR CLUSTERING

In this study, we focus of subpopulations that are based on location of the house and hot water consumption. A daily model forecasting error of internal temperatures for each floor in the house was calculated for each house, as mentioned in above section. Later, the individual error of homes was grouped into one of the subpopulations. KME, KMDTW, and Ward clustering were applied to the model forecasting error of each

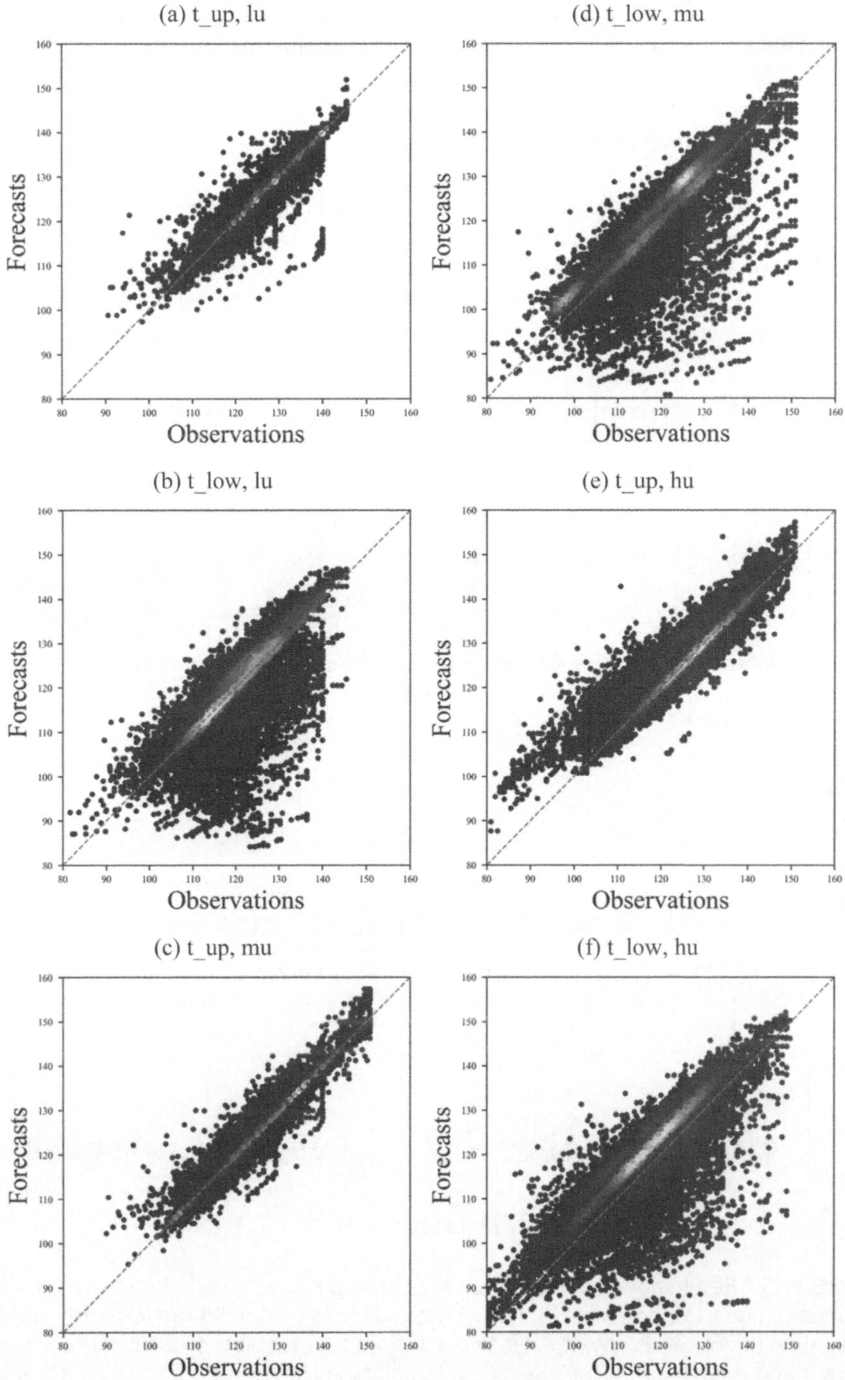

FIGURE 11.4 Scatter plot of forecasts and observations of water heaters with low ((a), (b)), medium ((c), (d)), and high ((e), (f)) water use. The dotted line is a 45-degree line.

subpopulation to differentiate between anomaly (high error) and non-anomaly (low error) instances. Results for error clustering for the subpopulations are provided below.

11.4.1 Location of the Homes

First, each house was divided into one of the two subpopulations, *corner* and *interior*, based on the location of the homes in the neighbourhood. For all floors, internal temperature error was calculated during the active control time period from 10:00 a.m. to 10:00 p.m. Three clustering algorithms were applied on the subpopulations and the results are presented in Figures 11.5–11.7. For the corner homes subpopulation, KME

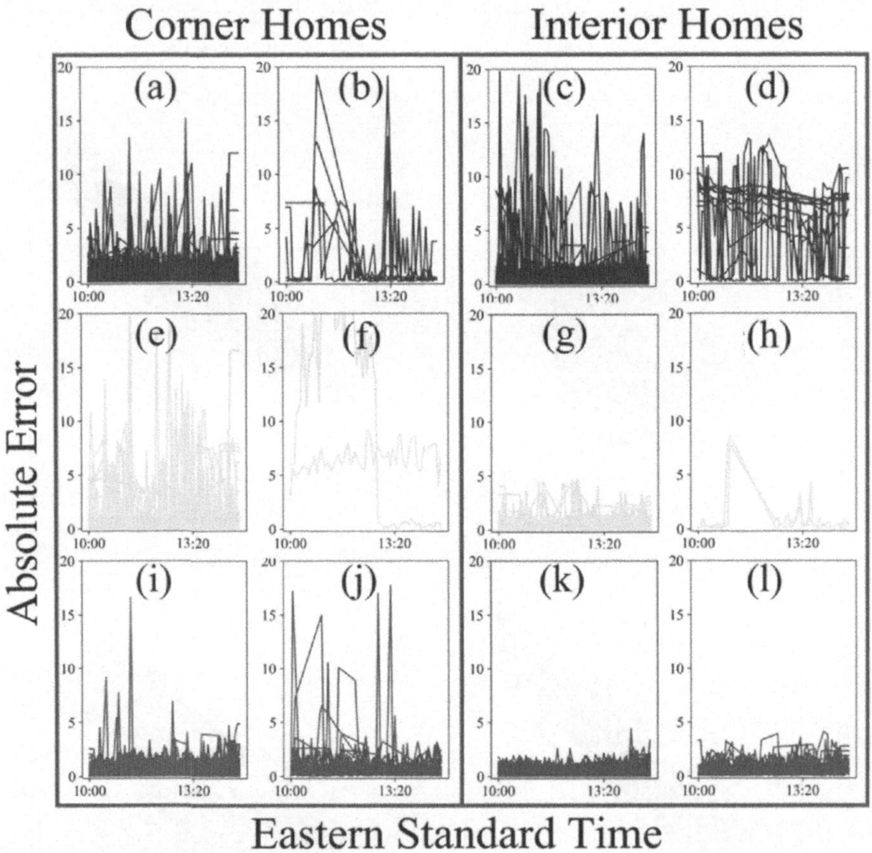

FIGURE 11.5 Results for two clusters approach using K-means with Euclidean distance of both interior and exterior homes for floor 1 (from (a) to (d)), floor 2 (from (e) to (h)), and floor 3 (from (i) to (l)). For each row, the left two and right two subplots on each side correspond to corner and interior houses, respectively. Subplots in the left column ((a), (e), (1), (c), (g), and (k)) and right column ((b), (f), (j), (d), (h), and (l)) within corner and interior homes represent clusters without (low error) and with anomalies (high error), respectively. The *y*-axis varies from 0 to 20 and the *x*-axis varies from 10 to 22 hours.

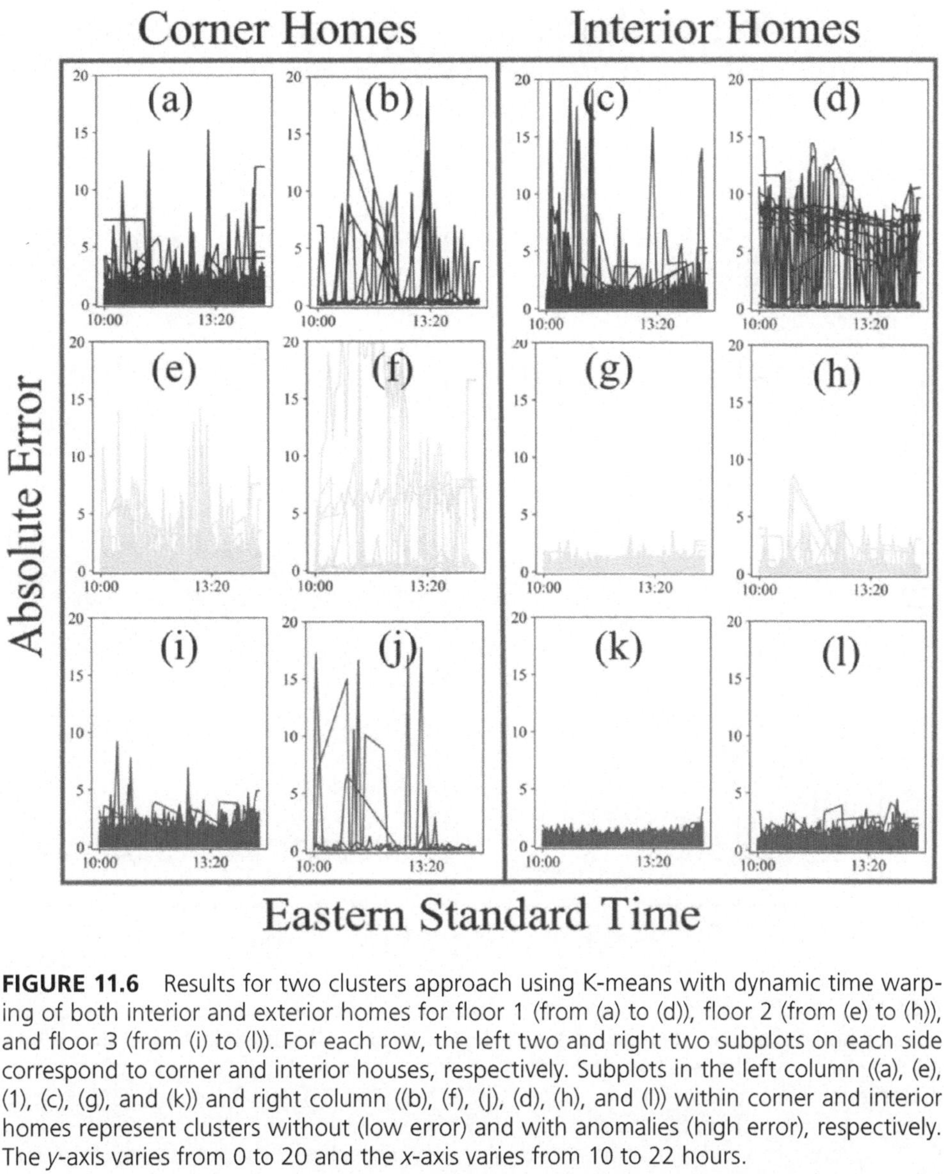

FIGURE 11.6 Results for two clusters approach using K-means with dynamic time warping of both interior and exterior homes for floor 1 (from (a) to (d)), floor 2 (from (e) to (h)), and floor 3 (from (i) to (l)). For each row, the left two and right two subplots on each side correspond to corner and interior houses, respectively. Subplots in the left column ((a), (e), (1), (c), (g), and (k)) and right column ((b), (f), (j), (d), (h), and (l)) within corner and interior homes represent clusters without (low error) and with anomalies (high error), respectively. The *y*-axis varies from 0 to 20 and the *x*-axis varies from 10 to 22 hours.

classified a lower number of error instances for floor 1 and floor 2 but not for floor 3 (Figure 11.5), whereas KMDTW and Ward clustering algorithms identified a lower number of error instances across all three floors (Figures 11.6 and 11.7). For the interior home subpopulations, in floor 2, only two instances were identified as anomalies by both KME and Ward clustering algorithms (Figures 11.5 and 11.7). In comparison, for floor 2, KMDTW resulted in a higher number of error instances or anomalies (Figure 11.6).

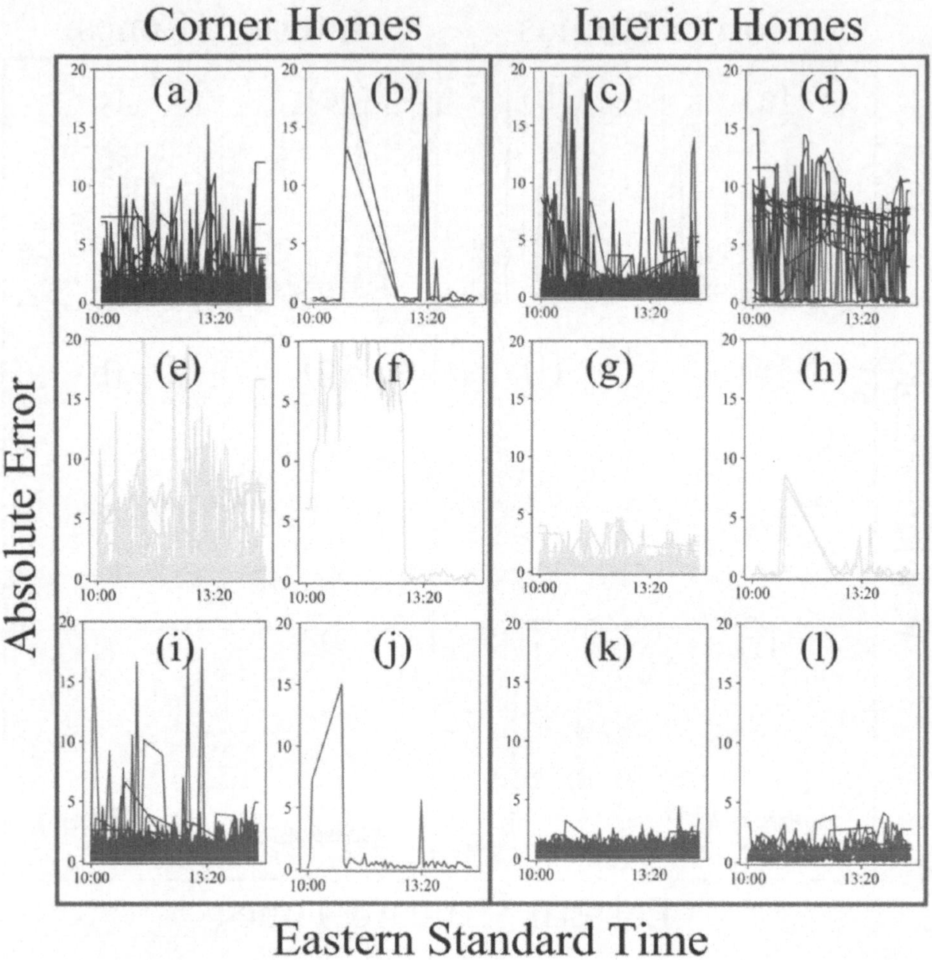

FIGURE 11.7 Results for two clusters approach using Ward clustering of both interior and exterior homes for floor 1 (from (a) to (d)), floor 2 (from (e) to (h)), and floor 3 (from (i) to (l)). For each row, the left two and right two subplots on each side correspond to corner and interior houses, respectively. Subplots in left column ((a), (e), (1), (c), (g), and (k)) and right column ((b), (f), (j), (d), (h), and (l)) within corner and interior homes represent clusters without (low error) and with anomalies (high error), respectively. The *y*-axis varies from 0 to 20 and the *x*-axis varies from 10 to 22 hours.

The above-mentioned results from two clustering approaches were generated with the assumption that all the instances can be divided into two groups: one group without anomalies (acceptable error) and another group that has anomalies (error above the tolerance). In all the scenarios, it is not necessary that this assumption holds true, especially in cases where there is an error of a different magnitude, such as high error and extremely high error (Figure 11.1b). To address this issue, previously, researchers used the elbow method and dendrograms to determine the optimal number of clusters for the

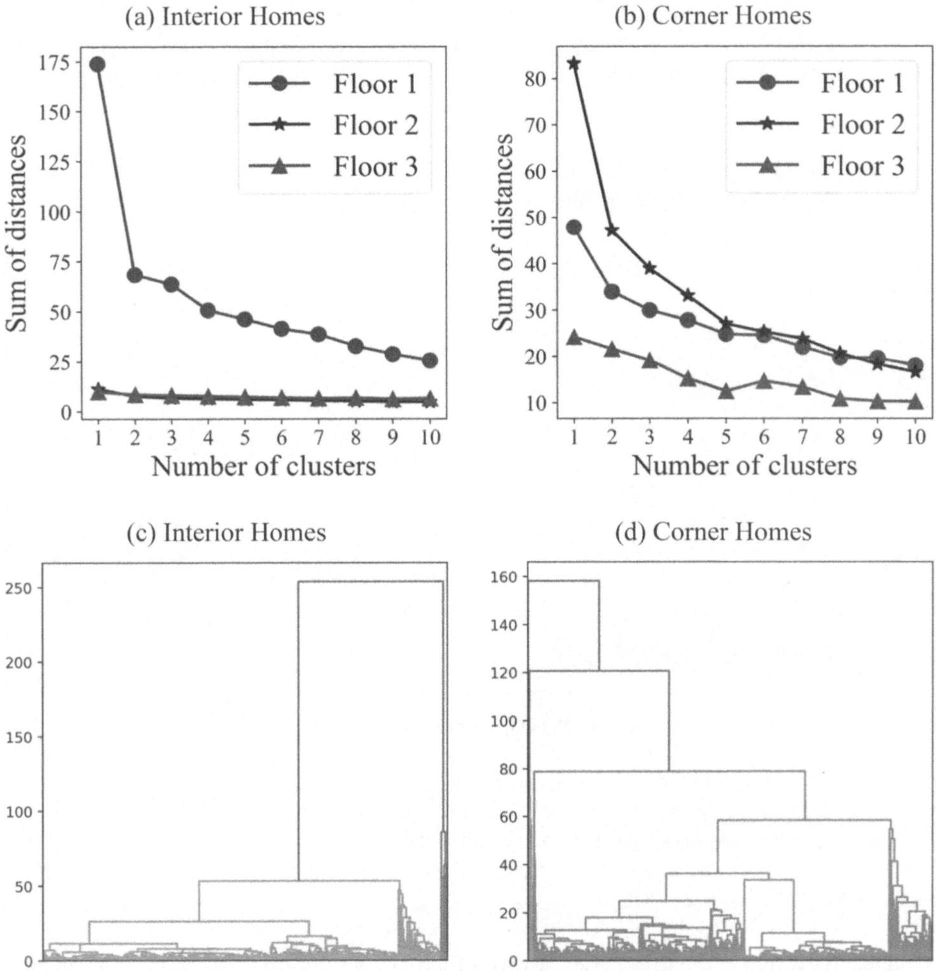

FIGURE 11.8 Clustering error by the number of clusters for three floors: (a) interior townhomes, (b) corner townhomes. Error dendrogram for (c) interior townhomes, (d) corner townhomes.

neighborhood. For interior homes, the distance between clusters declined at a slower rate when more than two clusters were used for clustering (Figure 11.8). However, for corner homes, the rate of decline in cluster distances slowed after five clusters. These results indicate that for the homes in the interior part of the neighborhood, two clusters are sufficient but may not be adequate for corner homes in the neighborhood. Put differently, two clusters may not be sufficient for both types of townhomes. To further investigate this idea, we classified corner homes into five categories. Compared to two clusters, five clusters resulted in a substantial increase in anomalies (cluster 2 to cluster 5) for floor 1 but not for floor 2 (Figures 11.5 and 11.9).

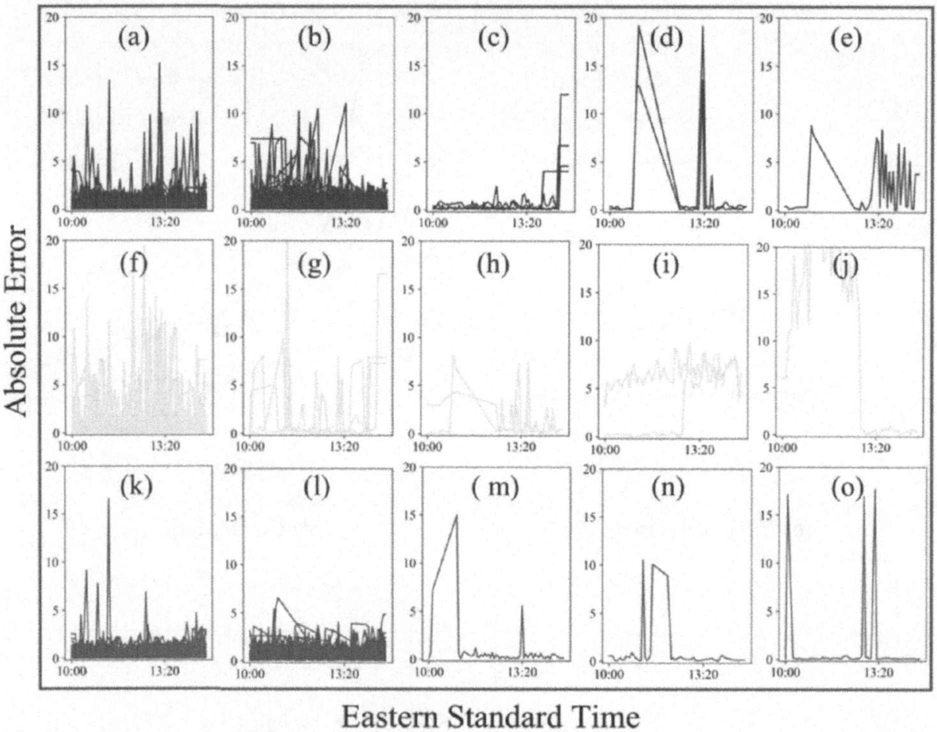

FIGURE 11.9 Five identified clusters using K-means with Euclidean distance for corner homes for floor 1 (from (a) to (e)), floor 2 (from (f) to (j)), and floor 3 (from (k) to (0)). For each, floor clusters are arranged from cluster 1 (left) to cluster 4 (right). Subplots in column 1 are associated with clusters without anomalies. The *y*-axis varies from 0 to 20 and the *x*-axis varies from 10 to 22 hours.

Due to overlap in the time series, it was difficult to get information, such as cluster size, so we provided additional results, including running time, cluster size (instances in each cluster), and other summary metrics in Tables 11.1 and 11.2 for corner and interior homes, respectively. For both types of townhomes and across all floors, Ward clustering ran quicker, followed by KME and by KMDTW algorithms. For corner homes, across all floors, the KMDTW algorithm identified the highest number of instances as anomalies, followed by KME and Ward clustering (Table 11.1). Similarly for interior homes, Ward was the quickest and KMDTW classified a higher time series into the anomaly group (Table 11.2).

11.4.2 Water Usage

Based on the consumption of the hot water, each house was classified into one of the three subpopulations: *high*, *low*, and *medium*. For each home, an upper (Temp_Up) and lower (Temp_Low) temperature error of the water tank was calculated. The elbow method was adopted to find the optimal number of clusters for the three subpopulations

TABLE 11.1 Summary of results for corner townhomes

		KME			*KMDTW*			*WARD*		
		FLOOR 1	*FLOOR 2*	*FLOOR 3*	*FLOOR 1*	*FLOOR 2*	*FLOOR 3*	*FLOOR 1*	*FLOOR 2*	*FLOOR 3*
Running Time (seconds)		0.132	0.1	0.1	24.024	10.778	16.956	0.015	0.007	0.009
Cluster Centre 1 (without anomalies)	mean	0.334	0.362	0.321	0.247	0.280	0.376	0.407	0.368	0.331
	median	0.329	0.366	0.326	0.207	0.215	0.313	0.408	0.366	0.332
	min	0.284	0.249	0.239	0.062	0.107	0.091	0.339	0.255	0.246
	max	0.406	0.456	0.391	1.339	1.156	1.315	0.487	0.472	0.421
	size	284	339	294	236	322	165	337	340	304
Cluster Centre 2	mean	0.883	2.574	0.710	0.615	0.501	0.173	2.147	1.921	0.717
	median	0.857	2.128	0.710	0.574	0.195	0.125	1.877	1.215	0.717
	min	0.590	0.067	0.507	0.156	0.074	0.061	0.144	0.139	0.473
	max	1.190	10.575	0.939	2.321	8.684	0.660	6.682	6.322	1.124
	size	56	3	61	96	16	141	4	5	52
Cluster Centre 3	mean	0.881	2.008	2.453	0.702	2.299	0.628	0.881	3.234	1.596
	median	0.301	1.581	0.402	0.232	2.035	0.537	0.301	0.889	0.259
	min	0.056	0.032	0.016	0.093	0.463	0.202	0.055	0.052	0.005
	max	6.796	6.203	14.961	7.765	6.727	2.949	6.796	16.578	10.530
	size	4	2	1	12	6	45	4	1	1
Cluster Centre 4	mean	3.466	4.766	1.596	3.870	2.934	0.839	3.466	6.535	1.152
	median	0.312	3.739	0.259	0.296	0.491	0.282	0.312	6.376	0.073
	min	0.039	1.586	0.005	0.041	0.040	0.060	0.039	3.127	0.001
	max	16.305	8.509	10.530	16.729	14.158	4.585	16.305	9.216	17.703
	size	2	2	1	2	2	4	2	1	1
Cluster Centre 5	mean	2.709	9.955	1.152	2.081	9.955	2.593	1.877	9.955	2.453
	median	1.294	11.346	0.073	0.273	11.346	0.305	0.260	11.346	0.403
	min	0.009	0.043	0.001	0.006	0.043	0.068	0.005	0.043	0.0157
	max	8.813	24.285	17.703	7.348	24.285	12.980	11.014	24.285	14.961
	Size	1	1	1	1	1	3	1	1	1

TABLE 11.2 Summary of results for interior townhomes

		KME			*KMDTW*			*WARD*		
		FLOOR 1	*FLOOR 2*	*FLOOR 3*	*FLOOR 1*	*FLOOR 2*	*FLOOR 3*	*FLOOR 1*	*FLOOR 2*	*FLOOR 3*
Running Time (seconds)		0.106	0.236	0.184	13.048	17.997	13.824	0.005	0.006	0.005
Cluster Centre 1 (without anomalies)	mean	0.332	0.281	0.299	0.249	0.225	0.248	0.313	0.281	0.324
	median	0.334	0.286	0.291	0.168	0.190	0.189	0.308	0.285	0.314
	min	0.261	0.217	0.250	0.078	0.077	0.079	0.258	0.217	0.280
	max	0.485	0.345	0.463	2.045	0.806	0.862	0.454	0.345	0.487
	size	285	296	258	282	282	237	283	296	284
Cluster Centre 2	mean	6.149	1.766	0.677	7.193	0.416	0.384	5.720	1.766	0.815
	median	6.065	0.448	0.669	7.023	0.226	0.218	5.736	0.448	0.780
	min	4.698	0.133	0.329	0.961	0.120	0.109	4.119	0.133	0.451
	max	9.019	8.213	0.912	10.812	5.097	2.301	8.014	8.218	1.242
	size	13	2	45	16	16	66	15	2	19

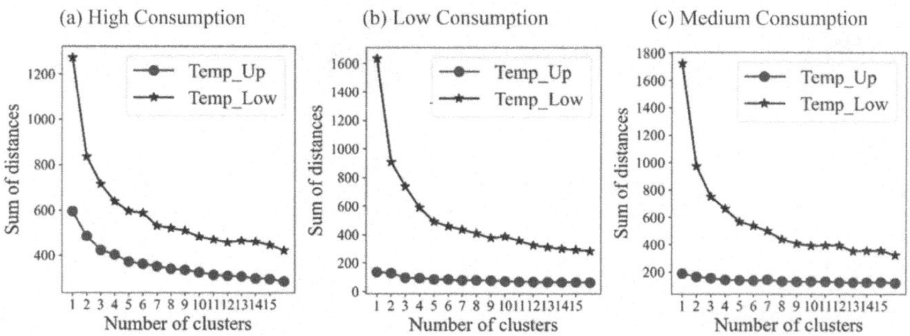

FIGURE 11.10 Clustering error by the number of clusters for upper and lower temperature of the water tank: (a) high water consumption homes, (b) low water consumption homes, and (c) medium water consumption homes.

(Figure 11.10). For all types of townhomes, the rate of decrease in distances was at a slower pace after two clusters, suggesting two was the optimal number of clusters. So, for each subpopulation and water heater temperature, we applied clustering separately using KME (Figure 11.11), KMDTW (Figure 11.12), and Ward clustering (Figure 11.13).

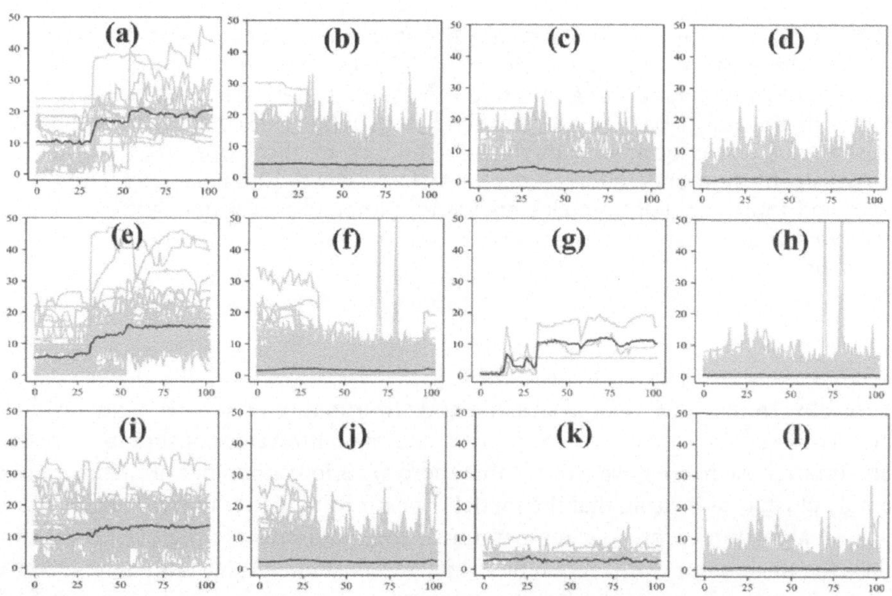

FIGURE 11.11 Results of K-means clustering of homes with high (from (a) to (d)), low (from (e) to (h)), and medium (from (i) to (l)) water usage. Subplots in columns one and two represent clusters associated with the temperature of the lower part of the tank and the subplots in columns three and four represent clusters associated with the temperature of the higher part of the tank. Subplots in the first and third columns are clusters with anomalies. Subplots in the second and fourth columns are clusters without anomalies. Cluster centres are represented with a thick line. The *y*-axis varies from 0 to 50 and the *x*-axis varies from 10 to 22 hours.

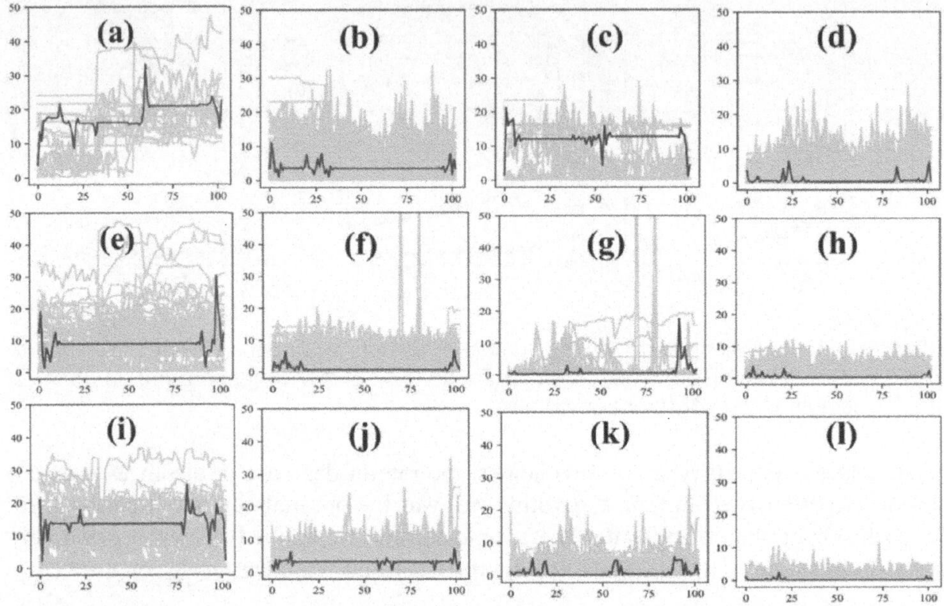

FIGURE 11.12 Results of K-means with dynamic time warping clustering of homes with high (from (a) to (d)), low (from (e) to (h)), and medium (from (i) to (l)) water usage. Subplots in columns one and two represent clusters associated with the temperature of the lower part of the tank and the subplots in columns three and four represent clusters associated with the temperature of the higher part of the tank. Subplots in the first and third columns are clusters with anomalies. Subplots in the second and fourth columns are clusters without anomalies. Cluster centres are represented with a thcik line. The *y*-axis varies from 0 to 50 and the *x*-axis varies from 10 to 22 hours.

The results from these approaches along with running time are summarised in Tables 11.3–11.5 for homes with high, low, and medium water consumption, respectively. From these results, we can see that there are a lower number of instances in an anomaly cluster, indicating that the model is performing better in most cases. Across the three algorithms, for both temperatures, the magnitude of the cluster centre is higher for the cluster with anomalies than the cluster without anomalies. Across all the subpopulations, for the lower temperature of the tank, KMDTW identified a higher number of instances as anomalies, whereas, for the upper temperature of the tank, KMDTW identified highest anomalies for homes with low and medium water usage but not for high water usage homes. For all the subpopulations, anomaly cluster centre magnitude is lower for the temperature of the upper portion of the water tank. Similar to the previous subpopulation, Ward ran quicker than the other two algorithms.

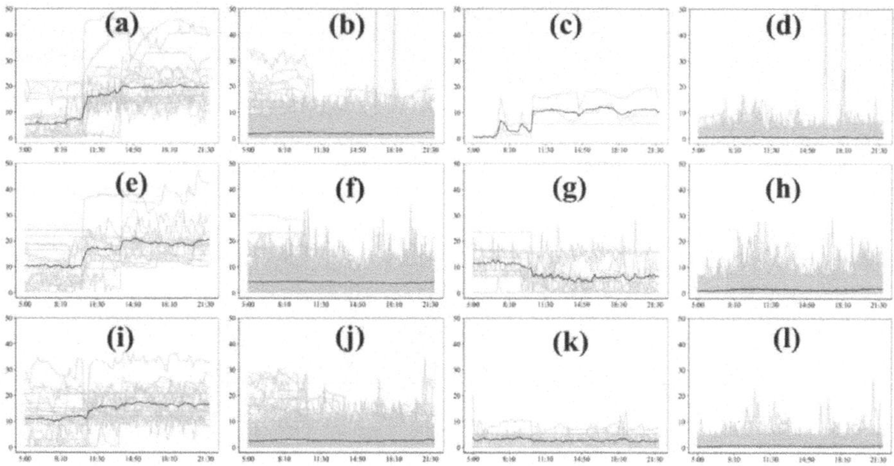

FIGURE 11.13 Results of Ward with dynamic time warping clustering of homes with high (from (a) to (d)), low (from (e) to (h)), and medium (from (i) to (l)) water usage. Subplots in columns one and two represent clusters associated with the temperature of the lower part of the tank and subplots in columns three and four represent clusters associated with the temperature of the higher part of the tank. Subplots in the first and third columns are clusters with anomalies. Subplots in the second and fourth columns are clusters without anomalies. Cluster centres are represented with a purple line. The *y*-axis varies from 0 to 50 and the *x*-axis varies from 10 to 22 hours.

TABLE 11.3 Summary of clustering results for low water usage homes

ALGORITHM	*VARIABLE*	*RUNNING TIME (SECONDS)*	*IDENTIFIED CLUSTER CENTRE WITH ANOMALIES*					*IDENTIFIED CLUSTER CENTRE WITHOUT ANOMALIES*				
			MEAN	*MEDIAN*	*MIN*	*MAX*	*SIZE*	*MEAN*	*MEDIAN*	*MIN*	*MAX*	*SIZE*
KME	t_low	0.703	11.795	13.276	5.428	15.977	45	1.710	1.680	1.399	2.103	675
	t_up	0.388	8.082	10.231	0.275	12.030	3	0.621	0.609	0.506	0.811	717
KMDTW	t_low	56.660	9.160	8.912	1.240	30.245	62	0.983	0.625	0.304	6.727	658
	t_up	46.376	0.901	0.456	0.134	17.654	22	0.363	0.135	0.124	3.645	698
Ward	t_low	0.031	14.66	17.770	5.167	20.718	24	1.915	1.893	1.677	2.295	696
	t_up	0.030	8.082	10.231	0.275	12.030	3	0.621	0.609	0.506	0.811	717

TABLE 11.4 Summary of clustering results for high water usage homes

ALGORITHM	*VARIABLE*	*RUNNING TIME (SECONDS)*	*IDENTIFIED CLUSTER CENTRE WITH ANOMALIES*					*IDENTIFIED CLUSTER CENTRE WITHOUT ANOMALIES*				
			MEAN	*MEDIAN*	*MIN*	*MAX*	*SIZE*	*MEAN*	*MEDIAN*	*MIN*	*MAX*	*SIZE*
KME	t_low	0.267	15.929	17.169	9.522	20.901	17	4.025	3.994	3.711	4.379	605
	t_up	0.389	3.688	3.619	2.863	4.901	106	0.968	0.956	0.608	1.352	516
KMDTW	t_low	37.260	18.0290	17.508	3.774	33.213	18	3.686	3.452	0.997	11.247	604
	t_up	34.468	12.29	12.729	1.166	21.211	17	0.737	0.303	0.167	6.290	605
Ward	t_low	0.015	15.929	17.169	9.522	20.901	17	4.025	3.994	3.711	4.379	605
	t_up	0.037	7.619	6.406	4.038	12.679	14	1.289	1.286	0.856	1.665	608

TABLE 11.5 Summary of clustering results for medium water usage homes

ALGORITHM	VARIABLE	RUNNING TIME (SECONDS)	IDENTIFIED CLUSTER CENTRE WITH ANOMALIES					IDENTIFIED CLUSTER CENTRE WITHOUT ANOMALIES				
			MEAN	MEDIAN	MIN	MAX	SIZE	MEAN	MEDIAN	MIN	MAX	SIZE
KME	t_low	0.345	11.964	12.760	9.169	13.643	42	2.441	2.395	2.136	2.915	449
	t_up	0.260	2.950	2.912	2.196	4.274	22	0.659	0.666	0.466	0.935	469
KMDTW	t_low	26.222	13.820	13.502	2.120	25.514	50	3.195	3.290	0.614	7.270	441
	t_up	14.635	0.216	0.133	0.123	2.330	238	1.082	0.547	0.152	5.786	253
Ward	t_low	0.015	14.474	15.819	9.938	17.482	24	2.679	2.629	2.377	3.142	467
	t_up	0.016	2.831	2.793	2.154	4.070	24	0.655	0.666	0.457	0.933	467

11.5 CONCLUSION

Previously anomaly detection was used to classify errors across all the townhomes. In the current book chapter, we performed anomaly detection on five subpopulations of townhomes based on the locations (corner and interior homes) and water consumption (high, low, and medium usage). The extended study concluded that error detection and clustering help find and isolate both unsystematic errors caused by outlier user behaviours or external disturbances and systematic errors caused by unobserved factors outside of utility control. We tested the proposed methodology on real-world data coming from a residential neighbourhood in Atlanta, GA. The suggested approach was efficient in clustering townhomes with high errors within a few seconds.

These findings provide an optimistic picture for the future of model predictive controllers in residential neighbourhoods. They indicate that a utility can quickly and cheaply find and isolate whole subpopulations of devices that happened to have a short-term or a long-term deviation from the predicted load. The next task is to correct these errors in a similarly efficient way, potentially allowing controllers to be simplified and become more adaptive to various needs in the industry.

ACKNOWLEDGEMENTS

This manuscript has been authored by UT-Battelle, LLC under Contract No. DE-AC05-00OR22725 with the U.S. Department of Energy. The United States Government retains and the publisher, by accepting the article for publication, acknowledges that the United States Government retains a non-exclusive, paid-up, irrevocable, world-wide license to publish or reproduce the published form of this manuscript, or allow others to do so, for United States Government purposes. The

Department of Energy will provide public access to these results of federally sponsored research in accordance with the DOE Public Access Plan (http://energy.gov/downloads/doe-public-access-plan).

REFERENCES

1. Cooper, D., & Cronje, W., "Autonomous Water Heater Control for Load Regulation on Smart Grids," *2016 IEEE International Energy Conference (ENERGYCON)*, Leuven, Belgium, 2016, pp. 1–6, doi: 10.1109/ENERGYCON.2016.7514084.
2. Starke, M. *et al.*, "Real-Time MPC for Residential Building Water Heater Systems to Support the Electric Grid," *2020 IEEE Power & Energy Society Innovative Smart Grid Technologies Conference (ISGT)*, Washington, DC, USA, 2020, pp. 1–5, doi: 10.1109/ISGT45199.2020.9087716.
3. Cui, B., Joe, J., Munk, J., Sun, J., & Kuruganti, T., "Load flexibility analysis of residential HVAC and water heating and commercial refrigeration" (No. ORNL/SPR-2019/1210). Oak Ridge National Lab.(ORNL), Oak Ridge, TN (United States), 2019.
4. Tsybina, E., Lebakula, V., Hill, J., Munk, J., & Zandi, H., "Structural Differences between Morning and Evening Peak in Optimized Water Heaters," *2024 IEEE Power & Energy Society Innovative Smart Grid Technologies Conference (ISGT)*, Washington, DC, USA, 2024, pp. 1–5, doi: 10.1109/ISGT59692.2024.10454236.
5. Lebakula, V., Zandi, H., Winstead, C., Tsybina, E., Kuruganti, T., & Hill, J., "Analysis of Building Model Forecasts using Autonomous HVAC Optimization System for Residential Neighborhood," *2023 IEEE Energy Conversion Congress and Exposition (ECCE)*, Nashville, TN, USA, 2023, pp. 1218–1224, doi: 10.1109/ECCE53617.2023.10362041.
6. Tsybina, E. *et al.*, "Peak Reduction Using Mode Adjustment of Heat Pump Water Heaters in a Residential Neighborhood," *2023 IEEE PES Innovative Smart Grid Technologies Latin America (ISGT-LA)*, San Juan, PR, USA, 2023, pp. 455–459, doi: 10.1109/ISGT-LA56058.2023.10328314.
7. Tsybina, E., Lebakula, V., Hill, J., Munk, J., & Zandi, H., "Using Synchronization as an Indicator of Controllability in a Fleet of Water Heaters," *2023 North American Power Symposium (NAPS)*, Asheville, NC, USA, 2023, pp. 1–6, doi: 10.1109/NAPS58826.2023.10318550.
8. Agah, N., Tsybina, E., Lebakula, V., Hill, J., Munk, J., & Zandi, H., "The Empirical Effect of Fleet Optimization on Synchronization and Rebound Effects in Heat Pump Water Heaters," *2023 IEEE PES Innovative Smart Grid Technologies Latin America (ISGT-LA)*, San Juan, PR, USA, 2023, pp. 460–464, doi: 10.1109/ISGT-LA56058.2023.10328308.
9. Tsybina, E., Lebakula, V., Hill, J., Munk, J., & Zandi, H., "Impact of Control on Availability and Cycling of Residential HVAC in a Real-World Experiment," *2023 IEEE PES Innovative Smart Grid Technologies Latin America (ISGT-LA)*, San Juan, PR, USA, 2023, pp. 450–454, doi: 10.1109/ISGT-LA56058.2023.10328300.
10. Tsybina, E., Lebakula, V., Hill, J., Munk, J., & Zandi, H., "Experimental Evidence on Latency in a Fleet of Controllable Water Heaters," *2023 IEEE Green Energy and Smart Systems Conference (IGESSC)*, Long Beach, CA, USA, 2023, pp. 1–5, doi: 10.1109/IGESSC59090.2023.10321753.
11. Kazmi, H., Mehmood, F., Lodeweyckx, S., & Driesen, J.. "Gigawatt-hour scale savings on a budget of zero: Deep reinforcement learning based optimal control of hot water systems," *Energy, 144*, 2018, pp. 159–168.

12. Ruelens, F., Claessens, B. J., Quaiyum, S., De Schutter, B., Babuška, R., & Belmans, R. "Reinforcement learning applied to an electric water heater: From theory to practice," *IEEE Transactions on Smart Grid*, *9*(4), 2018, pp. 3792–3800, doi: 10.1109/TSG.2016.2640184.
13. Al-jabery, K., Wunsch, D. C., Xiong, J., & Shi, Y., "A Novel Grid Load Management Technique Using Electric Water Heaters and Q-Learning," *2014 IEEE International Conference on Smart Grid Communications (SmartGridComm)*, Venice, Italy, 2014, pp. 776–781, doi: 10.1109/SmartGridComm.2014.7007742
14. Amasyali, K., Munk, J., Kurte, K., Kuruganti, T., & Zandi, H., "Deep reinforcement learning for autonomous water heater control," *Buildings*, *11*(11), 2021, p. 548.
15. Amasyali, K., Kurte, K., Zandi, H., Munk, J., Kotevska, O., & Smith, R., "Double Deep Q-Networks for Optimizing Electricity Cost of a Water Heater," *2021 IEEE Power & Energy Society Innovative Smart Grid Technologies Conference (ISGT)*, Washington, DC, USA, 2021, pp. 1–5, doi: 10.1109/ISGT49243.2021.9372205.
16. Firmino, P. R. A., de Mattos Neto, P. S., & Ferreira, T. A.. "Error modeling approach to improve time series forecasters," *Neurocomputing*, *153*, 2015, pp. 242–254.
17. Zeiml, S., Seiler, U., Altendorfer, K., & Felberbauer, T., "Simulation Evaluation of Automated Forecast Error Correction Based on Mean Percentage Error," *2020 Winter Simulation Conference (WSC)*, Orlando, FL, USA, 2020, pp. 1572–1583, doi: 10.1109/WSC48552.2020.9384055.
18. Lebakula, V., Tsybina, E., Amasyali, K., Hill, J., Munk, J., & Zandi, H., "Anomaly detection for MPC forecast in Fleet of Water Heaters," *2023 International Conference on Machine Learning and Applications (ICMLA)*, Jacksonville, FL, USA, 2023, pp. 249–256, doi: 10.1109/ICMLA58977.2023.00042.
19. Lebakula, V., Tsybina, E., Hill, J., Munk, J., & Zandi, H., "Autonomous Anomaly Detection for MPC Forecasts of HVAC Systems in Residential Communities," *2024 IEEE Power & Energy Society Innovative Smart Grid Technologies Conference (ISGT)*, Washington, DC, USA, 2024, pp. 1–5, doi: 10.1109/ISGT59692.2024.10454188.
20. Bock, H. H., 2007. Clustering methods: A history of k-means algorithms. In *Selected contributions in data analysis and classification*, Springer, pp. 161–172.
21. Ward, J. Jr. "Hierarchical grouping to optimize an objective function," *J. Amsr. Statis. Assoc*, *58*, 1963, pp. 236–244.
22. Zhang, Z., Tavenard, R., Bailly, A., Tang, X., Tang, P., & Corpetti, T., "Dynamic time warping under limited warping path length," *Information Sciences*, *393*, 2017, pp. 91–107.
23. Tsybina, E., Winstead, C., Hill, J., & Zandi, H., "Findings from Design and Operation of Connected Neighborhoods," *2023 IEEE Power & Energy Society General Meeting (PESGM)*, Orlando, FL, USA, 2023, pp. 1–5, doi: 10.1109/PESGM52003.2023.10253137.

12 A Hybrid Physics-Informed Neural Network

SEIRD Model for Forecasting COVID-19 Intensive Care Unit Demand in England

Michael Ajao-Olarinoye, Vasile Palade, Fei He, Petra A Wark, Zindoga Mukandavire, and Seyed Mousavi

12.1 INTRODUCTION

The SARS-CoV-2 virus caused COVID-19, a severe respiratory infectious disease with a high mortality rate, that triggered a global pandemic [1]. COVID-19 poses a complex and unprecedented challenge to public health and society, requiring scientific efforts to understand, model, diagnose, and control the disease. A significant challenge in mitigating this outbreak is the ability of individuals to transmit the virus during a period called the incubation period, with some remaining asymptomatic throughout the infection [2]. This complicates the detection and isolation of cases, as the disease can

DOI: 10.1201/9781003570882-15

facilitate presymptomatic and asymptomatic transmission. Consequently, it is imperative to meticulously monitor observed and unobserved infections and rigorously assess their epidemiological and social impacts.

The COVID-19 pandemic has restricted healthcare systems around the world, which requires them to implement effective resource allocation and intervention strategies to mitigate the impact of the virus [3]. Within healthcare systems, resource allocation includes both pharmaceutical interventions, such as medications and vaccines, and nonpharmaceutical interventions, including ICU beds, medical equipment, and healthcare personnel, in order to ensure effective and efficient provision of medical services [4, 5]. The usage of these resources is studied and modelled using mathematical models, statistical models, and machine learning, including deep learning algorithms [6–9].

Epidemiological modelling [10] and forecasting the demand for healthcare resources [7] are essential to inform allocation decisions. These models use data on various factors, such as prevalence of diseases, demographics of the population, transmission dynamics, and healthcare capacity, to predict the demand for different types of resources. Researchers have developed various mechanistic models to assist decision makers in designing effective public health policies. These models take advantage of statistics and epidemiology, such as the susceptible, infected, and recovered (SIR) model developed by Kermack and McKendrick in 1927 [11]. Subsequent refinements, including the susceptible, exposed, infected, and recovered (SEIR) and susceptible, infected, and susceptible (SIS) models, have been instrumental in analysing epidemic dynamics and evaluating the impact on public health infrastructures.

Compartmental models are mathematical models used to study an epidemic, represented by a system of ordinary differential equations (ODEs), which can be classified as deterministic or stochastic. Deterministic models can simulate a general scenario of disease propagation, whereas stochastic models are adept at simulating the spread of disease within small or subgroup populations and predicting potential outcomes by incorporating randomness in the simulation. These simulations are mostly solved using numerical methods, where the relationship between the state variables in the compartmental model and its parameters helps to understand the dynamics of the system of equations.

Statistical and machine learning (ML) models use data to significantly enhance short-term forecasting and long-term predictions. Deep learning models, such as RNNs [12], have been used to accurately analyse time series data to find patterns in infectious disease data, such as influenza or COVID-19, predicting the demand for beds and mechanical ventilators during a pandemic [13]. These models can identify important input variables, such as infection rates, mobility data, and public health interventions, that significantly impact predictions. Additionally, deep learning models can capture underlying nonlinear dynamics in the data, offering insights into complex interactions within the system. However, since these methods do not explicitly capture the mechanistic dynamics of epidemics, they may not provide comprehensive information on how the epidemic will evolve over longer periods.

Due to recent advances in using data-driven ML frameworks to solve physics problems, these have been significantly influenced by the work of Raissi et al. [14]. In their 2019 paper, they introduced the concept of physics-informed neural network (PINN), which integrates the underlying physical laws described by partial differential equations

(PDE) directly into the learning process. This innovative approach ensures that the ML models not only fit the data but also integrate the governing physical principles. Researchers have demonstrated that the use of PINNs can be applicable to different fields, including fluid dynamics, quantum mechanics, structural analysis, and epidemiological modelling. These simulations are calibrated with data collected from disease surveillance, such as those presented later. In this study, we focus on data from the United Kingdom, specifically the regions of the National Health Service (NHS) of England (Figure 12.1).

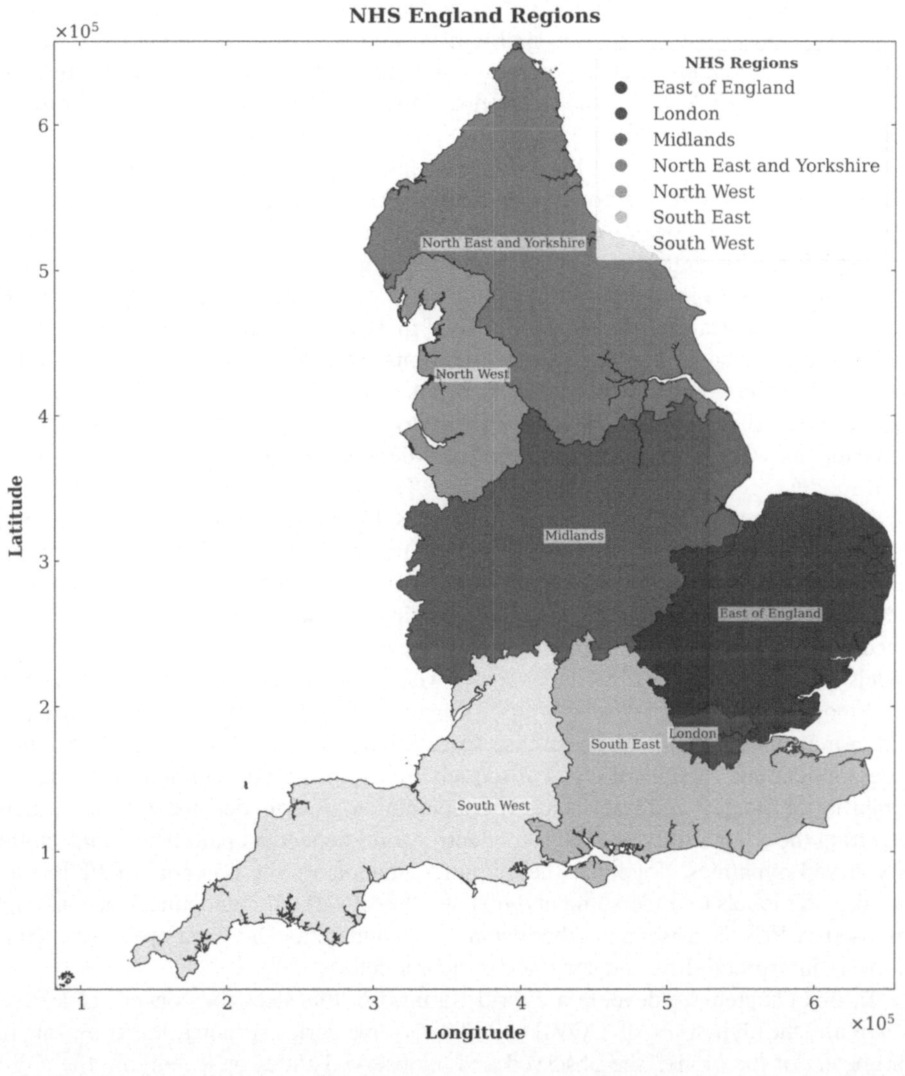

FIGURE 12.1 NHS England Regions Map showing the geographical distribution of the National Health Service (NHS) Regions in England. The data used to create this map were obtained from the Office for National Statistics[1] using matplotlib and geopandas in Python.

Significant contributions have been made in the application of PINNs to epidemiological modelling, demonstrating their ability to reconcile data-driven approaches with traditional model-based simulations [13–15]. Amaral et al. [16] leveraged data-driven methodologies to predict COVID-19 trends in São Paulo and Brazil, demonstrating how machine learning models can enhance pandemic forecasting and support public health decision-making. Similarly, Cuomo et al. [17] have introduced innovative approaches to modelling infectious diseases by incorporating the dynamics of known diseases directly into the learning process. This integration allows for more accurate predictions and estimates of dynamic parameters, adapting to the evolving nature of pandemics. These hybrid models have shown promise in forecasting disease spread and resource demand, informing public health responses, and policy making. The integration of PINNs with epidemiological models offers a powerful tool to understand the dynamics of infectious diseases, such as COVID-19, and to develop effective strategies to control and mitigate their impact, even with limited data.

PINNs have been applied for parameter and state estimation in epidemiological models, particularly in the context of the COVID-19 pandemic. Han et al. focused on the estimation of constant parameters within traditional epidemiological models, such as SIR, SEIR, and SEIRS. Their work highlighted the ability of PINNs to accurately estimate both model parameters and state variables. Nguyen et al. [18] and Hu et al. [19] demonstrated the efficacy of PINN in estimating unknown parameters and determining the basic number of reproductions, which is crucial to understanding the spread of the virus. Furthermore, Berkhahn and Ehrhardt [20] and Grimm et al. [21] extended this application using PINNs to model infection and hospitalisation scenarios, as well as to understand the time-dependent contact rate in the SIR and SEIR models.

Subsequent advances were made by Oluwasakin and Khaliq [22] and Torku et al. [23], who implemented deep learning techniques on data sets that comprised daily COVID-19 infection counts to forecast the incidence of Omicron variant infections and to evaluate the effectiveness of COVID-19 vaccines, respectively. Ning et al. [24] introduced a PINN framework designed to enhance parameter estimation in compartmental models by incorporating time-dependent variables. Ogueda et al. [25] improved the SIRD model to incorporate transportation dynamics to analyse the impact between population movement on infectious disease spread. Kharazmi et al. [26] further extended these applications to epidemiological models by exploring the identifiability and predictability of integer- and fractional-order epidemiological models using PINNs, demonstrating the adaptability of PINNs to identify time-dependent parameters and capture unobserved dynamics. Together, these studies underscore the potential of PINNs and data-driven models to understand and manage the COVID-19 pandemic. The challenge proposed by PINNs, as seen in other papers, is to understand how to improve the generalisation, interpretability, and accuracy of the models.

In this chapter, we develop a hybrid framework that uses the concept of a PINN to capture the dynamics of COVID-19. This framework estimates the time-varying parameters of the model, the observed and unobserved states by leveraging the ability of the network to learn from data, and integrate physical laws directly into the learning process of a modified SEIRD model. Furthermore, we use of the strength of RNNs to combine the data from a PINN and the lagged covariates of the target variables. This approach is used to forecast the demand for the ICU, specifically the daily number of mechanical ventilator beds occupied by patients with COVID-19.

This chapter introduces a robust, data-driven epidemiological compartmental model $\left(SEI_aI_sHCRD\right)$ that incorporates factors such as reinfection, asymptomatic carriers, symptomatic patients, hospitalised individuals, and critically ill patients requiring mechanical ventilation. It demonstrates the effectiveness of PINNs in estimating the model's time-varying parameters, observed and unobserved states, thereby providing accurate predictions of disease spread and healthcare resource demand. Furthermore, the chapter integrates the output of PINNs with lagged covariates of the COVID-19 target variables to forecast the utilisation of hospital resources, particularly ICU beds, in England. The performance of the proposed model is evaluated using various deep learning models, including RNNs, to predict the demand for healthcare resources, specifically ICU beds and mechanical ventilators, during the COVID-19 pandemic.

12.2 METHODS

We develop a novel hybrid approach to model and forecast the dynamics of infectious diseases and the demand for healthcare resources. Our methodology integrates the PINN framework with traditional epidemiological models to capture the dynamics of infectious diseases, estimate key parameters, and unobserved states in the model. We then used the output of the PINN model with various lagged variables of ICU beds to forecast the demand for healthcare resources, using various deep learning models. The following sections describe the SEIRD model, the PINN model, and the deep learning models used in this study.

12.2.1 Data Collection and Preprocessing

We collected daily records of infections, deaths, cumulative cases, and hospital activity data from the COVID-19 NHS hospital activity webpage [27] and Google Open Data [28]. The dataset spans from May 2020 to December 2023 and focuses on data from England. The dataset contains multiple features of daily records, selected only for England for this study. Due to the nature of the data, a 7-day moving average was applied to the data to reduce noise by smoothing the data and subsequently using it to train the proposed hybrid framework, as exemplified by the model training architecture depicted in Figure 12.4. Forecasting demand for ICU beds in the NHS regions is crucial for effective resource allocation and management, particularly during the COVID-19 pandemic (Figure 12.2).

The insights gathered from this analysis show that the number of new admissions and ventilator beds occupied was high in the early stages of the pandemic, especially during the lockdown periods. However, as the pandemic progressed, the number of new admissions and ventilator beds occupied decreased, indicating a fluctuation in the data.

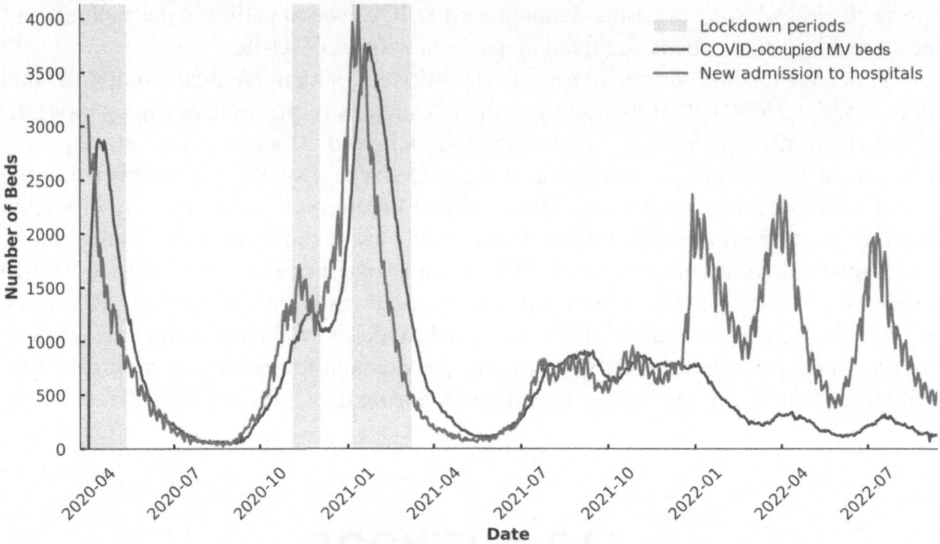

FIGURE 12.2 COVID-19 hospitalisation and mechanical ventilation bed use in England. This graph shows the number of COVID-19 occupied mechanical ventilator (MV) beds and new hospital admissions in England from April 2020 to July 2022. The shaded areas indicate lockdown periods. Peaks in MV bed usage and hospital admissions correspond to waves of COVID-19 infections, highlighting the impact of lockdown measures on reducing hospital resource demand.

This relationship is crucial to understanding the impact of new admissions on demand for ventilator beds and resource planning.

The PINN modelling approach incorporates key epidemiological variables, including new daily symptomatic infections (I_s), hospital admissions (H), critical care mechanical ventilator (C) and deaths (D). For data variables, I_s, H, D which are calculated from the cumulative infected cases, hospital admissions, and deaths, respectively, we used the following equations to estimate daily values from the cumulative data. We let C_t represent the cumulative value of the data on day t. The daily value D_t is given by:

$$D_t = C_t - C_{t-1} \tag{12.1}$$

This calculation was done for the daily new symptomatic infections (I_s), daily new hospitals admissions (H), and daily new deaths (D) data.

12.2.2 SEIRD-Based Compartmental Models for Epidemiological Dynamics

We present an enhanced model that extends the traditional deterministic frameworks of SIR, SEIR, and SEIRD in infectious disease modelling by incorporating compartments for hospital admissions and critical care. This enhancement aims to better represent the complexity of disease progression and the demand for healthcare resources. Research indicates that both asymptomatic and symptomatic individuals can transmit the virus. In particular, the infectious period for asymptomatic individuals is longer than for symptomatic individuals, and the transition from symptomatic infection to hospitalisation is crucial to understanding the disease spread. Hospitalisation dynamics are vital to assess healthcare resource needs, including ICU beds and mechanical ventilators. During the COVID-19 pandemic, managing hospital admissions and ensuring the availability of mechanical ventilators were essential for patient care.

COVID-19, which is a respiratory disease, can cause severe respiratory distress that requires mechanical ventilation. Therefore, the demand for ventilators in the ICU serves as a critical indicator of the severity of the pandemic and the strain of the health system.

This model integrates these considerations into the SEIRD framework, dividing the population into susceptible (S), exposed (E), asymptomatic (I_a), symptomatic (I_s), hospital admissions (H), critical care on mechanical ventilator (C), recovered (R) and death (D) compartments. The model is governed by a system of ODEs that describe the flow of individuals between compartments.

$$
\begin{aligned}
\frac{dS}{dt} &= -\beta \frac{S(I_s + I_a)}{N} + \eta R \\
\frac{dE}{dt} &= \beta \frac{S(I_s + I_a)}{N} - \alpha E \\
\frac{dI_s}{dt} &= \alpha \rho E - d_s I_s \\
\frac{dI_a}{dt} &= \alpha (1-\rho) E - d_a I_a \\
\frac{dH}{dt} &= d_s \omega I_s - d_H H - \mu H \\
\frac{dC}{dt} &= d_H (1-\omega) H - \gamma_c C - \delta_c C \\
\frac{dR}{dt} &= d_s (1-\omega) I_s + d_a I_a + d_H (1-\mu) H + \gamma_c C - \eta R \\
\frac{dD}{dt} &= \mu H + \delta_c C
\end{aligned}
\tag{12.2}
$$

where N is the total population, β is the transmission rate, η is the rate of return to the susceptible compartment, ω is the hospitalisation ratio, μ is the death rate from hospital, γ_c is the recovery rate from critical care, δ_c is the death rate in critical care, ρ is the proportion of symptomatic infections, α is the incubation period, d_s is the infectious period for symptomatic individuals, d_a is the infectious period for asymptomatic individuals, and d_H is the hospitalisation days. The model is subject to the initial conditions of the states of the compartmental model $S(t_0), E(t_0), I_s(t_0), I_a(t_0), H(t_0), C(t_0), R(t_0)$

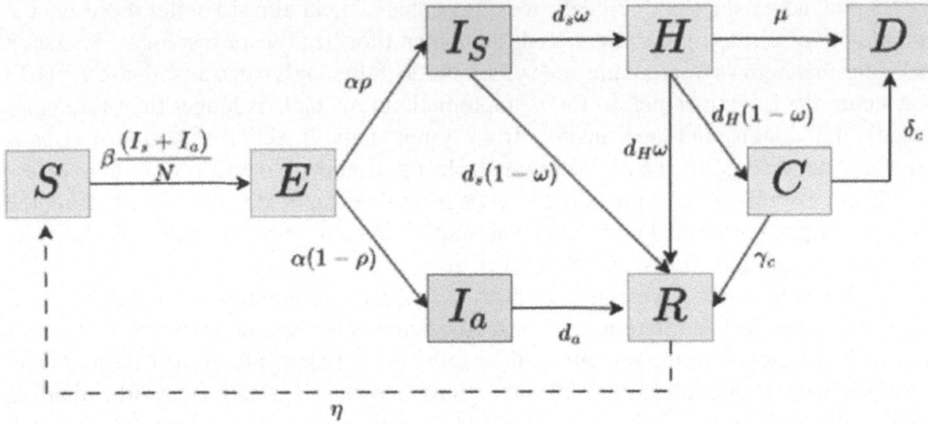

FIGURE 12.3 Compartmental flow of the SEIRD model for disease transmission dynamics.

and $D(t_0)$, and N is the sum of all the states of the epidemiological model, as shown in Equation 12.2.

$$S(t)+E(t)+I_a(t)+I_s(t)+H(t)+C(t)+R(t)+D(t)=N,\forall t\geq t_0 \tag{12.3}$$

where t_0 is the initial time. The model captures the dynamics of disease transmission, progression, hospitalisation, recovery, and mortality, providing information on disease spread and the demand for healthcare resources. The compartmental flow of the model is shown in Figure 12.3, which depicts the transitions between the different compartments and the corresponding transmission, hospitalisation, recovery, and mortality rates.

12.2.3 Physics-Informed Neural Network in Epidemiology Modelling

PINNs are a class of deep learning models that integrate neural networks with physical laws to solve both forward and inverse problems across various fields, including fluid dynamics, quantum mechanics, and epidemiology. PINNs have been used to estimate unknown parameters (constant or time-varying) and unobserved states in epidemiological models by incorporating the dynamics of known diseases directly into the learning process. This is achieved by including the physics of dynamics in the loss function, with differential equations representing known physical laws integrated directly into the neural network architecture. This approach leverages both data and epidemiological dynamics, encoded through ODEs, to train the neural networks. A typical PINN architecture aims to approximate the state variables of an epidemiological model.

Our architecture involves two neural networks, one for the estimation of time-varying parameters and one for the estimation of state; both neural networks are trained

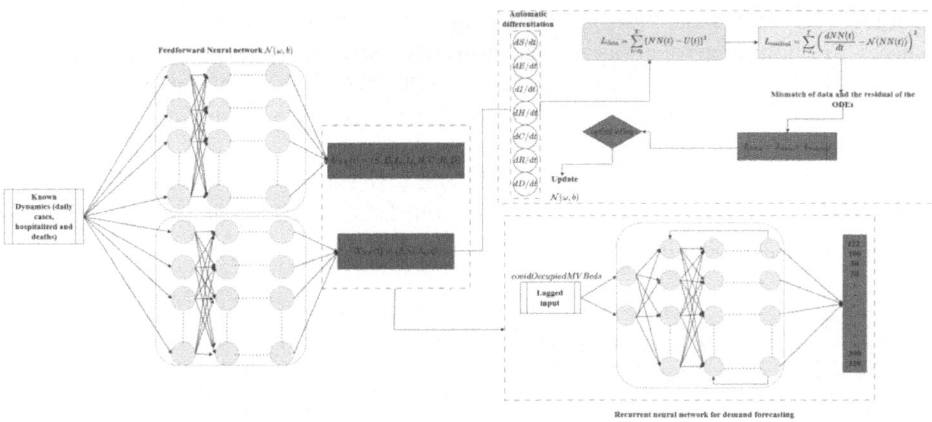

FIGURE 12.4 Architecture of the PINN and RNN framework for epidemiological modelling and demand forecasting.

simultaneously. The state estimation network, denoted as $U_{NN}(t)$ in Figure 12.4, is a feedforward network (FFN) incorporating a residual connection block derived from the ResNet architecture, with a hyperbolic tangent function (tanh) activation function for each layer and a linear output. The predefined constant parameters are initialised randomly, with each layer employing a sigmoid activation function.

This architecture approximates the solution to the state variable of the epidemiological model. The residual connection helps to mitigate the problem of vanishing gradients in the state estimation neural network. The time-varying parameter network $X_{NN}(t)$, uses an FFN with a tanh activation function for each layer. Each time-varying parameter passes through a sigmoid activation function for its output. The neural network is designed to initialise and learn all the parameters defined. Both networks are trained simultaneously to estimate the states and parameters of the model. The objective is to minimise the residuals of the ODEs that govern the dynamics of the SEIRD model.

$$U_{NN}(t) = [S(t), E(t), I_s(t), I_a(t), H(t), C(t), R(t), D(t)] \tag{12.4}$$

where $U_{NN}(t)$ represents different compartments of the model such as susceptible, exposed, symptomatic infected, asymptomatic infected, hospitalised, critical, recovered, and deceased populations. The parameter estimation network, denoted as $X_{NN}(t)$ estimates each of the time varying parameters of the epidemiological model.

$$X_{NN}(t) = [\beta(t), \gamma_c(t), \delta_c(t), \eta(t), \omega(t), \mu(t)] \tag{12.5}$$

where $\beta(t), \gamma_c(t), \delta_c(t)$, $\omega(t), \mu(t)$ and $\eta(t)$ are parameters that evolve over time. The model is governed by the following equation, which captures the dynamics of the state variables:

$$\frac{dU}{dt} + \mathcal{N}[U(t), \lambda] = 0 \tag{12.6}$$

where $\frac{dU}{dt}$ represents the time derivative of the state variables U, and $\mathcal{N}[U(t),\lambda]$ denotes the nonlinear operator representing the epidemiological dynamics defined by the ODEs, and λ is a set of learned model parameter. The training of PINNs involves minimising a loss function composed of two parts: the data fidelity loss and the physics-based residual loss. The data fidelity loss L_{data}, ensures that the model output matches observed epidemiological data, using the available data of new daily symptomatic infections (I_s), hospital admissions (H), critical care mechanical ventilator usage (C) and deaths (D).

$$L_{\text{data}} = \frac{1}{N}\sum_{i=0}^{N}\left(\left(I_s^{\text{pred}} - I_s^{\text{data}}\right)^2 + \left(H^{\text{pred}} - H^{\text{data}}\right)^2 + \left(C^{\text{pred}} - C^{\text{data}}\right)^2 + \left(D^{\text{pred}} - D^{\text{data}}\right)^2\right), \quad (12.7)$$

and a physics-based residual loss, L_{residual}, is derived from the physical laws governing the epidemiological model and ensures that the neural network predictions adhere to these laws:

$$L_{\text{residual}} = \frac{1}{N}\sum_{i=0}^{N}\left(\left(\frac{dS^{\text{pred}}}{dt} - \mathcal{N}_S\left[S^{\text{pred}},\lambda\right]\right)^2 + \left(\frac{dE^{\text{pred}}}{dt} - \mathcal{N}_{\mathcal{E}}\left[E^{\text{pred}},\lambda\right]\right)^2 + \ldots + \left(\frac{dD^{\text{pred}}}{dt} - \mathcal{N}_{\mathcal{D}}\left[D^{\text{pred}},\lambda\right]\right)^2\right) \quad (12.8)$$

where $\frac{dU^{\text{pred}}}{dt}$ represents the time derivative of the neural network approximation U^{pred} for each state in the model, and $\mathcal{N}[U^{\text{pred}};\lambda]$ denotes the right-hand side of the ODE, which is the residual of the solution of the model. The derivative of U^{pred} is computed by calculating the gradient with automatic differentiation. The residual loss ensures that the neural network accurately captures the dynamics of the epidemiological model. By integrating the dynamics dictated by the ODEs into the neural network through automatic differentiation, PINNs ensure that both the network weights and the model parameters are optimised to faithfully reproduce both the observed data and the model dynamics. The overall optimisation problem of the loss function is given by:

$$L_{\text{total}} = L_{\text{data}} + L_{\text{residual}} \quad (12.9)$$

This enables simultaneous learning of network parameters and epidemiological dynamics, thus improving model prediction and understanding. The training architecture of the model is shown in Figure 12.4. The objective is to minimise the residuals of the ODEs that govern the dynamics of the SEIRD model. In the diagram below, Y in the data represents the shorts form for the available data and X in the residual loss represents the predicted state variable for each compartment. Also N_X in the figure is the corresponding compartmental model equations which depends on the predicted states X^{pred} and the parameters λ^{pred} at time t_i.

12.2.4 Deep Learning Models for Forecasting

In particular, models such as RNN, LSTM, and gated recurrent unit (GRU) have been widely used in time series forecasting for epidemiological modelling, effectively capturing temporal dependencies in disease spread. [13]. Time series data are a sequence of data points collected or recorded at successive points in time, typically at uniform intervals. These models are particularly well suited for time series analysis because they are designed to capture temporal dependencies and patterns within the data. Given the sequential nature of our data and the selected target variable, we treated the problem as a many-to-one time series forecasting task. Specifically, we selected the daily use of ventilators in England from COVID-19 hospital activity data as the target variable.

12.2.4.1 Recurrent neural networks

In Rumelhart et al. [29], the backpropagation algorithm was introduced, a technique used to train neural networks such as FFNs and RNNs. The backpropagation algorithm is a supervised learning method that uses gradient descent to minimise the error between the predicted and actual outputs of the network. By iteratively adjusting the weights of the network in the direction that reduces the error, backpropagation enables the network to learn complex patterns from the data. Unlike FFNs, RNNs are particularly suitable for handling sequential data.

Let the input to the RNN at time t be x_t, and let the hidden state of the previous timestep be H_{t-1}. The updated equations are given by:

$$A_t = W \cdot H_{t-1} + U \cdot x_t + b_1 \tag{12.10}$$

$$H_t = \tanh(A_t) \tag{12.11}$$

$$o_t = V \cdot H_t + b_2 \tag{12.12}$$

where W, U, and V are the weight matrices, b_1 and b_2 are the biases and *tanh* is the activation function. The output o_t is then passed through a SoftMax function to obtain the final output. The RNN is trained using the backpropagation through time (BPTT) algorithm, which is a variant of the backpropagation algorithm used to train RNNs.

Traditional RNNs are known to have vanishing and exploding gradients problems, which hamper their ability to learn long-term dependencies. To address this, Hochreiter and Schmidhuber [12] developed LSTM networks in 1997, which have memory cells and gates to control the flow of information. This design ensures that gradients are propagated stably, making LSTMs especially effective for learning long-term dependencies in data.

For an input x_t at time t and the previous hidden state H_{t-1}, LSTMs operate through three key gates:

- Input Gate (I_t)

$$I_t = \sigma(W_{xi}x_t + W_{hi}H_{t-1} + b_i) \tag{12.13}$$

- Forget Gate (F_t)

$$F_t = \sigma\left(W_{xf}x_t + W_{hf}H_{t-1} + b_f\right) \tag{12.14}$$

- Output Gate (O_t)

$$O_t = \sigma\left(W_{xo}x_t + W_{ho}H_{t-1} + b_o\right) \tag{12.15}$$

These gates work together to update the memory cell as:

$$C_t = F_t \odot C_{t-1} + I_t \odot \tanh\left(W_{xc}x_t + W_{hc}H_{t-1} + b_c\right) \tag{12.16}$$

and to compute the hidden state as:

$$H_t = O_t \odot \tanh\left(C_t\right) \tag{12.17}$$

The LSTM architecture contains weight matrices W, biases b, sigmoid function σ, and element-wise multiplication $\odot$. The GRU proposed by Cho et al. [30] in 2014 is based on the LSTM algorithm that was created as a simplified version that was still capable of managing long-term temporal dependencies. The GRU combines the forget and input gates into a single update gate, and the cell state and hidden state are merged into a single state. The GRU architecture is characterised by the following equations:

$$Z_t = \sigma\left(W_{xz}x_t + W_{hz}H_{t-1} + b_z\right) \tag{12.18}$$

$$R_t = \sigma\left(W_{xr}x_t + W_{hr}H_{t-1} + b_r\right) \tag{12.19}$$

$$H_t = \left(1 - Z_t\right) \odot H_{t-1} + Z_t \odot \tanh\left(W_{xh}x_t + W_{hh}\left(R_t \odot H_{t-1}\right) + b_h\right) \tag{12.20}$$

where Z_t is the update gate, R_t is the reset gate and H_t is the hidden state at time t. The GRU architecture contains weight matrices W, biases b, sigmoid function σ, and element-wise multiplication $\odot$. bidirectional versions of the models extend the standard unidirectional architecture by processing the input sequence in both directions. This allows the model to use data from both past and future states, improving its ability to capture long-term dependencies.

12.2.4.2 Sequence-to-sequence models

Sequence-to-sequence (Seq2Seq) models, which are characterised by their encoder-decoder architecture, have been highly successful in a variety of tasks, such as machine translation and time series forecasting. This approach was first popularised in the work of Sutskever et al. [31] and Cho et al. [30]. The fundamental principle underlying the Seq2Seq model involves encoding sequential data into a latent representation space that is later decoded to produce the target output sequence. This latent representation effectively encapsulates the inherent patterns and structures within the data, analogous to the way humans discern patterns and features in intricate entities.

A variety of techniques can be used to construct a Seq2Seq model. In this study, we used a variety of combinations of encoders consisting of RNN, LSTM, and GRU with a fully configured network decoder architecture. Either the fully connected layer can use the latest hidden state from the encoder to predict the desired output, or all the hidden states can be flattened into a single long vector and used to predict the output. The Seq2Seq model is trained using the Adam optimiser, which is a variant of the stochastic gradient descent algorithm that computes adaptive learning rates for each parameter. The model is trained using the mean squared error (MSE) loss function, which calculates the average squared difference between the predicted and actual values. During experimentation with respect to the decoder aspect of the Seq2Seq model, a fully connected (FC) network was used for our forecasting problem.

12.2.4.3 Hyperparameter optimisation using simulated annealing

We need to optimally determine which combination of parameters would provide the minimum scoring validation mean absolute error (MAE) when training ML models. Hyperparameter tuning is essential for optimising the performance of ML models, and given the vast hyperparameter space, efficient optimisation techniques are imperative. In this work, we employ simulated annealing (SA) [32], a probabilistic optimisation method inspired by the annealing process in material science, to fine-tune the hyperparameters of the models [33]. SA is particularly suited for hyperparameter tuning due to its ability to explore the hyperparameter space and avoid local minima.

In this instance, multiple objectives are passed to the SA algorithm for hyperparameter tuning, such as the one discussed by Fischetti and Stringher [34]. The algorithm then iteratively adjusts the hyperparameters to minimise the validation loss, which is the MAE in this case. By accepting worse solutions with a probability that decreases over time, the algorithm effectively explores the hyperparameter space and converges towards the optimal solution.

The process of SA for hyperparameter tuning involves several steps. The inputs to the algorithm include the validation loss function, the neighbour function, the cooling schedule, the initial hyperparameters, the initial temperature, and the number of iterations. The algorithm starts with an initial set of hyperparameters and computes the initial validation loss. It then iteratively selects a neighbour of the current hyperparameters and evaluates the new validation loss. If the new hyperparameters result in a lower validation loss, or satisfy the probabilistic acceptance criterion, they are accepted as the new set of hyperparameters. This iterative process continues for a specified number of iterations, at the end of which the algorithm returns the optimal hyperparameters that minimise the validation loss function.

The cooling schedule, typically a factor by which the temperature is reduced, plays a crucial role in the algorithm's performance. The temperature controls the likelihood of accepting worse solutions, allowing the algorithm to escape local minima and explore the hyperparameter space more effectively. As the temperature decreases, the algorithm becomes more conservative in accepting worse solutions, gradually converging towards the optimal hyperparameters.

Algorithm 1: Hyperparameter Optimisation with SA

Input: $f(x)$ - validation loss function, $N(x)$ - neighbour function, $T(t)$ - cooling schedule, x_0 - initial hyperparameters, T_0 - initial temperature, k - iterations
Output: x_{best} - optimal hyperparameters.

Function: HYPERPARAMETER_TUNING_SA (f, N, T_0, x_0, k)
  Initialize $x \leftarrow x_0$, $f_{\text{best}} \leftarrow f(x_0)$, $x_{\text{best}} \leftarrow x_0$, $T \leftarrow T_0$
  for $t = 1$ **to** k **do**
    $x' \sim N(x)$
    $\Delta f \leftarrow f(x') - f(x)$
    if $\Delta f < 0$ *or* $\text{rand}(0,1) < \exp(-\Delta f / T)$ ***then***
      $x \leftarrow x'$
      if $f(x') < f_{\text{best}}$ ***then***
        $f_{\text{best}} \leftarrow f(x')$
        $\text{x}_{\text{best}} \leftarrow \text{x}'$
      end if
    end if
    $T \leftarrow \alpha T$
  end for
  return $\mathbf{x}_{\text{best}}$
end Function

12.3 RESULTS

In this section, we present the results of our study, evaluating the performance of our proposed hybrid framework. This includes the accuracy of the PINN model for estimating the time-varying parameters and unobserved states in the SEIRD model, as well as the effectiveness of various deep learning models for forecasting the demand for ICU beds in England.

12.3.1 Experimentation Setup

The initial values of the observed states I_s, H, C and D were obtained from the dataset, while the initial values of the unobserved states S, E, I_a and R were set to 0. The first network, used to estimate the states, consists of five hidden layers, each containing 20 neurons. Each layer uses the tanh as the activation function, with a sigmoid activation function in the final layer. The second neural network is constructed as a fully connected feedforward architecture, featuring three hidden layers with 20 neurons each. Every layer, including the final one designated to predict the model's time-varying parameters, adopts the tanh activation function. A linear output layer with a sigmoid activation function is employed to ensure that the parameters' output remains positive. Both networks were trained simultaneously to estimate the states and parameters of the model and to minimise the residuals of the ODEs that govern the dynamics of the SEIRD model, initialised with random weights using the Xavier initialisation method [35]. This technique ensures that the variance of the inputs and outputs of each layer remains constant, which helps to avoid the issues of vanishing and exploding gradients that can significantly hinder the training of deep neural networks.

The PINN model was trained using the Adam optimiser with a learning rate of 1×10^{-4}. The Adam optimiser, proposed by Kingma and Ba [36], is a popular method for stochastic optimisation in machine learning. The training spanned 50,000 epochs, with early stopping based on a patience of 200 epochs in the training loss. The learning rate was reduced using a step size of 80% every 5,000 epochs. The entire training was completed in 50,000 epochs with all training data, incorporating early stopping to prevent overfitting and to finalise the training promptly, allowing predictions to be made.

For the RNN forecasting aspect, the lagged values of (C) were included to capture the temporal differences in the data. The lags 1, 2, 3, 4, 5, 7, and 21 were chosen because the data showed a corresponding correlation with these lags. Data analysis was carried out, with forecasts of 3, 7, and 21 days, to determine the optimal lag. The models were trained using a single step forecasting strategy with a horizon of 1.

Data regularisation was performed on the preprocessed variables before training using MinMax normalisation, expressed by the following equation:

$$x_{\text{norm}} = \frac{x - x_{\text{min}}}{x_{\text{max}} - x_{\text{min}}} \tag{12.21}$$

The experiments were carried out on a high-performance computing (HPC) machine with an Intel(R) Xeon(R) Gold 5217 CPU and a Nvidia Quadro RTX 8000 based on a Linux platform. The code was written in Python, using major libraries such as PyTorch and PyTorch Lightning for model training and evaluation. The models were trained with Adam Optimiser with a learning rate of 1×10^{-3} and a batch size of 32. Training spanned a minimum of 5 epochs and a maximum of 100 epochs, with early stopping based on patience of 20 epochs while monitoring validation loss. Table 12.1 represent the hyperparameter search space for simulated annealing algorithm to tune the RNN models that we are experimenting on with the data from the PINN.

12.3.2 Evaluation of the Models

The performance of the forecasting models and the PINN model was evaluated using the following key metrics. The MAE is a measure of the average absolute difference between the predicted and observed values. The mean absolute percentage error (MAPE) is a measure of the average percentage difference between the predicted and observed values. The normalised root mean squared error (NRMSE) is a measure of the root mean squared error normalised by the range of the observed values. The mean absolute scaled error (MASE) is a measure of the average absolute difference between

TABLE 12.1 Hyperparameters search space for the simulated annealing algorithm experiment

HYPERPARAMETER	*SEARCH SPACE*
RNN encoder type	GRU, LSTM, RNN
Hidden size	32, 128
Number of layers	5, 100
Bidirectionality	True, False
The decoder uses all hidden layers	True, False

the predicted and observed values, scaled by the average absolute difference of the naive forecast. The metrics are defined as follows:

$$\text{MAE} = \frac{1}{N}\sum_{i=1}^{N}\left|f_i - y_i\right|, \tag{12.22}$$

$$\text{MAPE} = \frac{1}{N}\sum_{i=1}^{N}\left|\frac{f_i - y_i}{y_i}\right| \times 100, \tag{12.23}$$

$$\text{NRMSE} = \frac{\sqrt{\frac{1}{N}\Sigma_{i=1}^{N}\left(f_i - y_i\right)^2}}{y_{\max} - y_{\min}}, \tag{12.24}$$

$$\text{MASE} = \frac{\frac{1}{N}\Sigma_{i=1}^{N}\left|f_i - y_i\right|}{\frac{1}{N}\Sigma_{i=1}^{N}\left|y_i - y_{\text{naive}}\right|} \tag{12.25}$$

In these equations, f_i is the predicted value, y_i is the observed value, N is the number of samples, $y_{\max}$ and $y_{\min}$ are the maximum and minimum observed values, and y_{naive} is the naive prediction. The metrics were used to evaluate the performance of the models in forecasting the demand for ICU beds in England and estimating the time-varying parameters of the SEIRD model.

12.3.3 Parameter and State Estimation Using PINNs

The PINN model was trained to estimate the time-varying parameters and unobserved states of the SEIRD model. Figure 12.5 shows the processed data in blue and the corresponding prediction in red, demonstrating that the model can effectively simulate the data using the SEIRD-PINNs based methodology. The close alignment of the observed and predicted values across all four graphs in Figure 12.5 indicates a high degree of accuracy in the model's predictions. This high correlation between observed and predicted

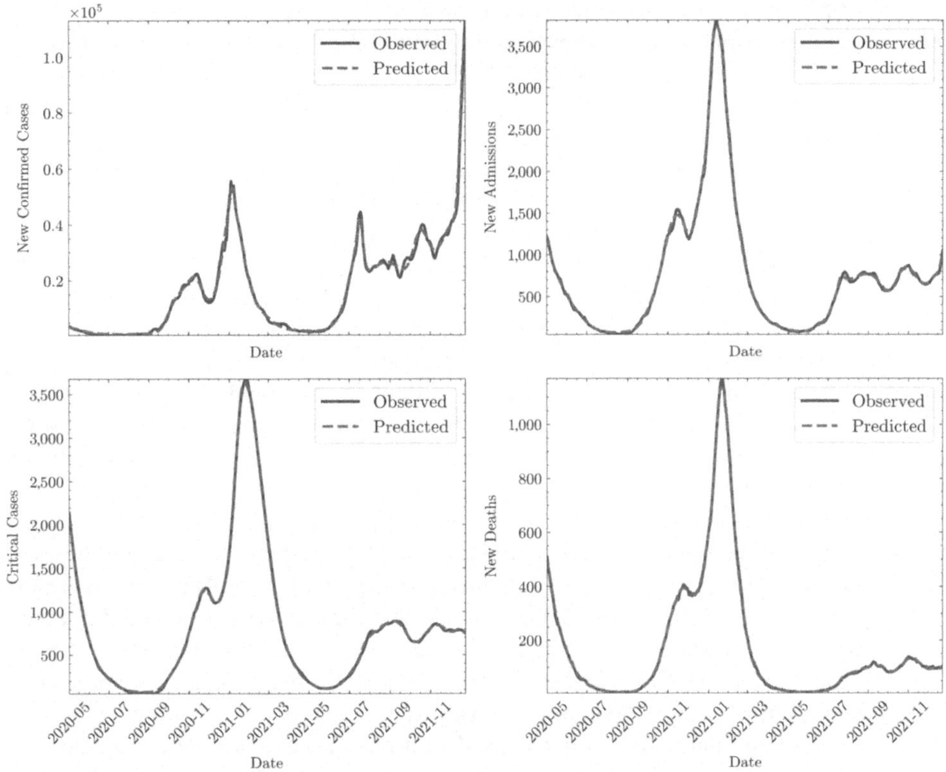

FIGURE 12.5 Comparison of observed versus predicted values for ICU bed demand. These graphs compare the observed data with the prediction of the model for key COVID-19 metrics. (a) New confirmed cases, (b) new admissions, (c) critical cases, and (d) new deaths. The close alignment of observed and predicted data validates the accuracy in forecasting the pandemic's impact of the pandemic on healthcare systems.

values for new confirmed cases, new admissions, critical cases, and new deaths demonstrates the robustness of the PINN model in handling real-world epidemiological data.

The model was also able to handle the time-varying parameter estimation. By studying the β parameters, we can see that the transmission rate was higher in the beginning part of the data, which was in the early stages of the pandemic, and gradually shows the peak and deep, which was due to all the lockdown and quarantining periods that were implemented by the government. This trend corresponds to the early phases of the COVID-19 pandemic, when transmission was high due to lack of interventions and public awareness. The gradual decline and stabilisation $\beta(t)$ reflect the impact of public health measures such as social distancing, mask mandates, and vaccination campaigns. During lockdown periods, a significant reduction is observed in new hospital admissions and MV bed usage, highlighting the effectiveness of these measures in curbing infection rates and reducing the strain of the healthcare system. The model accurately captures these dynamics, reflecting the reduction in $\beta(t)$ and $\omega(t)$ during these periods. Following the lift of lockdowns, there is an initial increase in hospital admissions and critical bed usage, reflecting the lagged effect of relaxed restrictions on infection rates. The model predictions show a corresponding increase in $\beta(t)$ and ω(t) indicating the rise in transmission and hospitalisation rates.

The recovery rate $\gamma_c(t)$ showed notable fluctuations, which can be correlated with changes in clinical treatments and healthcare protocols. Peaks in $\gamma_c(t)$align with periods where there might have been advancements in treatment options or increased availability of medical resources, leading to improved recovery rates from critical care. The death rate $\delta_c(t)$ demonstrated fluctuations consistent with the waves of the pandemic. High $\delta_c(t)$ values during peak infection periods reflect the increased strain on healthcare systems and the severity of the disease. The observed trends indicate periods of high mortality in critical care units, which underscores the critical phases of the pandemic. The reinfection rate $\eta(t)$ experienced a significant dip followed by recovery, potentially indicating the effect of immunity buildup in the population and the impact of vaccination. The initial drop in $\eta(t)$ could be due to the effectiveness of the first wave of vaccinations and natural immunity, while the subsequent increase may be attributed to the emergence of new variants or waning immunity. The hospitalisation ratio $\omega(t)$ showed periodic peaks and troughs, reflecting the dynamic nature of hospital admissions throughout the pandemic. High values of $\omega(t)$ correspond to surges in hospital admissions during major waves, while the troughs align with periods of reduced transmission and effective public health interventions. The death rate $\mu(t)$ indicated a declining trend with occasional spikes. This trend is consistent with overall improvements in clinical care and patient management. The spikes in $\mu(t)$ during certain periods might reflect overwhelming hospital conditions during peak waves, leading to higher mortality rates. The observed data align with the predicted model outcomes, demonstrating the model's ability to accurately capture and forecast the demand for hospital resources in response to infection waves. The fluctuations in $\delta_c(t)$ and $\mu(t)$ reflect the observed real-world trends in mortality and healthcare system strain during peak infection periods (Table 12.2).

Table 12.3 summarises the performance of the PINN model for parameter and state estimation, indicating high precision in capturing the dynamics of the pandemic. These metrics show that the model effectively captures the temporal patterns in the data, crucial for accurate forecasting and resource planning. Figure 12.5 shows the comparison of the observed versus predicted values for the demand for ICU beds, indicating a close alignment and effective capture of dynamics. Figures 12.6 and 12.7 illustrate the output of the unobserved state and the time-varying parameters estimated by the PINN model, respectively.

TABLE 12.2 Evaluation metrics for states estimation

STATE	*NRMSE*	*MASE*	*MAE*
Infections	0.0128	31.28	833.66
Hospitalisations	0.0074	31.55	20.29
Critical	0.0064	37.60	17.45
Deaths	0.0064	33.69	4.91

TABLE 12.3 This table presents the MAPE to forecast COVID-19 infections, hospitalisations, critical cases, and deaths over 7, 14, 21, and 28 days in England, after training of the SEIRD-PINN model

STATE	*MAPE*			
	7 DAYS	*14 DAYS*	*21 DAYS*	*28 DAYS*
Infections	0.1517	0.0956	0.0920	0.0798
Hospitalisations	0.2831	0.1746	0.1281	0.1036
Critical	0.1116	0.0616	0.0434	0.0359
Deaths	0.1123	0.0677	0.0591	0.0469

The results of the time-varying parameters estimated by the PINN model show that the model effectively captures the dynamics of the pandemic. The results of these parameters reflect the expected behaviour studied in the literature; the same can be said for the unobserved states. The prediction of the next 7, 14, 21, and 28 days for the

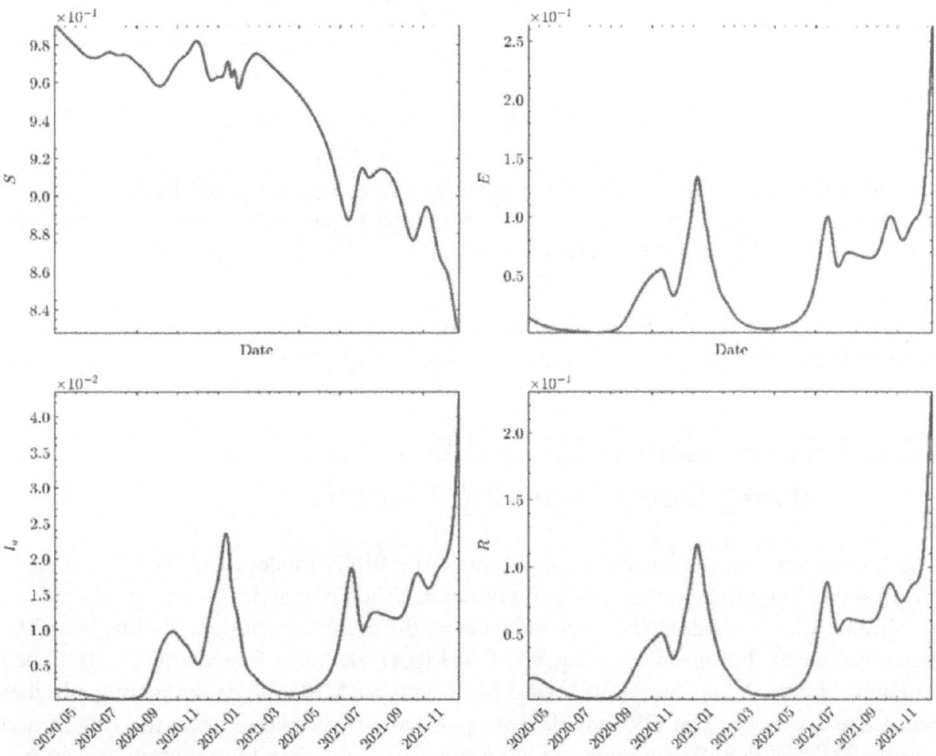

FIGURE 12.6 Unobserved state outputs estimated by the PINN model. The plots illustrate the evolution of the unobserved component of the model over time. These plots capture the dynamic changes in each compartment throughout the pandemic.

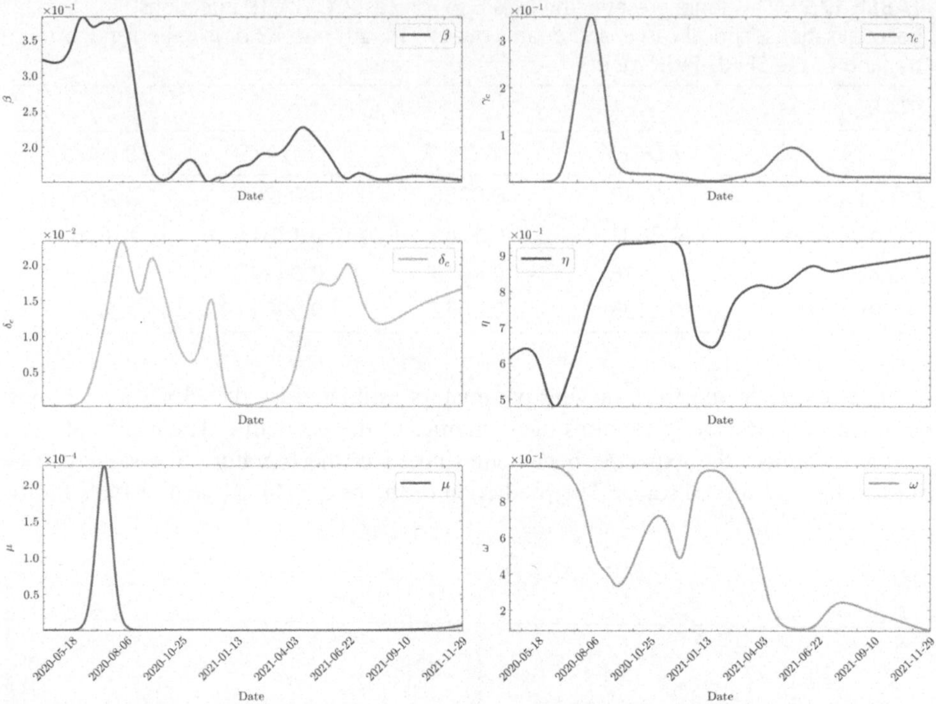

FIGURE 12.7 Time-varying parameters estimated by the PINN model. This figure shows the temporal evolution of various COVID-19-related parameters. Each plot reveals the dynamic nature of these parameters over time.

observed states of the model was also done to validate the effectiveness of PINNs for short-term predictions.

12.3.4 Forecasting ICU Bed Demand Using Deep Learning Models

The forecasting component used the output of the PINN model combined with lagged covariates of the target variable (C) to predict ICU bed demand.

Table 12.4 compares the performance of different algorithms, highlighting the effectiveness of the optimised Seq2Seq RNN (SA) model in forecasting the ICU bed demand. The model achieved the lowest MAE and NRMSE values, indicating a higher predictive accuracy. The GRU model also performed well, better than the LSTM and vanilla RNN models. The models, which were as a result of the hyperparameter optimisation using simulated annealing, performed well in capturing the temporal dependencies in the data, leading to more accurate forecasts.

Figure 12.8 compares the forecasts of different deep learning algorithms, demonstrating the effectiveness of the proposed hybrid framework in predicting ICU bed

TABLE 12.4 Performance metrics of different deep learning forecasting algorithms

ALGORITHM	*MAE*	*MAPE*	*NRMSE*	*MASE*
RNN	31.65	4.12%	0.1464	1.146
LSTM	22.36	2.92%	0.1092	0.8094
GRU	17.42	2.29%	0.0835	0.6305
Optimised GRA (SA)	21.97	8.86%	0.0245	0.6569
Seq2Seq LSTM	29.55	3.85%	0.1415	1.070
Optimised Seq2Seq RNN (SA)	16.73	2.20%	0.0819	0.6056

demand during the COVID-19 pandemic. The results indicate that our proposed hybrid framework is highly effective in capturing the dynamics of COVID-19 and forecasting the demand for healthcare resources. Integration of physical laws into the learning process ensures accurate parameter and state estimation, which is critical to understanding the spread of the virus and its impact on healthcare resources.

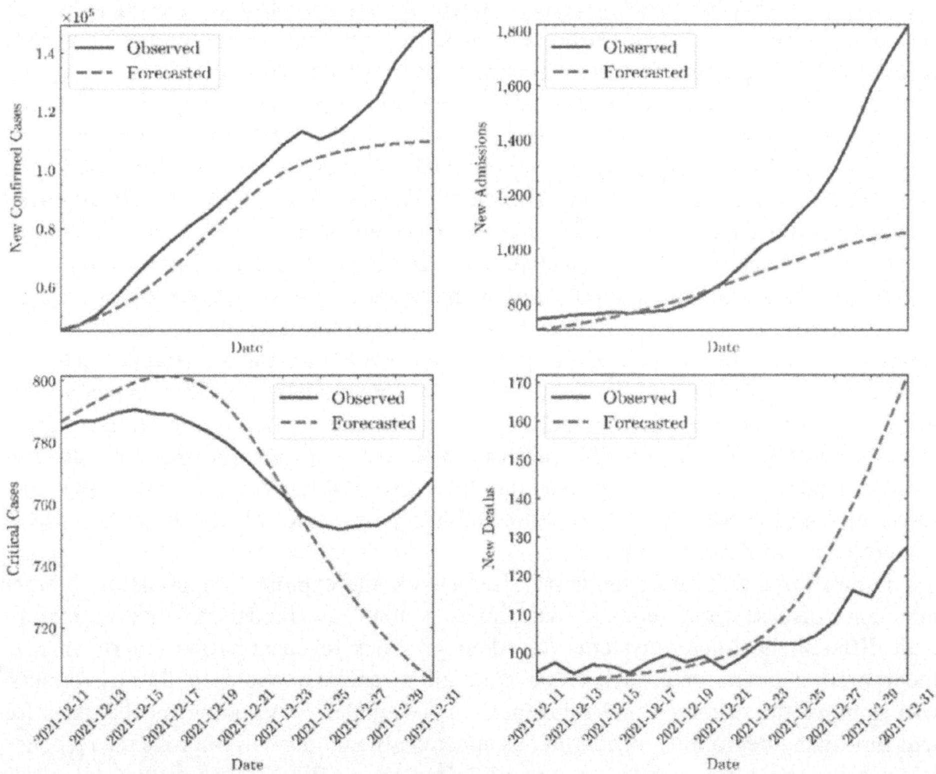

FIGURE 12.8 PINN 21-day forecast of the observed states of the model after training. future trends for key COVID-19 metrics. (a) New confirmed cases, (b) new admissions, (c) critical cases, and (d) new deaths. The observed data and forecasted trends highlight potential future scenarios, assisting in preparation and resource allocation.

Among the deep learning models evaluated, the optimised Seq2Seq RNN (SA) model showed the best performance, with the lowest MAE and NRMSE values. This suggests that the Seq2Seq architecture, optimised using SA, is particularly well suited for this forecasting task due to its ability to capture complex temporal patterns and dependencies in the data. The application of simulated annealing for hyperparameter tuning ensures that the model parameters are optimised for the best possible performance.

12.4 DISCUSSION AND CONCLUSION

This chapter has put forward a hybrid framework for modelling and forecasting COVID-19 dynamics and ICU bed demand in England. Through the integration of the output from SEIRD-PINNs with RNNs models, we can provide a way of simulating data and forecasting future trend. We leveraged both data-driven and mechanistic approaches to capture the pandemic's complexity in PINNs. Among the employed models, Optimised Seq2Seq RNN (SA) performed the best, achieving the lowest MAE and NRMSE values due to its encoder-decoder structure and SA for hyperparameter tuning. The PINN model further improved predictive accuracy and provided insight into the dynamics of the pandemic. Although the Optimised Seq2Seq RNN (SA) outperformed other approaches employed, all models showed good predictive performance and interpretability. Vanilla RNN, despite the higher error metrics, offered valuable information on trends in ICU bed demand due to its simplicity. The LSTM and GRU models improved over the vanilla RNN by managing long-term dependencies (Figure 12.9). The optimised GRU model, though less effective than Seq2Seq RNN, showed improved performance due to hyperparameter optimisation, making it robust for time series forecasting.

Our hybrid framework combines PINNs and deep learning models to address the complexities of COVID-19. PINNs incorporate epidemiological dynamics into the learning process, leading to accurate and interpretable models, crucial to understanding the spread of infectious disease and the impacts of intervention. Deep learning models, especially the Seq2Seq RNN optimised with SA, excel at capturing complex temporal dependencies, essential for accurate forecasting of healthcare resource demand. However, training these models is computationally intensive, which limits their use in resource-constrained environments.

Future work will refine the hybrid framework and explore its application to other infectious diseases and regions. We aim to extend our models to diverse regions with different healthcare systems and demographics to ensure robust performance. Incorporating diverse data, such as social mobility, vaccination rates, and virus variants, will improve the accuracy and robustness of the model. Developing mechanisms for real-time data integration and continuous update of the model will improve the response to emerging trends and pandemic dynamics. Integrating PINNs ensures that the model predicts future states and provides interpretable insights into disease dynamics, crucial for public health decision making. Accurate forecasts of the demand for beds in the ICU can assist healthcare systems in resource allocation and planning, ensuring critical care availability during peak periods.

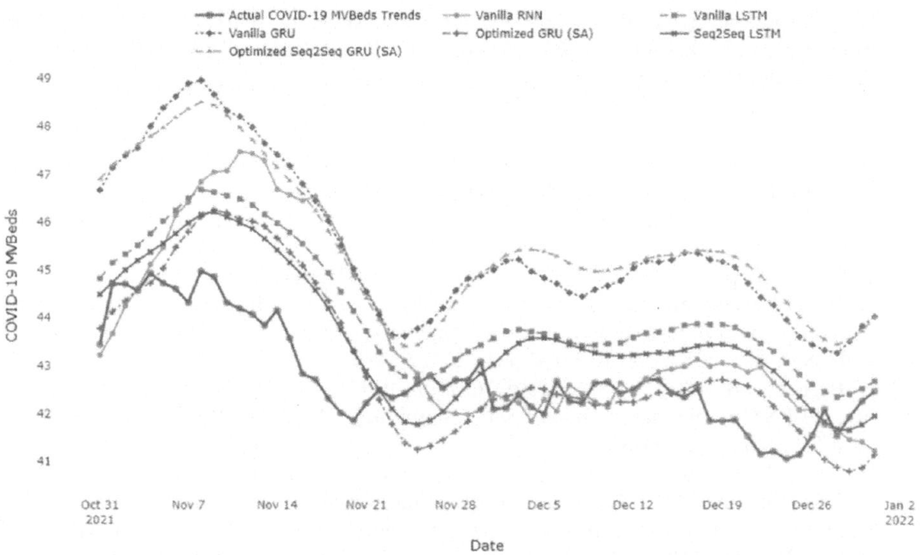

FIGURE 12.9 Comparison of COVID-19 ICU bed forecasts using various deep learning algorithms with data from England. This figure includes forecasts from the Vanilla RNN, LSTM, GRU, and Seq2Seq models, as well as their optimised versions with SA, highlighting the differences in predictive performance.

NOTE

1. Office for National Statistics licensed under the Open Government Licence v.3.0. website: NHS England (Regions) (July 2022) EN BFC - data.gov.uk

REFERENCES

1. R. Forman, R. Atun, M. McKee, and E. Mossialos, "12 lessons learned from the management of the coronavirus pandemic," *Health Policy*, vol. 124, no. 6, pp. 577–580, 2020, doi: 10.1016/j.healthpol.2020.05.008.
2. D. P. Oran, and E. J. Topol, "Prevalence of asymptomatic SARS-CoV-2 infection: A narrative review," *Annals of Internal Medicine*, vol. 173, no. 5, pp. 362–367, 2020, doi: 10.7326/M20-3012.
3. E. J. Emanuel *et al.*, "Fair allocation of scarce medical resources in the time of covid-19," *New England Journal of Medicine*, vol. 382, no. 21, pp. 2049–2055, May 2020, doi: 10.1056/NEJMsb2005114.
4. M. L. Brandeau, "Allocating resources to control infectious diseases," in Operations research and health care International Series in Operations Research & Management Science, *vol. 70, M. L. Brandeau, F. Sainfort, and W. P. Pierskalla, Eds. Springer, 2005, pp. 443–464. doi: 10.1007/1-4020-8066-2_17.*,

5. G. S. Zaric, and M. L. Brandeau, "Resource allocation for epidemic control over short time horizons," *Mathematical Biosciences*, vol. 171, no. 1, pp. 33–58, 2001, doi: 10.1016/S0025-5564(01)00050-5.
6. M. Ajao-Olarinoye, V. Palade, S. Mousavi, F. He, and P. A. Wark, "Deep learning based forecasting of COVID-19 hospitalisation in England: A comparative analysis," in *2023 International Conference on Machine Learning and Applications (ICMLA)*, Dec. 2023, pp. 1344–1349. doi: 10.1109/ICMLA58977.2023.00203.
7. M. I. Betti, A. H. Abouleish, V. Spofford, C. Peddigrew, A. Diener, and J. M. Heffernan, "COVID-19 vaccination and healthcare demand," *Bulletin of Mathematical Biology*, vol. 85, no. 5, p. 32, Mar. 2023, doi: 10.1007/s11538-023-01130-x.
8. H. Wu, K. Wang, and L. Xu, "How can age-based vaccine allocation strategies be optimized? A multi-objective optimization framework," *Frontiers in Public Health*, vol. 10, p. 934891, Sep. 2022, doi: 10.3389/fpubh.2022.934891.
9. D. Borges, and M. C. V. Nascimento, "COVID-19 ICU demand forecasting: A two-stage prophet-LSTM approach," *Applied Soft Computing*, vol. 125, p. 109181, Aug. 2022, doi: 10.1016/j.asoc.2022.109181.
10. E. Campillo-Funollet *et al.*, "Predicting and forecasting the impact of local outbreaks of COVID-19: Use of SEIR-D quantitative epidemiological modelling for healthcare demand and capacity," *International Journal of Epidemiology*, vol. 50, no. 4, pp. 1103–1113, Aug. 2021, doi: 10.1093/ije/dyab106.
11. W. O. Kermack, and A. G. McKendrick, "A contribution to the mathematical theory of epidemics," *Proceedings of the Royal Society of London. Series A, Containing Papers of a Mathematical and Physical Character*, vol. 115, no. 772, pp. 700–721, 1927.
12. S. Hochreiter, and J. Schmidhuber, "Long short-term memory," *Neural Computation*, vol. 9, no. 8, pp. 1735–1780, 1997, doi: 10.1162/neco.1997.9.8.1735.
13. F. Kamalov, K. Rajab, A. K. Cherukuri, A. Elnagar, and M. Safaraliev, "Deep learning for covid-19 forecasting: State-of-the-art review," *Neurocomputing*, vol. 511, pp. 142–154, Oct. 2022, doi: 10.1016/j.neucom.2022.09.005.
14. M. Raissi, P. Perdikaris, and G. E. Karniadakis, "Physics-informed neural networks: A deep learning framework for solving forward and inverse problems involving nonlinear partial differential equations," *Journal of Computational Physics*, vol. 378, pp. 686–707, Feb. 2019, doi: 10.1016/j.jcp.2018.10.045.
15. F. Amaral, W. Casaca, C. M. Oishi, and J. A. Cuminato, "Simulating immunization campaigns and vaccine protection against COVID-19 pandemic in Brazil," *IEEE Access*, vol. 9, pp. 126011–126022, 2021, doi: 10.1109/ACCESS.2021.3112036.
16. F. Amaral, W. Casaca, C. M. Oishi, and J. A. Cuminato, "Towards providing effective data-driven responses to predict the Covid-19 in São Paulo and Brazil," *Sensors*, vol. 21, no. 2, p. 540, 2021, doi: 10.3390/s21020540.
17. S. Cuomo, V. S. Di Cola, F. Giampaolo, G. Rozza, M. Raissi, and F. Piccialli, "Scientific machine learning through physics–informed neural networks: Where we are and what's next," *Journal of Scientific Computing*, vol. 92, no. 3, p. 88, Jul. 2022, doi: 10.1007/s10915-022-01939-z.
18. L. Nguyen, M. Raissi, and P. Seshaiyer, "Modeling, analysis and physics informed neural network approaches for studying the dynamics of COVID-19 involving human-human and human-pathogen interaction," *Computational and Mathematical Biophysics*, vol. 10, no. 1, pp. 1–17, 2022, doi: 10.1515/cmb-2022-0001.
19. H. Hu, C. M. Kennedy, P. G. Kevrekidis, and H.-K. Zhang, "A modified pinn approach for identifiable compartmental models in epidemiology with application to Covid-19," *Viruses*, vol. 14, no. 11, p. 2464, 2022, doi: 10.3390/v14112464.
20. S. Berkhahn, and M. Ehrhardt, "A physics-informed neural network to model COVID-19 infection and hospitalization scenarios," *Advances in Continuous and Discrete Models*, vol. 2022, no. 1, p. 61, 2022, doi: 10.1186/s13662-022-03733-5.

21. V. Grimm, A. Heinlein, A. Klawonn, M. Lanser, and J. Weber, "Estimating the time-dependent contact rate of SIR and SEIR models in mathematical epidemiology using physics-informed neural networks," *Electronic Transactions on Numerical Analysis*, vol. 56, pp. 1–27, 2022, doi: 10.1553/etna_vol56s1.
22. E. O. Oluwasakin, and A. Q. M. Khaliq, "Data-driven deep learning neural networks for predicting the number of individuals infected by COVID-19 omicron variant," *Epidemiologia*, vol. 4, no. 4, Art. no. 4, pp. 420–453, Dec. 2023, doi: 10.3390/epidemiologia4040037.
23. T. K. Torku, A. Q. M. Khaliq, and K. M. Furati, "Deep-data-driven neural networks for COVID-19 vaccine efficacy," *Epidemiologia*, vol. 2, no. 4, Art. no. 4, pp. 564–586, Dec. 2021, doi: 10.3390/epidemiologia2040039.
24. X. Ning, J. Guan, X.-A. Li, Y. Wei, and F. Chen, "Physics-informed neural networks integrating compartmental model for analyzing COVID-19 transmission dynamics," *Viruses*, vol. 15, no. 8, Art. no. 8, p. 1749, Aug. 2023, doi: 10.3390/v15081749.
25. A. Ogueda, E. G. Martinez, V. Arunachalam, and P. Seshaiyer, "Machine learning for predicting the dynamics of infectious diseases during travel through physics informed neural networks," *Journal of Machine Learning for Modeling and Computing*, 2023, [Online]. Available: https://api.semanticscholar.org/CorpusID:259876852
26. E. Kharazmi, M. Cai, X. Zheng, Z. Zhang, G. Lin, and G. E. Karniadakis, "Identifiability and predictability of integer- and fractional-order epidemiological models using physics-informed neural networks," *Nature Computational Science*, vol. 1, no. 11, Art. no. 11, pp. 744–753, Nov. 2021, doi: 10.1038/s43588-021-00158-0.
27. NHS England, "COVID-19 hospital activity." 2024. [Online]. Available: https://www.england.nhs.uk/statistics/statistical-work-areas/covid-19-hospital-activity/
28. G. LLC, "COVID-19 open data: A repository of COVID-19 related data." 2023. [Online]. Available: https://health.google.com/covid-19/open-data/raw-data
29. D. E. Rumelhart, G. E. Hinton, and R. J. Williams, "Learning representations by back-propagating errors," *Nature*, vol. 323, no. 6088, pp. 533–536, 1986, doi: 10.1038/323533a0.
30. K. Cho, B. Van Merriënboer, D. Bahdanau, and Y. Bengio, "On the properties of neural machine translation: Encoder-decoder approaches," *arXiv preprint arXiv:1409.1259,* 2014.
31. I. Sutskever, O. Vinyals, and Q. V. Le, "Sequence to sequence learning with neural networks," *Advances in Neural Information Processing Systems,* vol. 27, 2014, pp. 3104–3112.
32. P. J. Van Laarhoven, E. H. Aarts, P. J. van Laarhoven, and E. H. Aarts, *Simulated annealing*. Springer, 1987.
33. Y. Yoo, "Hyperparameter optimization of deep neural network using univariate dynamic encoding algorithm for searches," *Knowledge-Based Systems*, vol. 178, pp. 74–83, 2019, doi: 10.1016/j.knosys.2019.04.019.
34. M. Fischetti, and M. Stringher, "Embedded hyper-parameter tuning by Simulated Annealing," Jun. 04, 2019, *arXiv*: arXiv:1906.01504. doi: 10.48550/arXiv.1906.01504.
35. X. Glorot, and Y. Bengio, "Understanding the difficulty of training deep feedforward neural networks," in *Proceedings of the Thirteenth International Conference on Artificial Intelligence and Statistics*, JMLR Workshop and Conference Proceedings, Mar. 2010, pp. 249–256. Accessed: Aug. 01, 2024. [Online]. Available: https://proceedings.mlr.press/v9/glorot10a.html
36. D. P. Kingma, and J. Ba, "Adam: A method for stochastic optimization," in *International Conference on Learning Representations (ICLR)*, 2015.

PART FOUR

Deep Learning Methodological Approaches in Other Applications

13 A Novel Data Reduction Technique for Medicare Fraud Detection with Gaussian Mixture Models

John T. Hancock III and Taghi M. Khoshgoftaar

13.1 INTRODUCTION

Fraud within the Medicare program, a crucial public health insurance initiative of the U.S. federal government, represents a significant financial burden, incurring losses amounting to billions of dollars annually [1]. Fraudulent activities within the Medicare claims system facilitate illicit financial gains for unscrupulous providers. Given the substantial volume of claims processed, even a small fraction of fraudulent transactions can translate into a considerable financial loss. In the year 2019, the U.S. Department of Justice (DoJ) succeeded in recovering approximately $3 billion from such fraudulent practices [2]. However, the exact magnitude of losses incurred is somewhat obscured by the Centers for Medicare and Medicaid Services' (CMS) categorization of "improper payments," which encompasses both fraudulent and nonfraudulent erroneous payments [3].

DOI: 10.1201/9781003570882-17

Notably, the CMS reported improper payments in 2019 that exceeded the recovered amount by the DoJ by more than fivefold.

This situation unveils a substantial opportunity for the application of machine learning in reducing ambiguity and enhancing the precision of fraud detection. Advanced machine learning techniques could play a pivotal role in distinguishing fraudulent transactions from legitimate ones, thereby aiding the CMS in accurately determining the extent of fraudulent payments. Such refined detection capabilities would not only facilitate a more effective recovery of misappropriated funds but also serve as a valuable resource for law enforcement agencies in their efforts to combat financial fraud within the healthcare system.

Medicare insurance claims data offer a compelling domain for the application of classification techniques, particularly those adept at handling low-frequency events. The filing of fraudulent Medicare health insurance claims are low frequency, since fraudulently filed claims constitute a small fraction of total Medicare claims. Consequently, Medicare insurance claims data is a suitable subject for experiments with one-class classifiers. The data is similar to data in the true one-class scenario, since it is highly imbalanced; however, it is labeled so that we can evaluate the performance of the one-class classifiers. The true one-class scenario is extant when one knows there must be data of two types; however, only instances of one type are readily available for modeling.

The selection of the classifier type is primarily guided by the availability of labels within the dataset. In cases where labels are absent, unsupervised learning techniques come to the forefront as the most fitting approach. A quintessential example of such an unsupervised method is the isolation forest algorithm [4]. Conversely, when data from all classes are available and labeled, supervised machine learning classifiers are deemed more suitable. For an extensive overview of both supervised and unsupervised methods in the context of health insurance fraud detection, the reader is directed to [5]. One-class classifiers (OCCs), however, are particularly relevant in scenarios where there is an abundance of data from one class, but the other class is entirely absent. This situation is approximated by severe data imbalance. The datasets used in our study, which consist of highly imbalanced big Medicare data, serve as a prime example of such a one-class scenario approximation.

The expansive scale of data generated within the Medicare insurance system poses significant challenges typical of big data scenarios. A primary concern is the sheer volume of this data, which can significantly hinder the speed of model training. Consequently, there arises an imperative need for strategies that effectively reduce the scale of the data while concurrently preserving its essential characteristics. In our current study, we explore a data reduction technique that aims to minimize computational resource usage and reduce running times. Remarkably, this approach yields models that perform comparably to models constructed using the entire dataset.

To the best of our knowledge, we are the first to demonstrate and validate the efficacy of this specific data reduction technique. The innovation of our approach lies in how the data reduction method is combined with OCC techniques. Our extensive literature review underscores the uniqueness of our contribution; we found no existing study that implements the data reduction technique detailed in our research, with the exception of our previous study [6]. This study is an expansion of our previous study. This lack of previous related work highlights the novelty and potential impact of our data reduction technique in the field of fraud detection, particularly in contexts characterized by big data datasets.

Preliminary experiments with the same datasets used in this study revealed that one-class Gaussian mixture models (GMMs) exhibit superior performance when compared to one-class support vector machines (SVMs). This significant difference in performance led us to postpone further experimentation with one-class SVMs, primarily due to their relatively poor results. Our decision to focus on one-class GMMs is further supported by recent studies, which also report the superior efficacy of one-class GMMs over one-class SVMs [7, 8]. For those interested in a broader understanding of the applications of OCCs in big data application domains, we recommend the thorough review of the literature provided in [9].

In this chapter, we employ a data reduction technique to enhance the performance of one-class GMMs [10] in identifying fraudulent activities within Medicare Part D and Medicare Part B insurance claims data. This technique is designed to enable the training of one-class GMMs on a significantly reduced subset of the full dataset, thereby optimizing computational efficiency without compromising the effectiveness of fraud detection.

One of the sources we use in this study is data from Medicare Part D, hereafter referred to as the "Part D data." This data currently encompasses over 174 million records pertaining to provider prescription claims [11]. The Part D data contains information about healthcare providers, the pharmaceuticals prescribed by these providers, and the corresponding beneficiaries. A notable aspect of the Part D data is its growth trajectory, which is reflective of the increasing volume of insurance claims received by the CMS. Such growth is a characteristic of big data. The CMS has made this Part D data publicly accessible, thereby fostering an environment conducive to research, particularly in the critical area of fraud detection.

The second source for data which we use in this study is Medicare Part B data [12], which we will subsequently refer to as the "Part B data." The publicly available raw Part B data currently consists of 68 million records. Medicare Part B is an insurance plan that covers treatments and procedures that healthcare providers render to Medicare beneficiaries. It grows at a rate similar to the Part D data.

Our findings are pivotal in demonstrating that, with the datasets used in this study, it is feasible to train one-class GMMs on 80% less data without compromising performance, as measured by the area under the receiver operating characteristic curve (AUC) [13], and area under the precision recall curve (AUPRC) [14] metrics in the classification of Medicare Part B and Part D datasets. This significant data reduction translates to accelerated training times on large datasets, which is particularly beneficial in the context of big data. When classifying highly imbalanced data, using more data does not always guarantee better results since an overabundance of instances of the majority class may cause the classifier to become biased to the majority class during the training phase. The methodologies we propose highlight the potential of one-class GMMs in efficiently classifying highly imbalanced datasets, such as those encountered in Medicare fraud detection. For further insights into existing techniques addressing class imbalance in big data, we refer readers to the comprehensive survey by Leevy et al. [15]. The subsequent sections of this study cover the following subjects: related work, algorithms, methodology, results, analysis, and conclusions.

13.2 RELATED WORK

Our extensive literature review did not uncover any studies that apply OCCs and our data reduction technique to big data, with the exception of the study that this chapter expands on [6]. The existing research primarily explores novel sampling methods or ensembles of classifiers. We advocate that our approach, utilizing publicly available software for both classifiers and data reduction, offers greater reproducibility. Moreover, we combine these tools in an innovative way. Our findings suggest that our direction of research is pioneering in its focus on a data reduction technique specifically designed for the application of OCCs in the context of big data.

In 2022, Hayashi and Fujita's study on OCCs and imbalanced data [16] introduced an ensemble technique combining multiple OCCs trained on both majority and minority classes to classify various datasets. These datasets, however, are smaller when compared to the Medicare Part D and Part B datasets used in our research, with the largest dataset in their study comprising around 150,000 instances. Hayashi and Fujita compare the performance of their technique to binary-class classifiers (BCCs). We were not able to find where Hayashi and Fujita specify sample sizes in their study. We argue that their approach, training some OCCs on majority class samples and others on minority class samples, essentially aligns with BCC methodology. We find alignment with BCC methodology, since the complete ensemble is trained with instances of both classes. Our research, in contrast, focuses on the one-class GMM classifier trained exclusively on one class and addresses a much larger dataset with significant class imbalance.

Czarnowski introduces an ensemble method in [17] that segments a multiclass classification problem into distinct OCC tasks, using oversampling to boost classifier performance. This method's employment of OCCs and sampling is somewhat akin to our approach, but Czarnowski's technique diverges from our OCC strategy. In Czarnowski's approach, each OCC is tailored to different classes, requiring a data partitioning step into positive and negative examples for each class. This reliance on a partitioner for class separation implies that the system, in its entirety, is trained on more than a single class. This is another instance where the ensemble, as a whole, requires data of more than one class for training. In contrast, our study centers on a classifier trained exclusively on instances of a single class, with the premise that instances of other classes are unavailable during training.

Juszczak and Duin suggest selective sampling techniques to enhance OCCs' performance [18]. They focus on classification confidence, linked to the proximity of a sample to the decision boundary. In classification tasks involving a decision boundary, the proximity of a sample to this boundary inversely correlates with the confidence in its class membership. Samples situated closer to the boundary are typically associated with lower confidence regarding their classification. Categories of confidence include low confidence for both classes (ll), low target and high outlier confidence (lh), high target and low outlier confidence (hl), and high confidence for both (hh). Their method involves initial classification to set a boundary, selecting a confidence category, sampling and a human expert labeling, and then retraining the OCC with these augmented instances. They note better performance with lh and hh confidence regions, while ll and hl regions yield poorer results. Implementing this technique might be challenging due to the need for expert involvement in labeling. The expert may not be available, or may be costly and time-consuming to employ. The need for an expert in Juszczak and Duin could be considered a limitation, compared to our fully automated approach suitable for big data.

We were also able to discover recent studies that involve Medicare fraud detection. In this sense they are related works, however, they involve binary-class classification. Mayaki and Riveill [19] undertook a study similar to ours, involving BCCs. In their study, they compile data from the CMS spanning the years 2017 to 2019. This dataset was further enriched by labeling it with entries from the list of excluded individuals and entities (LEIE) [20]. In their work, they developed an innovative model, termed the multiple inputs neural network auto-encoder (MINN-AE), specifically designed to detect instances of Medicare fraud. A notable feature of MINN-AE is its auto-encoder component, which is based on long short-term memory (LSTM) networks. Auto-encoders, known for their efficacy in handling highly imbalanced data, have seen successful deployment in various domains [21]. The efficacy of MINN-AE was rigorously evaluated against a suite of established models, including logistic regression [22], random forest [23], XGBoost [24], and five other artificial neural network architectures. The metrics adopted for this evaluation encompassed precision, AUPRC, and the geometric mean. Mayaki and Riveill's empirical results are compelling, demonstrating that MINN-AE outperformed the other nine models in comparison. Mayaki and Riveill did not delve into detailed information regarding the experimental dataset, such as the number of instances and the feature set. Furthermore, Mayaki and Riveill did not explore applying OCCs to the task of fraud detection, as we do here.

Another recent study involving Medicare fraud detection is by Herland et al. [25]. Similar to our study, they investigate Medicare fraud detection utilizing datasets derived from the CMS's publicly available data. Their research encompasses several datasets: Medicare Physician and Other Practitioners (Part B) [26] for the years 2012–2015, Medicare Part D Prescribers (Part D) [11] for the years 2013–2015, and the Medicare Durable Medical Equipment, Prosthetics, Orthotics, and Supplies (DMEPOS) [27] for the same period. Furthermore, Herland et al. construct an integrated dataset, termed the Combined dataset, through the amalgamation of the Part B, Part D, and DMEPOS datasets. Similar to the data used in our study, Herland et al. have high imbalance in all four datasets. In terms of model development, Herland et al. explore the efficacy of logistic regression, random forest, and gradient boosting classifiers across these datasets. Their findings indicate that models built with the Combined dataset, particularly logistic regression, yield the best performance in fraud detection. Our study, however, diverges from Herland et al.'s approach in several respects. Notably, their research does not incorporate OCCs, and it does not cover the sampling technique we propose here.

Another related work that employs OCCs for classifying imbalanced data is the study by Hoang et al. [28]. The study involves the use of one-class SVMs in an ensemble model for anomaly detection in sensor data. Anomaly detection tasks implicitly involve working with imbalanced data. Since the study involves imbalanced data, and OCCs, it is related to this study. Hoang et al. utilize Shapley Additive exPlanations (SHAP) with their model for assessing feature importance. The researchers introduced an innovative model for the detection of anomalies in sensor data from industrial control systems. This model is an ensemble of various machine learning algorithms, including the one-class SVM. However, a divergence exists between their methodology and ours. Hoang et al. did not focus on a technique for data reduction with a single well-established one-class classification algorithm. Therefore, our work is distinct from Hoang et al.

"Data Reduction to Improve the Performance of One-Class Classifiers on Highly Imbalanced Big Data," by Hancock and Khoshgoftaar, is the basis of this study, which we were invited to expand upon [6]. In this work, we did the initial step of investigating the effect of applying sampling to highly imbalanced Medicare big data for use with one-class GMMs. We only used one dataset, the Medicare Part D data. We found that sampling as little as 5% of the original data yielded performance, in terms of AUC, that was not significantly different from using the entire dataset. By "significantly different" we mean that, at the $\alpha = 0.01$ significance level, there was no statistically significant difference in AUC scores of models built with the entire dataset, versus models built with a sample. For performance in terms of AUPRC, we found that models built with samples as small as 20% of the entire training data yielded AUPRC scores that were not significantly different from using the entire training data. In the present study, we expand the experiments performed in the previous study to encompass another highly imbalanced big Medicare data.

Our literature review highlights a notable gap in extant studies. Our initial study to combine OCCs with data reduction techniques for big data analysis is novel, and we expand on it here. Among the few related studies we have identified, we found one with the notable limitation of the necessity for expert intervention in data sampling. We found other studies which we posit do not conform to our more rigorous interpretation of one-class methodology. Furthermore, we noted additional studies in the healthcare fraud detection application domain, but they do not investigate the use of OCCs in conjunction with sampling techniques. We have developed and applied a unique data reduction technique that is especially adept at managing the challenges posed by severely imbalanced big data. We tailor OCC methodologies to the nuanced demands of big data applications, marking a significant advancement in the field by expanding our previous study.

13.3 ALGORITHMS

For this study, two algorithms play an important role. The primary algorithm is the one-class GMM, which is complemented by the second important component, calibration. Our methodology involves taking the raw output scores from the one-class GMM and applying calibration to these scores to transform them into probabilities. These probabilities are then utilized to compute AUC and AUPRC scores, serving as key metrics for evaluating the experimental outcomes of our study.

13.4 CALIBRATION

We found sigmoid calibration to be beneficial for calculating both the AUC and AUPRC scores when using a one-class GMM. We utilized the sklearn.mixture.GaussianMixture module from scikit-learn version 1.2.0 for our one-class GMM. During preliminary experiments with this implementation, we encountered a notable issue with the module's predict_proba method. The probabilities for class membership were all very near to 0.0 or 1.0. This extreme probability output led to minimal variation in precision and recall values across different decision threshold settings, resulting in precision-recall curves that lacked a sufficient number of points to provide meaningful insights.

To rectify the extreme values in the output of GMM, we implemented sigmoid calibration, which effectively transforms the raw GMM outputs into well-calibrated probability values [29]. More on calibration techniques can be found in [30, 31]. The sigmoid calibration process involves fitting a sigmoid function to map the raw scores from the GMM to probabilities, enabling a more gradual spread of precision and recall scores across different thresholds. Consequently, this allowed for the generation of a detailed and interpretable precision-recall curve, thereby resolving the issues with one-class GMM's predict_proba function outputs, and facilitating more meaningful AUC and AUPRC calculations.

13.5 ONE-CLASS GMM

In our study, the one-class GMM classifier, incorporated from the scikit-learn library [32], plays a crucial role. This classifier is based on the expectation-maximization (EM) algorithm, a method thoroughly discussed by Dempster et al. [33]. The EM algorithm is designed for finding optimal parameter values in probabilistic models. It operates through two iterative phases: expectation and maximization. In the expectation phase, model parameters are held constant to estimate the data probabilities. Subsequently, the maximization phase uses this data to refine the model parameters. This iterative process causes the model parameters to converge to their optimal values, thereby ensuring the robustness and reliability of the model.

The application of the EM algorithm is particularly effective in estimating the parameters of a one-class GMM. This process involves fitting a series of Gaussian distributions to a dataset for classification purposes. In the context of a one-class GMM, the training data is conceptualized as a mixture, essentially a summation, of various Gaussian distributions. Each Gaussian component within this mixture is designed to model distinct segments of the instances in the training dataset, thereby capturing the diverse characteristics present within the data.

The process of training a one-class GMM involves fitting a specified number of Gaussian components to the target data. Since we are discussing a one-class classifier, the training data is composed exclusively of instances from a single class. This fitting process entails determining the optimal mean and covariance parameters for each Gaussian distribution to best describe the dataset. When dealing with a dataset that has more than one dimension, the Gaussian components in the one-class GMM are multivariate Gaussian distributions. Here, the mean is represented as a vector, and the covariance is characterized by a matrix. The mean parameter specifies the central point of the distribution, while the covariance defines its width and overall shape. For univariate data, the covariance parameter simplifies to the variance of the data. This may aid one in conceptualizing the effect of the covariance matrix; the matrix dictates how the probability density of the Gaussian mixture spreads out in the space that contains the dataset.

Upon convergence of the EM algorithm, the trained one-class GMM can be utilized for classifying instances in the test dataset. Classification through one-class GMM involves evaluating the probability that a test instance could have been produced by the Gaussian distributions fit to the training data. If this likelihood falls below a predetermined threshold, it suggests that the test instance is not likely to have originated from the same distributions, indicating that it belongs to a different class than the one used in training. Preliminary experiments have shown that one-class GMMs trained on the majority class tend to outperform those trained on the minority class. Put another way, based on experience, we recommend training the one-class GMM on the majority class. This concludes our theoretical discussion of algorithms used in our study. In the next section on methodology, we describe how they are applied.

13.6 METHODOLOGY

Our methodology starts with dataset compilation. Datasets are compiled according to the technique described in [34]. Table 13.1 summarizes the key characteristics of the Part B and Part D datasets, after the entire process is complete.

We perform experiments with both Part D and Part B data. We use the same methodology with both datasets. In general terms, we conduct experiments by executing programs written in the Python programming language. The programs produce performance metrics which we then use for analysis.

We employ version 1.2.0 of scikit-learn, a widely used open-source Python library renowned for its extensive applications in machine learning and data analysis [32]. This library offers an array of algorithms and tools that facilitate the rapid development and assessment of models. Specifically, scikit-learn provides the necessary libraries

TABLE 13.1 Summary of Part B and Part D datasets

DATASET	*INSTANCE COUNT*	*FRAUDULENT*	*RATIO FRAUDULENT*
Part D	5344106	3700	0.07%
Part B	8669497	3954	0.05%

for implementing the one-class GMM and its calibration. For the one-class GMM, we adhere to the default settings in scikit-learn, with the exception of adjusting the n_components parameter to two, as determined by the optimal performance obtained in preliminary experiments using scikit-learn's GridSearchCV module.

Cross-validation is another important aspect of our methodology. We utilize scikit-learn's implementation of stratified k-fold cross-validation for assessing model performance. This technique involves dividing the dataset into k subsets (or folds) while ensuring the proportion of each class is consistent across all folds. During cross-validation, we perform k iterations, in each of which a different fold is reserved for testing, and the remaining k-1 folds are used for training. This method significantly reduces variability compared to a single train-test split, offering a more dependable and comprehensive evaluation of the model's performance.

In our experimental setup, we adopt a (k = 5) fivefold cross-validation method. From the training data, we remove instances of the minority class, keeping the test data intact. Depending on the experiment's parameters, we sample 5, 10, 15, 20, or 25% of the training data using the Python pandas module data frame object's sample function [35], with default seed values. As a benchmark, we also include experiments with 100% of the training data. Categorical attributes are processed using CatBoost encoding, similar to the method described in [36]. For detailed insights into CatBoost encoding, refer to [37], and for its applications in machine learning, see [38]. The one-class GMM is then fit to the training data, and subsequently, the entire test fold is classified using the trained model. The performance of these classifications is evaluated based on AUC and AUPRC metrics.

To the best of our knowledge, our study introduces a novel methodology in its application of the sampling technique with one-class GMMs. We believe this to be the first documentation of such a technique in conjunction with one-class GMMs. For comprehensive insights into established sampling techniques like random undersampling (RUS), please see [39]. It is important to distinguish our data reduction method from RUS. According to the definition used in the Python imbalanced-learning library imblearn [40], RUS involves discarding majority class instances until a specific minority-to-majority class ratio is achieved, retaining minority class instances. Conversely, in our approach for one-class classification, we first eliminate all instances of the minority class and then sample from the majority class.

In our experimental framework, we execute 10 iterations of fivefold cross-validation for each specified sampling level. This repeated iteration process is essential to mitigate variability and ensure a consistent and reliable estimate of the model's performance. By conducting multiple iterations, we aim to achieve performance evaluations for each sampling percentage that are robust and not influenced by anomalies in any particular data sample. Put another way, the multiple iterations protect against the scenario where we present results that are the outcome of a particularly lucky (or unlucky) partitioning of the data in fivefold cross-validation.

Our experimental design includes six sampling levels, and for each level, we conduct 10 iterations of fivefold cross-validation; hence, a total of 300 experiments. Therefore, we collect 300 AUC and AUPRC scores, providing a robust dataset for analysis. We then apply statistical tests to these experimental factors and outcomes. These tests are crucial in determining whether the sampling percentage used in training significantly impacts model performance metrics. The results of these tests are also instrumental in identifying the minimum amount of data required to train the models without compromising their performance. In the next section, we report the results of applying this experimental methodology.

13.7 RESULTS

Before delving into results, we explain the notation used in Tables 13.2–13.5 in this section. We represent the sample size by the number that is the result of dividing the sample size by the size of the entire dataset. For example, in the first row of Table 13.2, the figure 0.0500 means that the result pertains to experiments where the sample is 5% of the size of the entire dataset is used.

The first set of results we wish to present are for one-class GMM's performance in terms of AUPRC and AUC for various levels of the sampling factor applied to the Part D data. In Table 13.2, we observe that samples of sizes 20 and 25% of the entire training data yield performance higher than using the entire dataset.

TABLE 13.2 Mean AUPRC values by fraction for 10 iterations of fivefold cross-validation, for classifying the Part D dataset

FRACTION	*GMM SIGMOID*
0.0500	0.1343
0.1000	0.0000
0.1500	0.1401
0.2000	0.1442
0.2500	0.1443
1.0000	0.1437

TABLE 13.3 Mean AUC values by fraction for 10 iterations of fivefold cross-validation, for classifying the Part D dataset

FRACTION	*GMM SIGMOID*
0.0500	0.7280
0.1000	0.7298
0.1500	0.7285
0.2000	0.7291
0.2500	0.7275
1.0000	0.7257

TABLE 13.4 Mean AUPRC values by fraction for 10 iterations of fivefold cross-validation, for classifying the Part B dataset

FRACTION	*GMM SIGMOID*
0.0500	0.0019
0.1000	0.0018
0.1500	0.0017
0.2000	0.0018
0.2500	0.0018
1.0000	0.0016

TABLE 13.5 Mean AUC values by fraction for 10 iterations of fivefold cross-validation, for classifying the Part B dataset

FRACTION	*GMM SIGMOID*
0.0500	0.7042
0.1000	0.7214
0.1500	0.7121
0.2000	0.7031
0.2500	0.7210
1.0000	0.6178

Table 13.3 contains the AUC scores the one-class GMMs yield when classifying the Part D data. The trend of similar or better scores with smaller sample sizes is also apparent in Table 13.2.

The second set of results we wish to present are for classification of the Part B data. The results in Table 13.4 follow a pattern that is similar to what we see in the results for the Part D data. The AUPRC scores for fractions of sizes 20 and 25% of the data are higher than the AUPRC score one-class GMM yields when the entire dataset is used. Moreover, the mean AUPRC scores of models built with the other sample sizes of 5, 10, and 15% are also higher than the AUPRC score of the one-class GMM built with the entire dataset.

Finally, we present the mean AUC scores the one-class GMMs yield when classifying the Part B data. The similarity of the AUC scores in Tables 13.3 and 13.5, and the discrepancies of AUPRC scores in Tables 13.2 and 13.4 highlight how AUPRC can be a more informative metric in the classification of imbalanced big data. Both AUC and AUPRC involve the true positive rate. The key difference between the AUC and AUPRC metrics is that AUC involves the false positive rate, whereas AUPRC involves precision. The formula for the false positive rate is: $\frac{false\ positives}{true\ negatives + false\ positives}$

and the formula for precision is:

$$\frac{true\ positives}{true\ positives + false\ positives}$$

The difference between precision and false positive rate is that the false positive rate has true negatives in the denominator. In highly imbalanced big data, such as Part D and Part B data, a large number of true negatives can overwhelm the other terms. Precision does not suffer this weakness because it does not involve the true negative class.

The results in Tables 13.2–13.5 provide evidence that we can apply data sampling to one-class GMM's training data, and one-class GMM will yield performance similar to, or better than one-class GMM trained on the entire dataset. This is because we find models trained on a fraction of the training data yield similar or better AUC and AUPRC scores than models trained with all of the available training data.

Another result we wish to present is on the running time of one-class GMM when trained on sampled data versus the entire dataset. The data presented in Tables 13.2–13.5 indicate favorable variation in AUC or AUPRC scores when smaller samples of training data are used. This observation underscores the advantage of training models with the least amount of data required. The lower running times, coupled with similar or better AUC and AUPRC scores yielded by models trained with a fraction of the training data, are what determines the advantage. Typically, such a practice results in shorter training times for most classifiers. For example, training the one-class GMM with a 25% sample of the Part D data averaged around 5 minutes, compared to approximately 13 minutes when using the full dataset. We found similar speed-up for classifying Part B data as well. This significant reduction in training time suggests that as dataset sizes increase, employing our sampling technique could become essential. Moreover, in scenarios where the dataset size precludes the use of all available data, our findings support the strategy of training GMM on a subset of the data.

The presentation of AUC, AUPRC, and the discussion of running times concludes our presentation of results. In the next section, we move on to statistical analysis of the AUC and AUPRC scores reported in this section.

13.8 STATISTICAL ANALYSIS

This section contains a statistical analysis of the experimental outcomes depicted in Tables 13.2–3.5. This analysis aims to conclusively identify the smallest fraction of data that is sufficient for classifying Part D or Part B data in model building, ensuring performance on par with or superior to one-class GMM models trained on the entire dataset.

An initial step involves conducting an analysis of variance (ANOVA) [41] test to evaluate the impact of training data size on the AUC scores for one-class GMMs classifying Part D data. According to the results presented in Table 13.6, the proportion of majority class instances utilized in training the one-class GMM does not significantly

TABLE 13.6 ANOVA for percentage as a factor of one-class GMM's performance in terms of AUC in classifying the Part D dataset

	DF	*SUM SQ*	*MEAN SQ*	*F VALUE*	*PR(>F)*
Percentage	5.0000	0.0000	0.0000	0.7400	0.5912
Residuals	294.0000	0.0400	0.0000		

influence the AUC performance at the $\alpha = 0.01$ significance level. Consequently, it is feasible to train the one-class GMM with as little as a 5% sample of the Part D training data, achieving AUC performance equivalent to that obtained using the full set of majority class instances in the training data.

Next, we report the results of analysis of the AUPRC scores the one-class GMMs yield when classifying Part D data. The results of the ANOVA test, as indicated in Table 13.7, reveal that the portion of majority class instances used in training the one-class GMM significantly affects the AUPRC scores at an $\alpha = 0.01$ significance level. Consequently, to identify the optimal percentage of instances that leads to the best performance, we proceed to conduct Tukey's honestly significant difference (HSD) test [42], which is essential for determining the most effective sample size for model training. This test categorizes factors into groups, assigning them alphabetical labels that indicate their rank. Factors designated as "a" are correlated with the highest AUPRC scores, and this ranking proceeds in reverse order alphabetically with subsequent labels reflecting groups with decreasing AUPRC scores.

The outcome of the Tukey's HSD test, detailed in Table 13.8, reveals that utilizing merely 20% of the training data instances can achieve performance comparable to that achieved using the entire training dataset. Furthermore, it is observed that training the one-class GMM with 25% of the data positions first in Group "a," indicates that this fraction of data results in the highest mean AUPRC score for the model.

Our statistical analysis indicates that using only a fraction of the Part D dataset can result in models achieving performance comparable to those trained on the entire dataset, as measured by two different metrics. This finding is significant, as it suggests the possibility of constructing models that require less time for training due to the reduced volume of training data. Additionally, conducting the analysis across two distinct metrics adds a layer of robustness to our study, reinforcing the reliability of our conclusions.

Now we move on to perform a similar statistical analysis for the results of classifying the Part B data. The next analysis we wish to present is the result of the ANOVA test for the impact of the sample size on the one-class GMMs performance in terms of AUC. The result of the ANOVA test is reported in Table 13.9. The Pr(>F) value in the test result implies that the sample size has a statistically significant impact on AUC scores.

TABLE 13.7 ANOVA for percentage as a factor of one-class GMM's performance in terms of AUPRC in classifying the Part D dataset

	DF	*SUM SQ*	*MEAN SQ*	*F VALUE*	*PR(>F)*
Percentage	5.0000	0.0000	0.0000	3.9500	0.0017
Residuals	294.0000	0.0500	0.0000		

TABLE 13.8 HSD test groupings after ANOVA of AUPRC for the percentage factor for the Part D dataset

Group a consists of:	25, 20, 100
Group ab consists of:	10, 15
Group b consists of:	5

TABLE 13.9 ANOVA for fraction as a factor of performance in terms of AUC in classifying the Part B dataset

	DF	*SUM SQ*	*MEAN SQ*	*F VALUE*	*PR(>F)*
Fraction	5.0000	0.3900	0.0800	8.3400	*
Residuals	294.0000	2.7300	0.0100		

*indicates the value is less than 1×10.

Since the sample size has a statistically significant impact on the AUC score the one-class GMMs yield in classifying the Part B data, we conduct an HSD test to group the levels of the sample size factor. The HSD test result in Table 13.10 is that any sample size yields a better AUC score than using the entire dataset.

Next, we move on to test for the impact of the sample size on the one-class GMM's ability to classify the Part B data in terms of AUPRC scores. In order to determine whether the sample size has a statistically significant impact on AUPRC scores, we conduct an ANOVA test. The result of the ANOVA test is in Table 13.11. The Pr(>F) value reported in Table 13.11 means that the sample size does not have a significant influence over the AUPRC scores in one-class GMM's classification of the Part B data.

Since the result of the ANOVA test implies that the sample size does not have a statistically significant impact on AUPRC scores, this ends our statistical analysis. In conclusion, we find that applying sampling to the test data that we use to train one-class GMM to classify the Part D or Part B data either has no effect on AUC or AUPRC scores, or it yields an improvement. The HSD result for classifying the Part D data in Table 13.8 is the most restrictive. It implies that we can train one-class GMM on a sample of the training data as small as 20% of the original training data, and one-class GMM will yield performance similar to using all the training data. The key takeaway

TABLE 13.10 HSD test groupings after ANOVA of AUC for the fraction factor for the Part B dataset

Group a consists of:	0.10, 0.25, 0.15, 0.05, 0.2
Group b consists of:	1

TABLE 13.11 ANOVA for fraction as a factor of performance in terms of AUPRC in classifying the Part B dataset

	DF	*SUM SQ*	*MEAN SQ*	*F VALUE*	*PR(>F)*
Fraction	5.0000	0.0000	0.0000	2.0600	0.0699
Residuals	294.0000	0.0000	0.0000		

from the statistical analysis is proof that one can save time by training one-class GMM on samples of the training data. It is acceptable to take advantage of the time savings because the performance will be equivalent to or better than using the entire dataset.

13.9 CONCLUSIONS

In this study, we evaluated a novel data reduction technique aimed at enhancing the performance of one-class GMMs in the context of Medicare fraud detection. To our knowledge, this is the second exploration of data reduction methods specifically tailored to address the challenges posed by severely imbalanced big data in conjunction with OCCs. The first exploration was the study we expanded into this book chapter. Our findings demonstrate that reducing the training dataset to as little as 20% of its original size does not significantly affect the performance of one-class GMMs, as measured by both AUC and AUPRC scores. Notably, models built on 20% of the training data exhibit optimal performance. We make this claim based on the fact that, according to our statistical analysis, 20% is the smallest fraction of data that can be used to build models that yield performance that is not significantly different from models built with all of the training data. Moreover, since the models are built with a fraction of the data, there is a reduction in training time. Our results show this statement holds for two highly imbalanced big data datasets: Medicare Part D and Medicare Part B. These conclusions are bolstered by statistical tests, affirming the relationship between the size of the training data and experimental outcomes.

Our results for the Part B and Part D datasets show that sampling may have a significant impact on experimental outcomes in terms of AUC scores, but not a significant impact on AUPRC scores, or vice versa. The absence of an impact is nevertheless positive news. It implies we can train models with the smallest-sized sample of the training data used in experiments, resulting in the shortest training time, and still enjoy performance equivalent to training models with all the training data.

We would like to point out that our methodology can be used as a general technique in large-scale experiments with one-class GMMs to accelerate the pace of experimentation. One can apply our methodology to discover an optimal sample size of the training data to use, in terms of the metric one is interested in. In subsequent experiments, one-class GMMs may be trained on the same sample size of the training data, in less time, hence speeding up the pace of the subsequent experiments. Looking forward, we plan to extend the application of our data reduction approach for anomaly detection in a broader range of big data datasets.

REFERENCES

1. S. Zamost, and C. Brewer, "Inside the mind of criminals: How to brazenly steal $100 billion from Medicare and Medicaid." 2023. [Online]. Available: https://www.cnbc.com/2023/03/09/how-medicare-and-medicaid-fraud-became-a-100b-problem-for-the-us.html

2. Civil Division, U.S. Department of Justice, "Fraud statistics, overview." 2020. [Online]. Available: https://www.justice.gov/opa/press-release/file/1354316/download
3. Centers for Medicare and Medicaid Services, "2019 estimated improper payment rates for Centers for Medicare & Medicaid Services (CMS) programs." 2019. [Online]. Available: https://www.cms.gov/newsroom/fact-sheets/2019-estimated-improper-payment-rates-centers-medicare-medicaid-services-cms-programs
4. F. T. Liu, K. M. Ting, and Z.-H. Zhou, "Isolation forest," in *2008 Eighth IEEE International Conference on Data Mining*, IEEE, 2008, pp. 413–422.
5. R. Bauder, T. M. Khoshgoftaar, and N. Seliya, "A survey on the state of healthcare upcoding fraud analysis and detection," *Health Serv. Outcomes Res. Methodol.*, vol. 17, pp. 31–55, 2017.
6. J. Hancock, and T. M. Khoshgoftaar, "Data reduction to improve the performance of one-class classifiers on highly imbalanced big data," in *2023 22nd IEEE International Conference on Machine Learning and Applications (ICMLA)*, IEEE, 2023, pp. 465–471.
7. J. L. Leevy, J. Hancock, and T. M. Khoshgoftaar, "Comparative analysis of binary and one-class classification techniques for credit card fraud data," *J. Big Data*, vol. 10, no. 1, p. 118, 2023.
8. J. L. Leevy, J. Hancock, and T. M. Khoshgoftaar, "Assessing one-class and binary classification approaches for identifying Medicare fraud," in *2023 IEEE 24th International Conference on Information Reuse and Integration for Data Science (IRI)*, IEEE, 2023, pp. 267–272.
9. N. Seliya, A. A. Zadeh, and T. M. Khoshgoftaar, "A literature review on one-class classification and its potential applications in big data," *J. Big Data*, vol. 8, no. 1, pp. 1–31, 2021.
10. L. Kalantari, P. Gader, S. Graves, and S. A. Bohlman, "One-class gaussian process for possibilistic classification using imaging spectroscopy," *IEEE Geosci. Remote Sens. Lett.*, vol. 13, no. 7, pp. 967–971, 2016.
11. The Centers for Medicare and Medicaid Services, "Medicare part D prescribers – by provider and drug." 2021. [Online]. Available: https://data.cms.gov/provider-summary-by-type-of-service/medicare-part-d-prescribers/medicare-part-d-prescribers-by-provider-and-drug
12. The Centers for Medicare and Medicaid Services, "Medicare physician & other practitioners – by provider." 2021. [Online]. Available: https://data.cms.gov/provider-summary-by-type-of-service/medicare-physician-other-practitioners/medicare-physician-other-practitioners-by-provider
13. M. Bekkar, H. K. Djemaa, and T. A. Alitouche, "Evaluation measures for models assessment over imbalanced data sets," *J. Inf. Eng. Appl.*, vol. 3, no. 10, pp. 28–37, 2013.
14. K. Boyd, K. H. Eng, and C. D. Page, "Area under the precision-recall curve: point estimates and confidence intervals," in *Jt. Eur. Conf. Mach. Learn. Knowl. Discov. Databases*, pp. 451–466, 2013.
15. J. L. Leevy, T. M. Khoshgoftaar, R. A. Bauder, and N. Seliya, "A survey on addressing high-class imbalance in big data," *J. Big Data*, vol. 5, no. 1, pp. 1–30, 2018.
16. T. Hayashi, and H. Fujita, "One-class ensemble classifier for data imbalance problems," *Appl. Intell.*, vol. 52, no. 15, pp. 17073–17089, 2022.
17. I. Czarnowski, "Weighted ensemble with one-class classification and over-sampling and instance selection (WECOI): An approach for learning from imbalanced data streams," *J. Comput. Sci.*, vol. 61, p. 101614, 2022.
18. P. Juszczak, and R. P. Duin, "Uncertainty sampling methods for one-class classifiers," in *Proceedings of ICML-03, Workshop on Learning with Imbalanced Data Sets II*, CiteSeer, 2003, pp. 81–88.
19. M. Z. A. Mayaki, and M. Riveill, "Multiple inputs neural networks for fraud detection," in *2022 International Conference on Machine Learning, Control, and Robotics (MLCR)*, IEEE, 2022, pp. 8–13.

20. LEIE, "Office of inspector general LEIE downloadable databases." 2024. [Online]. Available: https://oig.hhs.gov/exclusions/exclusions_list.asp
21. Z. Salekshahrezaee, J. L. Leevy, and T. M. Khoshgoftaar, "A class-imbalanced study with feature extraction via PCA and convolutional autoencoder," in *2022 IEEE 23rd International Conference on Information Reuse and Integration for Data Science (IRI)*, IEEE, 2022, pp. 63–68.
22. S. L. Cessie, and J. C. V. Houwelingen, "Ridge estimators in logistic regression," *J. R. Stat. Soc. Ser. C Appl. Stat.*, vol. 41, no. 1, pp. 191–201, 1992.
23. L. Breiman, "Random forests," *Mach. Learn*, vol. 45, no. 1, pp. 5–32, 2001.
24. T. Chen, and C. Guestrin, "Xgboost: A scalable tree boosting system," in *Proc. 22nd ACM SIGKDD Int. Conf. Knowl. Discov. Data Min. - KDD '16*, 2016, [Online]. Available: http://dx.doi.org/10.1145/2939672.2939785
25. M. Herland, T. M. Khoshgoftaar, and R. A. Bauder, "Big data fraud detection using multiple Medicare data sources," *J. Big Data*, vol. 5, no. 1, pp. 1–21, 2018.
26. The Centers for Medicare and Medicaid Services, "Medicare physician & other practitioners – by provider and service." 2021. [Online]. Available: https://data.cms.gov/provider-summary-by-type-of-service/medicare-physician-other-practitioners/medicare-physician-other-practitioners-by-provider-and-service
27. The Centers for Medicare and Medicaid Services, "Medicare durable medical equipment, devices & supplies – by referring provider and service." 2021. [Online]. Available: https://data.cms.gov/provider-summary-by-type-of-service/medicare-durable-medical-equipment-devices-supplies/medicare-durable-medical-equipment-devices-supplies-by-referring-provider-and-service
28. D.T. Ha, N. X. Hoang, N. V. Hoang, N. H. Du, T. T. Huong, K. P. Tran "Explainable anomaly detection for industrial control system cybersecurity," *IFAC-Pap*, vol. 55, no. 10, pp. 1183–1188, 2022.
29. J. Platt, "Probabilistic outputs for support vector machines and comparisons to regularized likelihood methods," *Adv. Large Margin Classif.*, vol. 10, no. 3, pp. 61–74, 1999.
30. G. S. Morrison, "Tutorial on logistic-regression calibration and fusion: Converting a score to a likelihood ratio," *Aust. J. Forensic Sci.*, vol. 45, no. 2, pp. 173–197, 2013.
31. M. Kull, T. M. S. Filho, and P. Flach, "Beyond sigmoids: How to obtain well-calibrated probabilities from binary classifiers with beta calibration," *Electron. J. Stat.*, vol. 11, no. 2, pp. 5052–5080, 2017.
32. F. Pedregosa *et al.*, "Scikit-learn: Machine learning in python," *J. Mach. Learn. Res.*, vol. 12, pp. 2825–2830, 2011.
33. A. P. Dempster, N. M. Laird, and D. B. Rubin, "Maximum likelihood from incomplete data via the EM algorithm," *J. R. Stat. Soc. Ser. B Methodol.*, vol. 39, no. 1, pp. 1–22, 1977.
34. J. M. Johnson, and T. M. Khoshgoftaar, "Data-centric AI for healthcare fraud detection," *SN Comput. Sci.*, vol. 4, no. 4, p. 389, 2023.
35. W. McKinney, "Data structures for statistical computing in python," in *Proceedings of the 9th Python in Science Conference*, Austin, TX, 2010, pp. 51–56.
36. J. T. Hancock, T. M. Khoshgoftaar, and J. M. Johnson, "Evaluating classifier performance with highly imbalanced big data," *J. Big Data*, vol. 10, no. 1, p. 42, 2023.
37. L. Prokhorenkova, G. Gusev, A. Vorobev, A. V. Dorogush, and A. Gulin, "CatBoost: Unbiased boosting with categorical features," *Adv. Neural Inf. Process. Syst.*, vol. 31, pp. 1–11, 2018.
38. J. T. Hancock, and T. M. Khoshgoftaar, "CatBoost for big data: An interdisciplinary review," *J. Big Data*, vol. 7, no. 1, pp. 1–45, 2020.
39. J. M. Johnson, and T. M. Khoshgoftaar, "Deep learning and data sampling with imbalanced big data," in *2019 IEEE 20th International Conference on Information Reuse and Integration for Data Science (IRI)*, IEEE, 2019, pp. 175–183.

40. G. Lemaître, F. Nogueira, and C. K. Aridas, "imbalanced-learn: A python toolbox to tackle the curse of imbalanced datasets in machine learning," *J. Mach. Learn. Res.*, vol. 18, no. 17, pp. 1–5, 2017.
41. G. R. Iversen, and H. Norpoth, *Analysis of variance*. Newbury Park: SAGE, 1987.
42. J. W. Tukey, "Comparing individual means in the analysis of variance," *Biometrics*, vol. 5, no. 1, pp. 99–114, 1949.

14 Convolutional Recurrent Deep Q-Learning for Gas Source Localization with a Mobile Robot

Iliya Kulbaka, Ayan Dutta, Ladislau Bölöni,
O. Patrick Kreidl, and Swapnoneel Roy

14.1 INTRODUCTION

A mobile robot with a gas measurement sensor can measure gas concentrations at different locations in an unknown environment. The gas might be leaking from a bomb, drugs, or from a sub-sea pipeline [1–4]. It might be hazardous to send a human to find the gas source [5].

Therefore, using an autonomous mobile robot that can find the source of the gas leak would be a safer option [6, 7].

In this chapter, we explore the gas source localization (GSL) problem from a computational lens, introducing a novel learning algorithm that leverages gas concentration measurements from a robot's onboard sensor array. This approach is designed to be robust across different robot models, ensuring the broad applicability of our methodology. Traditional GSL research often assumes obstacle-free environments [8, 9], which does not reflect the complexity of real-world scenarios. In contrast, our algorithm is

DOI: 10.1201/9781003570882-18

uniquely capable of navigating environments with static obstacles, such as walls, which can significantly alter gas dispersion patterns.

To effectively address these challenges, we employ a deep Q-learning framework that processes the environmental state to guide the robot toward the gas source while minimizing travel distance. Our proposed neural network architecture comprises three convolutional layers paired with two stacked recurrent layers; specifically, long short-term memory (LSTM) modules [10]. The convolutional layers are tasked with extracting relevant spatial features, such as the concentrations of gas across the plane and the robot's previously visited locations, while the LSTM layers analyze the temporal sequence of movements and sensory readings to detect patterns that are critical for effective navigation and localization.

To further enhance our model's robustness and generalizability, we experimented with various neural network configurations. This testing approach involved implementing four distinct architectural setups to evaluate their performance in diverse environmental conditions. Each configuration varied in layer depth, and connectivity patterns to systematically assess how these factors influence the robot's ability to accurately pinpoint the gas source. This comprehensive testing not only validates the efficacy of our proposed solution but also provides insights into the adaptability of neural networks in complex GSL tasks.

We have trained the presented models for 30,000 episodes with random walls, random gas sources, and/or random robot start positions. Note that the robot start locations might have zero gas concentration levels. Once each model was trained, we deployed the trained model on 1,000 randomly generated test cases with various obstacles, robot start locations, and/or gas source configurations. Results show that our proposed technique achieves a high success rate in finding the gas sources in the test environments. An illustrative example of a few test cases at various checkpoints is presented in Figure 14.1. A preliminary version of this chapter has recently appeared as a conference paper at the 2023 International Conference on Machine Learning and Applications (ICMLA) [11].

14.2 RELATED WORK

GSL with an autonomous mobile robot is a practical problem as it can help save humans from hazardous exposures. A state-of-the-art review on robotic GSL can be found in [8]. From the algorithmic point of view, some of the earliest yet still relevant algorithms used chemical gradients to find the source [12]. Algorithms for both indoor and outdoor applications have been developed in the literature. Hutchinson et al. [13] evaluate the plume mapping capabilities of autonomous robots by comparing various mapping algorithms, such as Gaussian process regression, neural networks, and both polynomial and piecewise linear interpolation. Similar to our setting, where the robot (e.g., UAV) can move on a 2D plane in four directions, have been studied in [14]. The authors have proposed an informative path planning

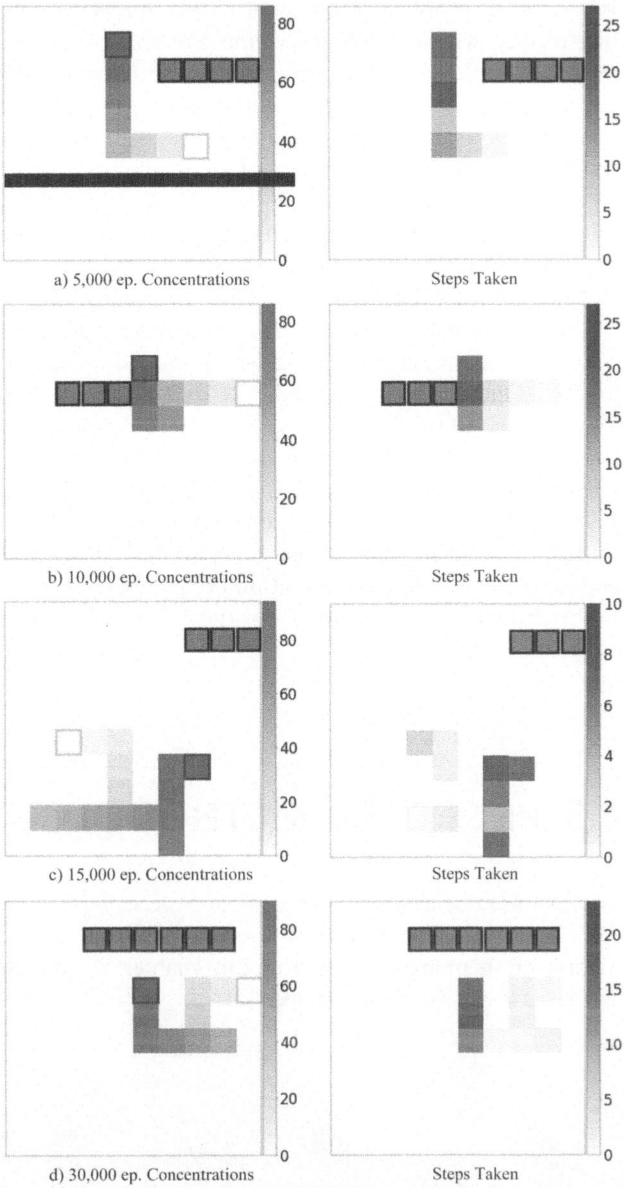

FIGURE 14.1 Gas source localization with a mobile robot using deep Q-learning. Four test samples are shown after (a) 5-, (b) 10-, (c) 15-, and (d) 30,000 episodes with random walls (black cells) present. The white cell with an outline indicates the starting position of the robot. The red cell indicates the source. Robot trajectories are shown with their corresponding measured gas concentration intensities (purple). One can observe that after only 5,000 episodes (top left), the robot takes a longer path to reach the source, whereas it takes the shortest path to reach the target when tested with a 30,000-episode-trained model. The robot's travel path is indicated by shades of orange, with deeper tones being the most recent steps and lighter tones being older steps.

approach to solve the GSL problem. Monroy et al. [15] presents a novel system for GSL using mobile robots, which integrates vision and chemical sensors to enhance detection accuracy in complex environments. In [9], the authors proposed a method that leverages the geometry of indoor environments for propagating local estimations to locate gas sources using a terrestrial mobile robot. Along a similar line, the authors in [16] have proposed the development of an estimation-based route planning (ERP) method to enable a wheeled mobile robot to trace chemical sources in outdoor environments using real-time data on wind conditions and chemical concentrations. In our paper, we imitate this from a machine learning perspective where the history of wind information and robot trajectories are memorized and leveraged using LSTM layers. LSTM has been used extensively in robotics for deep Q-learning applications [17, 18]. Recently, there have been GSL-related studies that used deep learning–based approaches. One of the closest studies to us is due to [1]. We model our gas distribution following their work. However, as suggested in some of the prior work, we incorporate the "memory" of the sensor measurements, unlike [1]. In [19], the proposed method outperforms traditional RL approaches by using domain knowledge to shape the reward mechanism and enhance observational data, which helps to reduce the learning time and improve decision accuracy. Unlike all of the relevant studies mentioned here, we utilize a combination of spatial as well as temporal deep machine learning architectures that learn from the prior gas concentrations, wind directions, and the robot's path.

14.3 PROPOSED METHODOLOGY

The problem model is the same as the conference paper version [11] as well as the base neural network architecture. A pictorial illustration is shown in Figure 14.2. For brevity, any further details on them are omitted from this chapter. For detailed descriptions about them, please refer to our conference paper [11].

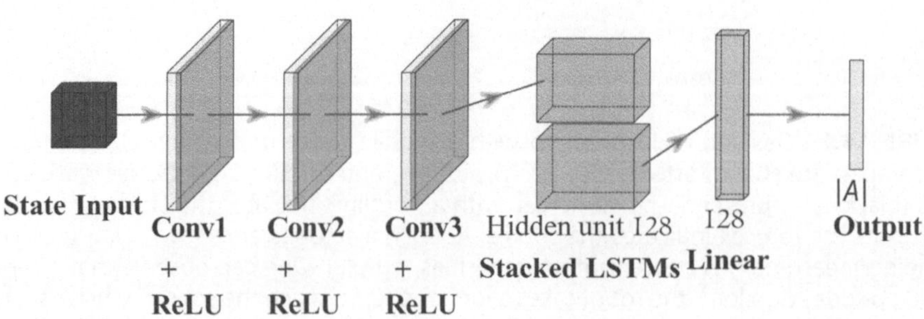

FIGURE 14.2 The proposed convolutional-recurrent neural network architecture.

14.3.1 Neural Network Architectures

We have compared our originally proposed network architecture in [11] to a set of other net- work architectures. These configurations are comprehensively outlined in Table 14.1, which pro- vides a summary of the structural differences among the models.

The network architectures assessed in our ablation studies include four distinct types, labeled as Type 1 through Type 4, each designed to test the effects of varying layer arrangements on the model's ability to process environmental information and pass through meaningful data effectively.

- **Type 1** architecture closely mirrors our proposed model, as described in Section 14.3.1, but with a notable reduction, featuring one fewer LSTM layer. This configuration was chosen to evaluate the impact of reducing temporal processing capacity while maintaining other parameters constant.
- **Type 2** configuration eschews convolutional neural network (CNN) layers entirely, relying solely on two stacked LSTM layers. This design tests the hypothesis that a purely temporal analysis might be sufficient in certain scenarios, particularly where spatial relationships are less critical or more uniform.
- **Type 3** maintains a balanced approach with an equal number of CNN and LSTM layers, specifically two of each. This symmetrical design is intended to assess how evenly distributed processing of spatial and temporal data affects the model's performance.
- **Type 4** consists of a single CNN layer coupled with two LSTM layers, providing a skewed emphasis toward temporal data processing over spatial data integration.

These architectural permutations were carefully constructed to explore how different combinations of CNN and LSTM layers influence the robot's capability to understand and interact with its environment. Each model variant adheres to the same underlying mechanics of deep learning, differing only in the layer configuration. This methodical variation allows us to isolate the effects of each architectural element on the overall effectiveness of the network, particularly in terms of how well information is processed and passed through the system. Training and testing methodologies follow from our conference paper version [11] and are not repeated here due to space constraints.

TABLE 14.1 Descriptions of networks used for performance comparison

LABEL	*CONFIGURATION*	*TRAINABLE PARAMETERS*
Our original (this chapter and [11])	3 CNN + 2 LSTM	238, 260
Type 1	3 CNN + 1 LSTM	106, 164
Type 2	0 CNN + 2 LSTM	403, 972
Type 3	2 CNN + 2 LSTM	334, 212
Type 4	1 CNN + 2 LSTM	495, 732

14.4 EXPERIMENTS AND RESULTS

14.4.1 Settings

The proposed technique is implemented in Python, with the use of the PyTorch library. The main parameters used in the experiments are listed in Table 14.2. The environment size is 10×10 m^2. Similar to our conference paper version [11], we have three environment configurations: (1) *All random*, (2) *No walls*, and (3) *No walls set robot position.* Similar to [11], we have compared against two baselines, namely (1) *Greedy* and (2) *Random*. For more details about these settings, please refer to [11].

14.4.2 Results

We are interested in investigating three main metrics to evaluate the performance of the neural network algorithm: reward (a higher reward indicates a shorter path found), steps (the lower value indicates a more efficient trajectory), and success rate (success percent in locating the gas source).

TABLE 14.2 List of parameters used in our experiments

PRANETERS	*VALUES*
State	4 × 10 × 10 tensor
Action	Up, Down, Left, Right
Number of training episodes	30, 000
Episode length	100
Priority replay memory size	20, 000
Mini-batch size	32
Sample length	20
Discount factor	0.90
Learning rate	0.0001
Target network update frequency	20
Epsilon decay type	Linear
Epsilon decay rate	0.0002
Epsilon start value	1.0
Epsilon end value	0.001
Loss function	Mean Square Error
Optimizer	Adam
Number of testing episodes	1,000

14.4.2.1 Training

The training results are presented in Figure 14.3. As can be seen, the proposed technique rapidly increases its efficiency concerning the episodic reward, which steadily goes up before converging and the number of steps gradually decreases before it almost converges. It can be seen that all networks regardless of configurations follow relatively

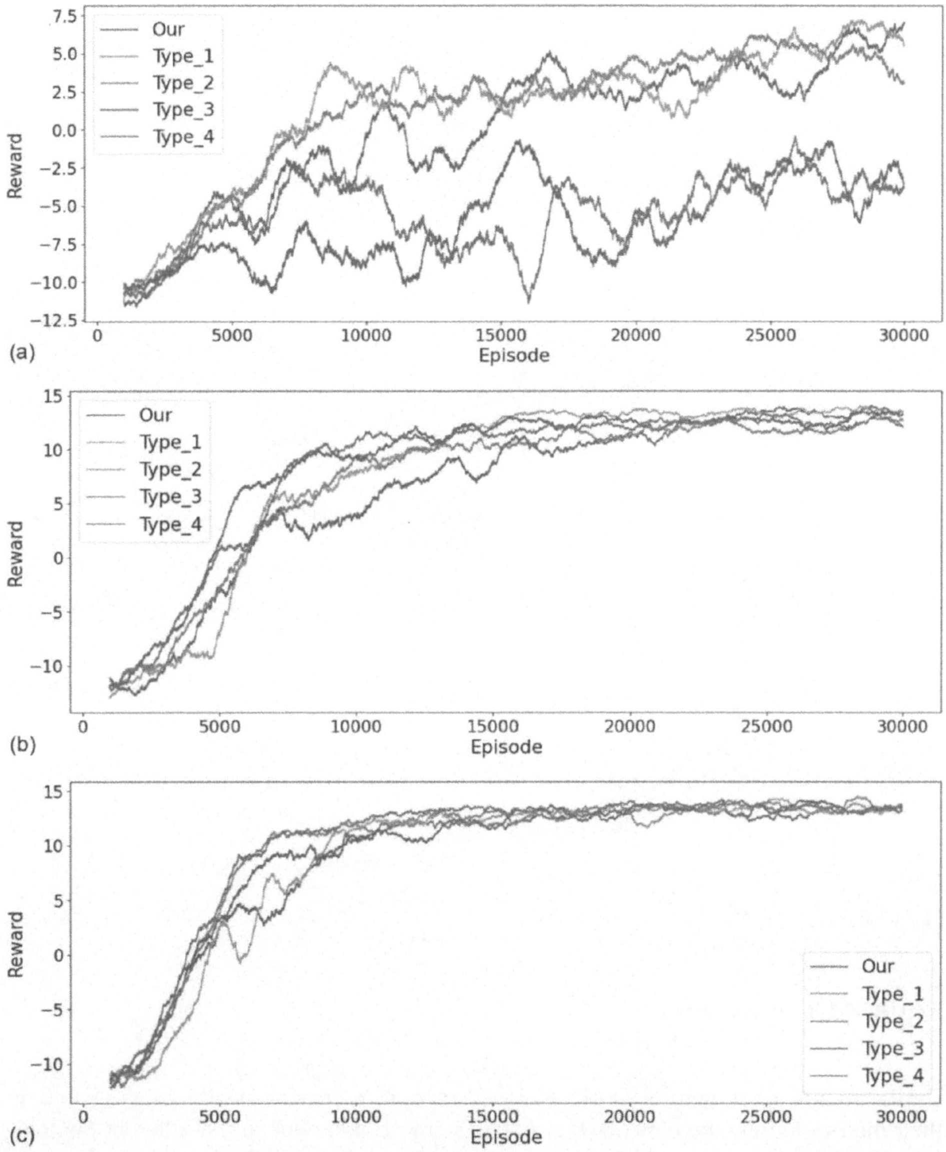

FIGURE 14.3 Training plots (1,000-episode rolling average) for reward and steps, covering all model types. All random, (b,e): No walls, (c,f): No walls, set robot position. (Continued)

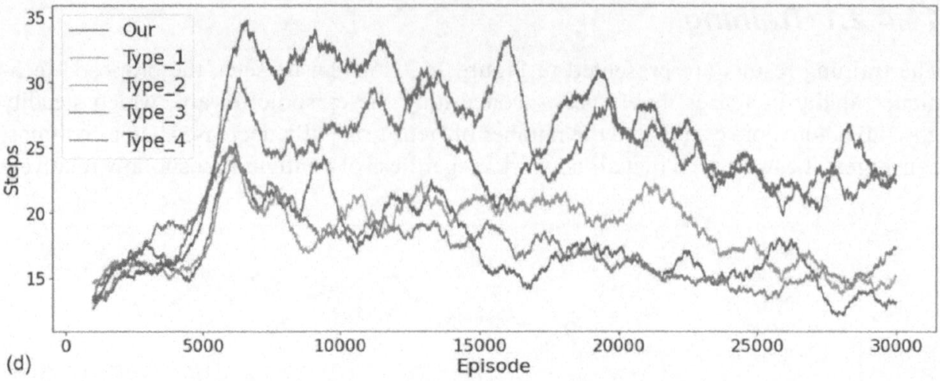

(d)

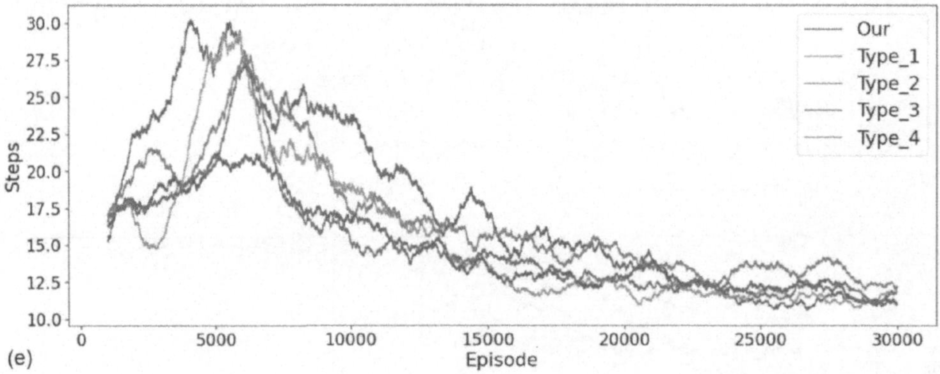

(e)

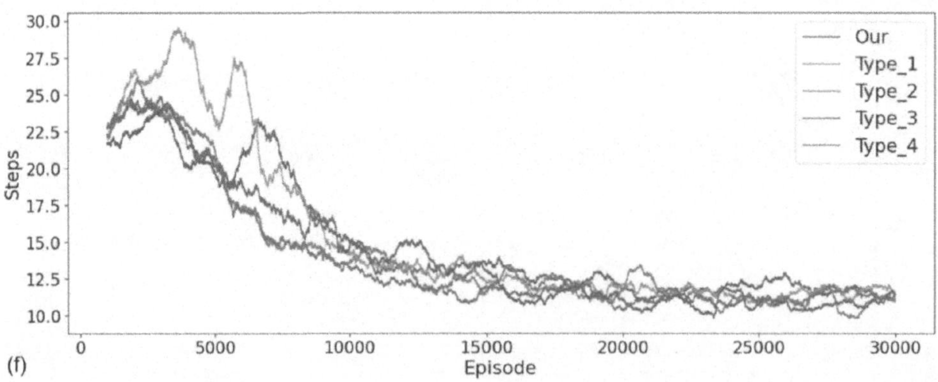

(f)

FIGURE 14.3 (Continued)

similar trends, as shown in Figure 14.3. All random scenarios are the most diverse in their metrics among the network types. This is most likely due to the inherent randomness of the environment, which impacted each network case differently. At around episode 5,500, ϵ value of 0.001, resulting in all of the actions taken past episode 5,500 to primary driven by the proposed neural network.

14.4.2.2 Testing

The testing results are presented in Figures 14.4–14.6. These plots show the improvements provided by our proposed framework over the baselines in the three main metrics. The data presented in Tables 14.3–14.5 show averages for testing performance in three testing environments along with standard deviation. Individually, the results can be misleading; for example, in Table 14.4, the random baseline and our proposed algorithm have similar average steps per environment configuration, however, this alone does not draw a meaningful conclusion. When combined with the reward and success rate tables (Tables 14.3 and 14.5), we can understand that the random movements might not have led the robot to the gas source. Overall, across all configurations, our proposed algorithm achieves a success rate of 91.67% in finding the gas source, whereas the random and the greedy baselines yield 22.33% and 44.33%, respectively.

14.4.2.3 Discussion of the ablation test results

Comparing our proposed model with the other four model types reveals a complex landscape of performance nuances that go beyond straightforward comparison metrics. Most importantly, the testing phase exposed significant disparities in their performance levels, indicating that similar training trajectories do not necessarily translate to comparable real-world efficacy.

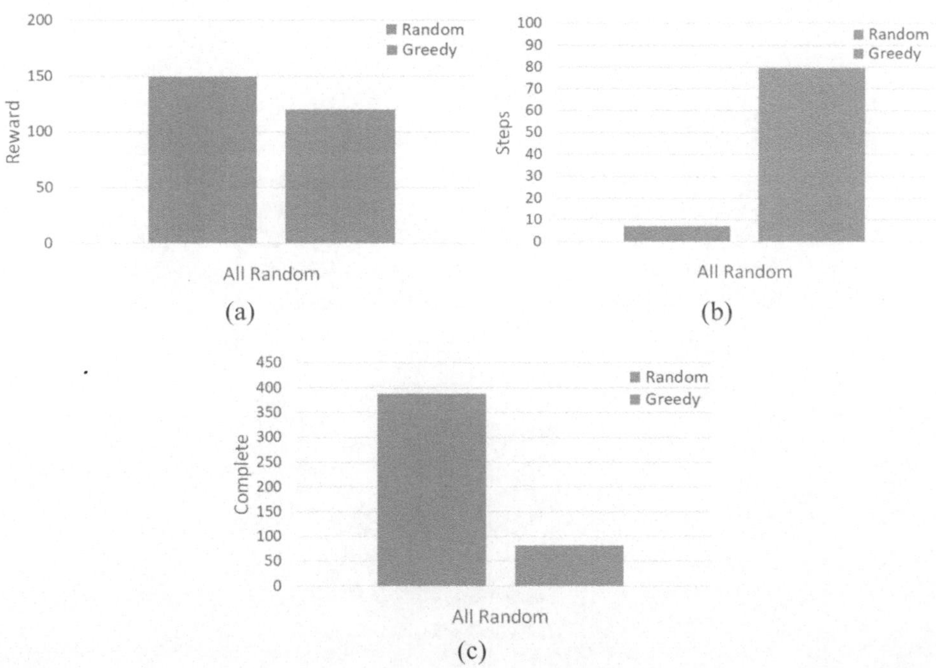

FIGURE 14.4 Improvement of performance metrics (%) by our model on the baselines: All random, (b,e): No walls, (c,f): No walls, set robot position.

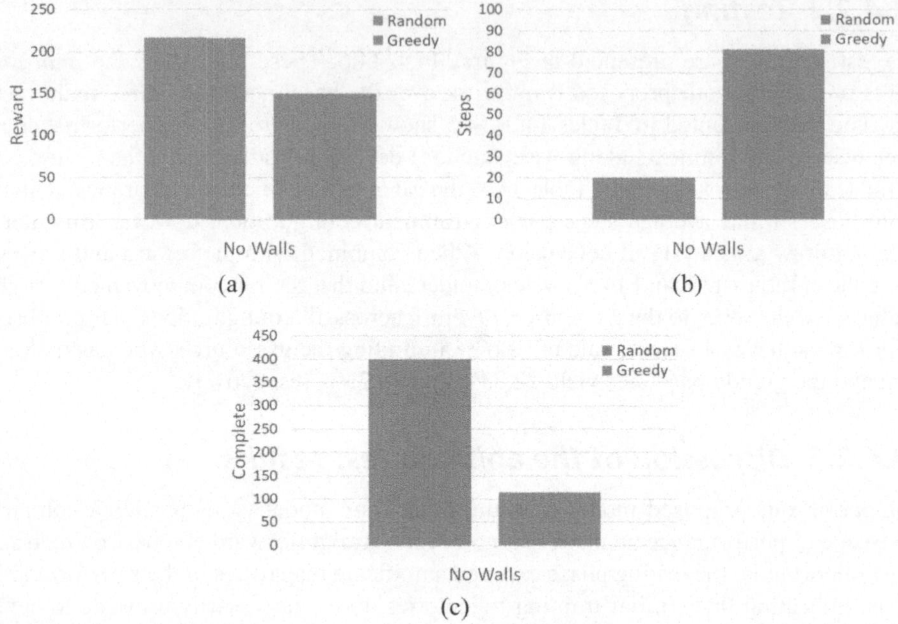

FIGURE 14.5 Improvement of performance metrics (%) by our model on the baselines: (a)–(c) No walls case.

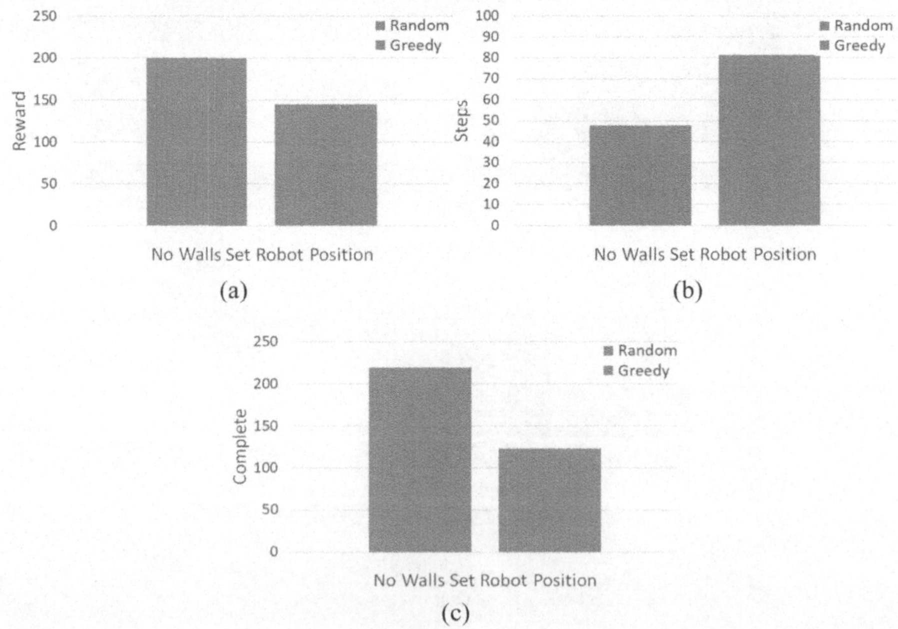

FIGURE 14.6 Improvement of performance metrics (%) by our model on the baselines: (a)–(c) Set robot position case.

TABLE 14.3 Average reward gained during 1,000-episode testing on three different configurations

	ALL RANDOM	*NO WALLS*	*SET ROBOT POSITION*
Random	−12 ± 13.31	−12 ± 14	−13 ± 18.57
Greedy	−30 ± 34.48	−28 ± 34.76	−29 ± 35.02
Type 1	**8 ± 17.54**	14 ± 7.49	14 ± 6.46
Type 2	2 ± 20.76	13 ± 10.32	14 ± 6.73
Type 3	−4 ± 25.91	13 ± 9.64	**14 ± 8.58**
Type 4	−4 ± 25.05	12 ± 10.54	13 ± 9.37
Ours	6 ± 16.90	**14 ± 8.40**	13 ± 9.71

TABLE 14.4 Average steps during 1,000-episode testing on three different configurations

	ALL RANDOM	*NO WALLS*	*SET ROBOT POSITION*
Random	14 ± 16.82	15 ± 17.64	23 ± 20.65
Greedy	64 ± 42.49	63 ± 42.68	64 ± 42.48
Type 1	14 ± 19.27	**11 ± 10.01**	11 ± 8.79
Type 2	18 ± 21.83	12 ± 12.80	**10 ± 8.91**
Type 3	24 ± 29.77	12 ± 11.45	11 ± 10.48
Type 4	23 ± 28.95	12 ± 12.64	11 ± 11.37
Ours	**13 ± 14.38**	14 ± 8.40	12 ± 11.72

TABLE 14.5 Average success rate (%) during 1,000-episode testing on three different configurations

	ALL RANDOM	*NO WALLS*	*SET ROBOT POSITION*
Random	16	20	31
Greedy	43	46	44
Type 1	**83**	99	**99**
Type 2	70	97	99
Type 3	60	97	98
Type 4	59	97	97
Ours	78	**99**	98

During the rigorous testing phase, particularly in the "All Random" environment configuration – known for its complexity and unpredictability – it became apparent that not all models performed equally. Specifically, Types 3 and 4 models exhibited notably inferior performance, which was a critical insight. These models, which featured fewer convolutional layers, struggled significantly. The reduction in CNN layers led to a deficient extraction of spatial features, which are crucial for the subsequent LSTM layers to process temporal dependencies effectively. This inadequate feature extraction compromised the overall information flow within the network, leading to a partial and often inaccurate understanding of the environment.

In stark contrast, the Type 2 model, which entirely lacked CNN layers and relied solely on stacked LSTM layers, performed remarkably well. Quantitatively, the Type 2 model outperformed Types 3 and 4 by substantial margins: it was 342%, 35%, and 15% better than Type 3, and 348%, 30%, and 17% better than Type 4 in terms of average reward, steps taken to completion, and completion rate, respectively, in the "All Random" configuration. This suggests that a purely temporal approach, unencumbered by potentially misleading spatial preprocessing, can be more effective in environments where spatial relationships are less predictable and more complex to model accurately. At the same time, this network has almost double the number of trainable parameters than our proposed neural network.

Interestingly, the Type 1 configuration, which reduced the LSTM component to just one layer, showed improved performance over our proposed multi-LSTM approach. The enhancements were noticeable: there was a 20% increase in average reward, a 9% increase in steps taken, and a 6% increase in completion rate compared to our model. This configuration suggests that a single LSTM layer might be more adept at processing direct outputs from the CNN layers. The potential redundancy of a second LSTM layer could lead to a dilution of effectiveness, where the "secondhand" interpretations from an additional LSTM layer do not contribute to, and might even detract from, the model's ability to integrate and act upon environmental data effectively. Interestingly, this network has the least number of trainable parameters. Therefore, its light weight is an added advantage.

When comparing results for less complex environment configurations – such as "No Walls" and "Set Robot" position – the models performed relatively similarly, achieving completion rates in the high nineties, with similar average reward and step metrics. This consistency suggests that the reduced complexity allowed the models to fully understand and respond to environmental patterns, underscoring the critical role of environmental context in model performance.

These findings underscore the importance of tailored neural network configurations to specific environmental challenges. The variations in performance across different model types highlight the need for a deliberate and context-aware approach in neural network design, especially in complex and dynamic settings such as those encountered in GSL tasks. Our analysis not only demonstrates the varied capabilities of different neural network architectures but also guides future designs and improvements in robotic sensory networks.

14.5 CONCLUSION AND FUTURE WORK

In this paper, we address the problem of localizing a gas source using a mobile robot equipped with sophisticated sensory tools including gas sensors (e.g., an electronic nose) and anemometers to measure both the concentration of gases and the wind parameters such as speed and direction. Our approach centers on a novel deep recurrent Q-learning model that autonomously develops a navigation policy through training.

The versatility and robustness of our trained model were rigorously evaluated in simulated environments characterized by variables such as random obstacle placements, varying starting positions for the robot, and unpredictable GSLs. Our experimental results demonstrate that the proposed deep recurrent Q-learning strategy not only consistently locates the gas source but also significantly surpasses traditional methods such as gradient ascent and random-walk strategies in terms of success rate.

A comparative analysis of different neural network configurations further enriched our understanding. Our primary model incorporated multiple convolutional and LSTM layers, facilitating detailed spatial and temporal analysis crucial for navigating and adapting to dynamic environmental conditions. We found that models with fewer CNN layers struggled, particularly in environments with high randomness, where comprehensive spatial processing is essential for successful localization. Conversely, the model devoid of CNN layers and relying solely on LSTM stacks excelled under the same conditions, suggesting that for certain environmental complexities, focusing on temporal data processing might be more effective.

Meanwhile, the model that simplified the LSTM structure to a single layer unexpectedly outperformed more complex configurations. This indicates that in scenarios where immediate sensor outputs are more relevant to decision making, a streamlined neural architecture can prevent the overfitting and redundancy that sometimes plague more intricate networks. Our ambitious future objective is to develop a universally robust deep reinforcement learning–enabled robotic system capable of effective GSL under any environmental conditions.

FUNDING

This work is supported in part by NSF CPS Grants #1932300 and #1931767.

REFERENCES

1. X. Chen, C. Fu, and J. Huang, “A deep Q-network for robotic odor/gas source localization: Modeling, measurement and comparative study,” *Measurement*, vol. 183, p. 109725, 2021.
2. M. Hutchinson, C. Liu, P. Thomas, and W.-H. Chen, “Unmanned aerial vehicle-based hazardous materials response: Information-theoretic hazardous source search and reconstruction,” *IEEE Robotics & Automation Magazine*, vol. 27, no. 3, pp. 108–119, 2019.
3. M. Mohan, T. Panwar, and M. Singh, “Development of dense gas dispersion model for emergency preparedness,” *Atmospheric Environment*, vol. 29, no. 16, pp. 2075–2087, 1995.

4. C. Liu, Y. Liao, S. Wang, and Y. Li, "Quantifying leakage and dispersion behaviors for sub-sea natural gas pipelines," *Ocean Engineering*, vol. 216, p. 108107, 2020.
5. J. Huo, M. Liu, K. A. Neusypin, H. Liu, M. Guo, and Y. Xiao, "Autonomous search of radioactive sources through mobile robots," *Sensors*, vol. 20, no. 12, p. 3461, 2020.
6. M. A. Arain, V. Hernandez Bennetts, E. Schaffernicht, and A. J. Lilienthal, "Sniffing out fugitive methane emissions: Autonomous remote gas inspection with a Mobile robot," *The International Journal of Robotics Research*, vol. 40, no. 4–5, pp. 782–814, 2021.
7. W. Rahmaniar, and A. Wicaksono, "Design and implementation of a mobile robot for carbon monoxide monitoring," *Journal of Robotics and Control (JRC)*, vol. 2, no. 1, pp. 1–6, 2021.
8. A. Francis, S. Li, C. Griffiths, and J. Sienz, "Gas source localization and mapping with mobile robots: A review," *Journal of Field Robotics*, vol. 39, no. 8, pp. 1341–1373, 2022.
9. P. Ojeda, J. Monroy, and J. Gonzalez-Jimenez, "Information-driven gas source localization exploiting gas and wind local measurements for autonomous mobile robots," *IEEE Robotics and Automation Letters*, vol. 6, no. 2, pp. 1320–1326, 2021.
10. S. Hochreiter, and J. Schmidhuber, "Long short-term memory," *Neural Computation*, vol. 9, no. 8, pp. 1735–1780, 1997.
11. I. Kulbaka, A. Dutta, L. Bölöni, O. P. Kreidl, and S. Roy, "CNN-LSTM-based deep recurrent q-learning for robotic gas source localization," in *2023 International Conference on Machine Learning and Applications (ICMLA)*, pp. 1060–1065, IEEE, 2023.
12. R. Rozas, J. Morales, and D. Vega, "Artificial smell detection for robotic navigation," in *Proc. of Fifth Int. Conf. on Advanced Robotics*, pp. 1730–1733, 1991.
13. M. Hutchinson, P. Ladosz, C. Liu, and W.-H. Chen, "Experimental assessment of plume mapping using point measurements from unmanned vehicles," in *Proc. of Int. Conf. on Robotics and Automation (ICRA-2019)*, pp. 7720–7726, 2019.
14. F. Rahbar, A. Marjovi, and A. Martinoli, "An algorithm for odor source localization based on source term estimation," in *Proc. of Int. Conf. on Robotics and Automation (ICRA-2019)*, pp. 973–979, 2019.
15. J. Monroy, J.-R. Ruiz-Sarmiento, F.-A. Moreno, F. Melendez-Fernandez, C. Galindo, and J. Gonzalez-Jimenez, "A semantic-based gas source localization with a mobile robot combining vision and chemical sensing," *Sensors*, vol. 18, no. 12, p. 4174, 2018.
16. J.-G. Li, M.-L. Cao, and Q.-H. Meng, "Chemical source searching by controlling a wheeled mobile robot to follow an online planned route in outdoor field environments," *Sensors*, vol. 19, no. 2, p. 426, 2019.
17. J. Orr, and A. Dutta, "Multi-agent deep reinforcement learning for multi-robot applications: A survey," *Sensors*, vol. 23, no. 7, p. 3625, 2023.
18. T. Said, J. Wolbert, S. Khodadadeh, A. Dutta, O. P. Kreidl, L. Bölöni, and S. Roy, "Multi-robot information sampling using deep mean field reinforcement learning," in *2021 IEEE International Conference on Systems, Man, and Cybernetics (SMC)*, pp. 1215–1220, IEEE, 2021.
19. T. Wiedemann, C. Vlaicu, J. Josifovski, and A. Viseras, "Robotic information gathering with reinforcement learning assisted by domain knowledge: An application to gas source localization," *IEEE Access*, vol. 9, pp. 13159–13172, 2021.

Conditioned Cycles in Sparse Data Domains

Applications to the Physical Sciences

15

Maria Barger, Randy Paffenroth, and Harsh Pathak

15.1 INTRODUCTION

This book chapter is an extended version of our previous paper, which was published at the IEEE ICMLA conference in December of 2023. Machine learning algorithms require large amounts of data to train, yet it is often expensive and time-consuming to gather data in the physical sciences. As such, many problems in machine learning for the physical sciences suffer from small training data.

Cycles – similar to those of CycleGANs [1–4] and in a paper from which we drew inspiration [5] – and conditioning have been shown to improve machine learning prediction accuracy [6–9], but less literature exists regarding the application of a combination of these methods to low-data problems within the physical sciences. In our work, we employ the idea of "conditioning," which is common in statistical literature [10], by enforcing that important physical parameters explicitly appear as components of the hidden layers of autoencoders. *The idea of conditioning the hidden layers of autoencoders with important physical parameters is one of the key contributions of our work and is detailed in this chapter.*

DOI: 10.1201/9781003570882-19

Another important element of our work is the utilization of iterations, where a single neural network is designed to have the same size input and output (regardless of the desired size of prediction compared to predictor) such that the neural network can be applied iteratively to its output multiple times and refine its prediction. This concept of iteration is an extension of our work with cycles and is inspired by both recurrent neural networks (RNNs) and the mathematical field of dynamical systems. We show that the use of iterations will often improve prediction performance on real-world problems.

In our work, we address two challenging low-data problems and measure performance of our regressors both with and without cycles, conditioning, and iterations. Much of our work in this chapter is inspired by and uses language based upon another paper [5]. Figure 15.1 shows the neural network structure we utilize in many of our experiments, highlighting the cyclic nature of our training process and the range conditions imposed upon our hidden layer. We also contribute the addition of "oracle" conditioning, a method of conditioning in training that allows for faster training convergence. Our ultimate goal is the production of neural networks capable of producing accurate testing predictions when only sparse real-world data is available.

Our first real-world application concerns the source localization of over-water electromagnetic (EM) signals collected by a passive receiver. Similar work has been produced as in our previous paper [11], but our work herein is generalizable to other domains. While we discuss our application to contextualize the results shown in this chapter, it is important to note that these concepts and loss functions can be applied across many problems in the physical sciences with specific physical constraints. Our

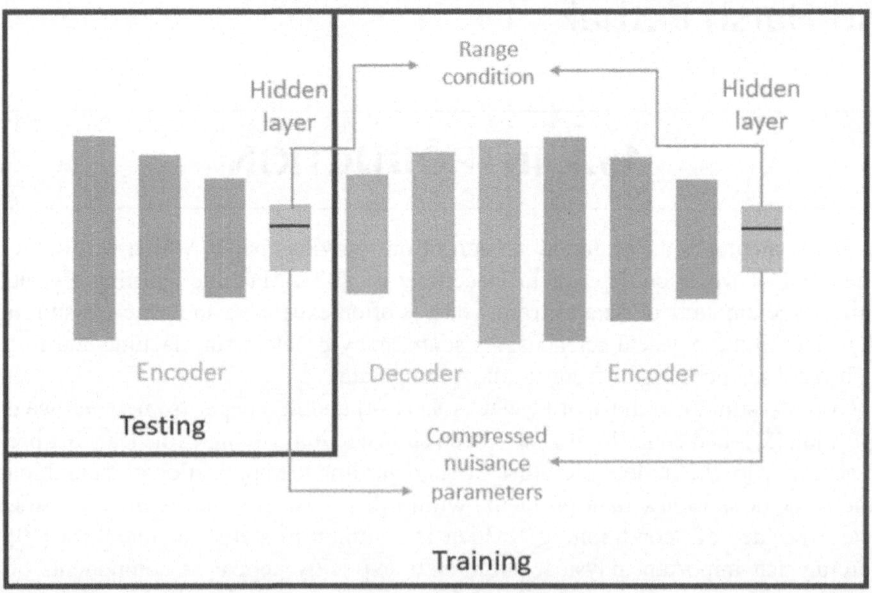

FIGURE 15.1 The novel neural network structure we propose. We implement cycles with oracle conditioning in the hidden layer to improve final performance. Note that cycles are implemented in the training phase but not the testing phase.

loss functions are written to be generalizable to such problems. To demonstrate the generalizability of our loss functions and theory, we also include results based upon a lunar lander simulation. When analyzing the lunar lander data, we attempt to distinguish two piloting neural networks from each other based only upon the trajectory coordinates.

15.2 OUR APPLICATION

We focus on two applications: EM source localization and synthetic lunar lander pattern of life analysis.

15.2.1 Electromagnetic Source Localization

Our first real-world application concerns determining the distance between an over-water EM signal emitter and a passive receiver, based only on the received signal. This problem is not solvable *a priori* but can be solved when the EM signals are trapped by layers of humidity known as evaporation ducts [12]. Without evaporation ducts, we cannot distinguish between a strong signal emitted far from the receiver and a weak signal emitted near the receiver. The existence of evaporation ducts causes EM signals to propagate farther than the horizon and impacts the signal propagation such that similar signals emitted from different distances are distinguishable. Figure 15.2 demonstrates an EM signal propagating in an evaporation duct.

Our datasets consist of EM signals paired with corresponding ranges. As input, neural networks in this chapter utilize the EM signals and produce predictions of the range from emitter to receiver. Figure 15.3 shows the 10 m tall array used to collect the EM signals in our dataset. The array collects 20 signal measurements in 3.5 seconds, yielding 20 signals which correspond to the same range in the dataset.

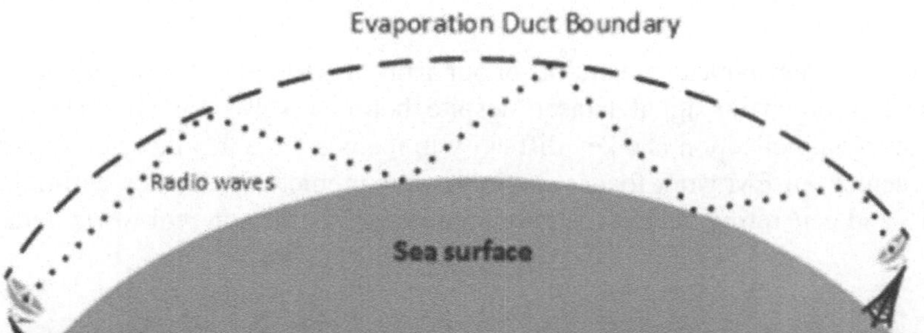

FIGURE 15.2 A radio signal propagating through an evaporation duct [13]. The duct traps EM signals, ensuring they propagate father than normal and allowing us to differentiate between similar signals emitted from different distances.

FIGURE 15.3 An image of the 10 m tall array used in our data collection [14]. The array holds 40 sensors, each collecting 20 measurements in 3.5 seconds, and must be erected by a crane. The mechanical difficulty of data collection limits our collection capabilities and consequently the size of our data.

Due to the mechanical constraints of our array, it is time-consuming and financially expensive to collect our signal datasets despite their relatively small size. The expensive nature of signal collection is a key difficulty in many source localization problems. As such, much of our EM work focuses on increasing prediction accuracy within low-data domains and generating data to increase prediction accuracy on real-world data.

15.2.2 Lunar Lander Pattern of Life Analysis

Our second application involves work analyzing unmanned aerial vehicle (UAV) trajectories. Our pattern of life analysis is meant to serve as a proxy for real-world drone trajectory analysis. The ultimate question we wish to answer in our UAV work is: based upon only trajectory data, can our machine learning models accurately differentiate between two different piloting softwares? As we do not currently have access to real-world drone data, we have utilized trajectories we generated via the HuggingFace Lunar Lander setting in the OpenAI gym [15]. The lunar lander problem is meant to serve as a proxy for real-world drone trajectories and future work within the project will utilize real-world data for the classification task.

To accomplish our pattern of life analysis task, we make use of two open-source models trained to pilot the lunar landers [16, 17]. We refer to these models as "Company A" and "Company B" in our analysis. Both models are able to pilot the lunar landers from varying starting points in randomized terrains and land safely. To a human eye, it is difficult to distinguish between trajectories from Company A and Company B. As such, this task contains a level of ambiguity and will be moderately nontrivial for a neural network to solve.

15.3 OUR CONTRIBUTIONS

Our key contributions are:

- A novel neural network structure for addressing small data problems in the physical sciences, as shown in Figure 15.1,
- Application of "oracle" conditioning for cyclic autoencoder training to accelerate and regularize the training process,
- Application of iterative neural networks to improve prediction performance,
- Demonstration of these ideas in a challenging over-water EM source localization problem and trajectory pattern of life analysis, and
- An example of how such methods can be leveraged in a wide range of small data problems in the physical sciences.

15.4 RELEVANT CONCEPTS

Herein we introduce six important concepts for our experiments which have produced noteworthy testing performance: autoencoders [18], cycles [4], iteration [19], continuation [20], conditioning [6], and oracle conditioning. Our work is motivated by the ultimate goal of improved prediction accuracy in low-data environments. This chapter demonstrates the effectiveness of our methods at improving prediction accuracy in low-data domains.

As a basis for comparison and to demonstrate the benefits of cycles and oracle conditioning, we provide an example of a "vanilla" regression network adapted from a previous paper [11] lacking both cycles and conditioning. The regressor is constructed with the same architecture as the multitask encoders utilized in our cycles and conditioning experiments to ensure that the computational differences arise from different training methodology rather than architectural differences between the neural networks.

15.4.1 Autoencoders

Autoencoders are neural networks composed of two parts: an encoder and a decoder. A piece of data will be used as input for the encoder, which compresses the data into a lower dimension known as the hidden layer of the autoencoder. This compressed data is then used as input for the decoder which returns the data to its original size. While autoencoders are often used as denoising neural networks, one may also use the decoder of an autoencoder as a data generator [18].

15.4.2 Cycles

Cycles [4] are regularizers applied for better predictions with small datasets and sparse data. Our cycles are modeled after the cycles in another paper [4], commonly known as CycleGAN. Similar cycles have been used to improve training efficiency and denoise data [9] and frame prediction in videos [8]. We implement cycles by developing an encoder-decoder-encoder structure where the first and second encoders are the same neural network.

By employing cycles, we create multitask encoders and evaluate multiple performance metrics in our loss function simultaneously. Figure 15.1 illustrates the structure of our cycle neural networks.

15.4.3 Iterations

Our work with cycles led us to consider iterating neural networks more broadly. Inspired by dynamical systems, RNNs, and the "stable diffusion" models which have become increasingly popular in the field of machine learning, we construct what we consider "iterative neural networks": neural networks which repeat the cycle multiple times instead of once. The connection between traditional neural network perspectives, dynamical systems, and RNNs is explored further in extant literature [19, 21–23]. We consider an *n-iterative neural network* to be a neural network whose architecture iterates n times for some integer n. Figure 15.4 shows an example of a three-iterative neural network.

There are many possible loss functions for an iterative neural network. One can consider utilizing only the final prediction outcome in the loss function, somewhat akin to the process of reinforcement learning. Alternately, one can consider utilizing the sum of all predictions at the interim and final steps. Finally, one can consider utilizing a weighted sum of all predictions where less weight is assigned to earlier predictions in the iterative process while more weight is assigned to later predictions. We have implemented all three loss functions and have found the best results when utilizing the third

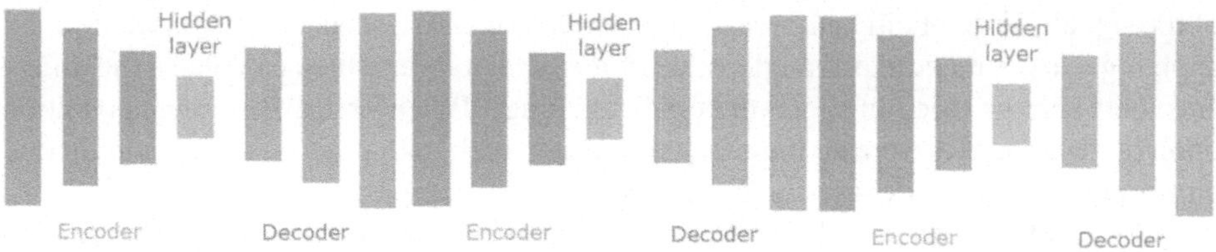

FIGURE 15.4 An image of a three-iterative neural network. While the neural network shown in **Figure 15.1** can be considered a "1.5-iterative neural network" we do not consider it to be a true iterative neural network as the entire architecture does not iterate. The current image is a three-iterative neural network as the entire neural network architecture iterates three times.

possible function. As such, the third loss function will be the only iterative loss function discussed in depth in this chapter.

15.4.4 Continuation

Generally, "continuation" methods refer to methods of extending a manifold or moving from a simple manifold to the complicated manifold of interest to find the global minimum while avoiding suboptimal local minima [20, 24]. In our work, we utilize the spirit of this "continuation" term and that of natural parameter continuation [25] and use it to refer to moving from a simple loss function to a more complicated loss function. When training our iterative neural networks, we utilize continuation in our loss function by dividing our training epochs into sections determined by the number of final iterations we wish our neural network to use. For example, for a three-iterative neural network training for n total epochs, we would train using only the first iteration's loss for the first $\lfloor \frac{n}{3} \rfloor$ epochs. From epochs $\lfloor \frac{n}{3} \rfloor + 1$ to $\lfloor \frac{2n}{3} \rfloor$, we train using the first and second iterations' loss terms. For epochs $\lfloor \frac{2n}{3} \rfloor + 1$ and beyond, we train with loss terms calculated based upon all three iterations. These continuation loss functions can easily be adapted for neural networks with more or less iterations by simply altering the fraction denominator and ensuring the iterations are trained sequentially.

In our work, we have found that continuation is a powerful tool which significantly improves performance and allows for convergence in cases where predictions do not converge without continuation.

15.4.5 Conditioning

Conditioning is a well-known technique for training neural networks for data generation [26]. We use the term "conditioning" to refer to a method inspired by conditional probabilities in classical statistics. In many problems in the physical sciences, there are numerical constraints important to the dataset. In our EM example, our dataset is composed of signals and ranges where the ranges provide a constraint upon the appearance of the signal. We explicitly enforce a range prediction into our encoder's hidden layer by applying a loss function to a single value of the hidden layer which calculates the difference between the true range and the predicted range. By enforcing this range condition upon our encoders, we create multitask encoders which both produce a range prediction and compress real-world signals for the decoder to decompress.

Conditioning can also be used to improve performance when enforcing restrictions upon neural network hidden layers [6, 7, 27, 28]. Due to the periodic nature of EM signals and the nature of our dataset itself, where 20 signals correspond to the same range, conditioning is useful for unique signal reconstruction from the hidden layers. We consider our hidden layers to be a combination of the conditioning parameter and a compression of the nuisance parameters for each signal. Without conditioning, our neural networks would not necessarily produce useful signals for given ranges. Similarly, our neural networks would be unable to reconstruct multiple signals for a single given condition without the compression of the nuisance parameters. By combining conditioning and the compression of nuisance parameters, we are able to reconstruct multiple physically possible signals at any given range.

15.4.6 Oracle Conditioning

Oracle conditioning is similar to standard conditioning except that in training it incorporates the true value of the conditions rather than relying on the encoder to produce good condition predictions (i.e., an "oracle" provides the correct condition to the decoder in training). The input to the decoder for oracle conditioning will be the compressed form of the encoder's input and a true condition distinct from the condition prediction produced by the encoder. While this is an example of data snooping [29, 30] in training, we do not utilize oracle conditioning in testing, thereby avoiding data snooping during our final evaluations.

Theoretically, standard conditioning will converge to approximately the same performance as oracle conditioning with sufficient training. Oracle conditioning removes the reliance upon the encoder producing good condition predictions, especially at early stages in training when the encoder's performance is poor. This leads to better test performance with shorter training times.

15.5 ELECTROMAGNETIC EXPERIMENTAL DESIGN

To artificially create a lower-data section in our EM training dataset, we select a 10 km segment of our dataset and remove 90% of the training data. Removing the data in such a way leaves a mere 300 training points in the 10 km data gap. As such, our training data consists of roughly 10,000 training points outside of the 10 km data gap and only

300 points within the gap. Our testing data consists of 2,500 points outside of the gap and 2,700 points within the gap.

The 10 km low-data gap is designed to simulate a low-data environment such as one where sensors intermittently fail or signals are only intermittently transmitted.

To ensure fair comparisons between the five experiments, we train our neural networks for 110,000 epochs each.

By implementing cycles and iterations, we can evaluate three factors simultaneously: our encoder's range prediction performance from real-world signals, our decoder's reconstruction of compressed real-world signals, and our encoder's range prediction performance based upon these signal reconstructions. When writing loss functions, we separate our evaluation factors into three terms, labeled A, B, and C. Term A always represents the initial prediction loss of the multitask encoder's prediction based upon the real-world signal. Term B always represents the decoder's reconstruction loss based upon the encoder's compressed form of the signal and some form of conditioning upon range. Term C always represents the encoder's final range prediction loss based upon the decoder's signal reconstruction.

15.5.1 Experiment 1: Unconditioned Regressors

The neural network architecture for our unconditioned regressor is adapted from the best-performing regressor in our previous work [11]. We modify the architecture such that the regressor becomes a regressor-encoder, producing a compressed form of nuisance paramters as well as a regression prediction. Let $\mathbf{x}_i$ be a real-world signal with $\tilde{\mathbf{x}}_i$ as the same signal with added noise. Let $E(\tilde{\mathbf{x}}_i)$ be our regressor-encoder. E produces two outputs, called $\hat{\mathbf{z}}_i$ and $\hat{\mathbf{c}}_i$, which are the compressed nuisance parameters and the range condition, respectively. Our unconditioned regression loss function is the mean-squared error (MSE) loss between the true range and the regressor's predicted range, shown in Equation 15.1.

$$
\begin{aligned}
&Loss = \frac{1}{n}\sum_{i=1}^{n} \\
&A : \left\|\mathbf{y}_i - \hat{\mathbf{c}}_i\right\|_F^2
\end{aligned}
\qquad (15.1)
$$

Note that we utilize the squared Frobenius norm, denoted $\| \mathbf{x} \|_F^2$ in our loss function.

The latent spaces of noncyclic, unconditioned neural networks often display useful information regarding the data encodings. Our unconditioned neural networks often produce hidden layers, such as in Figure 15.5, where multiple ranges could be assigned to a point in the hidden layer based upon differing metrics. Regressors with hidden layers lacking a well-defined method of assigning range to encoded points make avoidable mistakes in their predictions. While some regressors produce hidden layers which respect the importance of range, these demonstrate an unsupervised learning of range. To ensure our hidden layers represent range reliably, we introduce

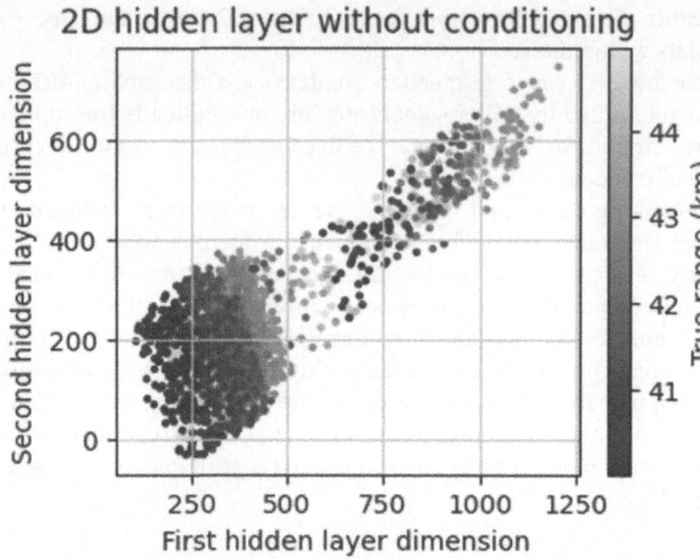

FIGURE 15.5 Compressed nuisance parameters of the training data when processed by the unconditioned regressor, colored by true range. This neural network and neural networks with similar hidden layers, where there are multiple methods of defining range within the hidden layer, will make avoidable mistakes when predicting ranges based upon signals.

the loss function in Experiment 2, which enforces a supervised learning of range in the hidden layer.

15.5.2 Experiment 2: Conditioned Cycles

Our second experiment utilizes cycles and conditioning methods. We create a range-predicting encoder E and decoder D and then we train a cycle with the loss function shown in Equation 15.2.

Our loss function has the primary benefit of allowing us to train a range predictor and a data reconstructor simultaneously. The multitask encoder produces a meaningful latent space for signal reconstruction while also predicting range. The conditional decoder is trained to decode signals from this meaningful hidden layer with conditioning on range and can serve as a traditional decoder or as a conditional data generator.

In the cyclic conditioning experiment and future experiments, our loss function includes two distinct conditioning predictions. We denote these conditioning predictions as $\hat{\mathbf{c}}_{i,1}$ and $\hat{\mathbf{c}}_{i,2}$ for the first and second condition predictions, respectively. Additionally, in the current and future experiments, we use the encoder's two outputs as input to the decoder in training. To emphasize the conditioning aspect of our work, we write the decoder's output as $D(\hat{\mathbf{z}}_i, \hat{\mathbf{c}}_i)$. Let each λ_k where k ranges from 1 to the number of terms in the loss function be scaling factors for each term in the loss to ensure that each term begins training at values roughly equivalent to each other. Our loss function becomes Equation 15.2.

$$
\begin{aligned}
&Loss = \frac{1}{n}\sum_{i=1}^{n} \\
&A: \lambda_1 \left\| \mathbf{y}_i - \hat{\mathbf{c}}_{i,1} \right\|_F^2 + \\
&B: \lambda_2 \left\| \mathbf{x}_i - D\left(\hat{\mathbf{z}}_i, \hat{\mathbf{c}}_{i,1}\right) \right\|_F^2 + \\
&Loss = \frac{1}{n}\sum_{i=1}^{n}
\end{aligned}
\qquad (15.2)
$$

With cyclic conditioning, we see a hidden layer more suitable for EM source localization. Figure 15.6 shows a hidden layer of a conditioned cyclic autoencoder, colored by the true range of the input signal. By explicitly including range prediction in our loss function, we enforce restrictions on the hidden layers of our models, which now show a

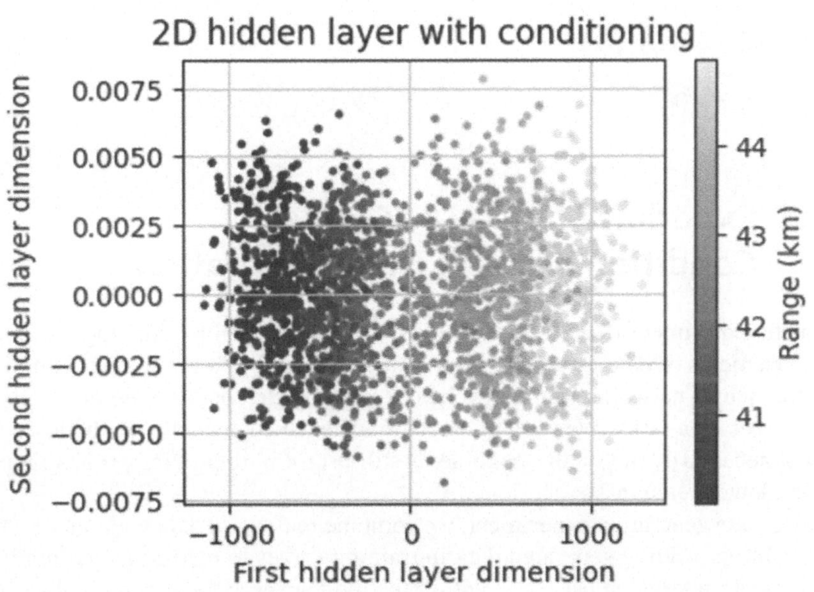

FIGURE 15.6 Training data after being compressed by a conditioned regressor, colored by a true range. This hidden layer is more appropriate for EM source localization than that shown in **Figure 15.5** as the conditioned hidden layer yields a well-defined method of determining range. Note that this 2D hidden layer has a dimension corresponding to range (shown as the *x*-axis) and a dimension corresponding to nuisance parameters (shown as the *y*-axis). The encoding of nuisance parameters in this way is necessary due to the 20 signals corresponding to each range in our dataset.

smooth gradient with respect to range. Such hidden layers yield a well-defined method of determining the range condition and reduce avoidable mistakes in range predictions based upon given signals.

15.5.3 Experiment 3: Cyclic Oracle Conditioning

Our third experiment utilizes cyclic oracle conditioning. We use a similar loss function to that of Experiment 2; however, we replace the encoder's range prediction in the loss function with the true range to induce oracle conditioning. The loss function for our cyclic oracle conditioning can be seen in Equation 15.3. Note that oracle conditioning is distinct from standard conditioning in that the input to the decoder is both the $\hat{\mathbf{z}}_i$ term produced by the encoder and the true condition.

$$
\begin{aligned}
Loss &= \frac{1}{n}\sum_{i=1}^{n} \\
A &: \lambda_1 \left\| \mathbf{y}_i - \hat{\mathbf{c}}_{i,1} \right\|_F^2 + \\
B &: \lambda_2 \left\| \mathbf{x}_i - D\left(\hat{\mathbf{z}}_i, \mathbf{y}_i\right) \right\|_F^2 + \\
C &: \lambda_3 \left\| \mathbf{y}_i - \hat{\mathbf{c}}_{i,2} \right\|_F^2
\end{aligned}
\tag{15.3}
$$

15.5.4 Experiment 4: Cyclic Oracle Conditioning with Data Generation

Our fourth experiment demonstrates the strength of oracle conditioning while incorporating elements of data generation in training. A benefit of our previous experiments is that the neural networks trained cyclically with oracle conditioning are well suited for data generation tasks. We wish to create models which generate synthetic data that, while not necessarily physically possible, is still useful in improving prediction performance in downstream tasks.

In our data generation experiment, we combine real-world data with data generated during training. Adding generated data in training serves to regularize our predictions and enforce an internal consistency within the hidden space for generated data as well as for real-world data. We combine the real-world and synthetic data in each epoch to create a roughly equal distribution of real to synthetic data within our sparse-data "gap" but also to have a 10-to-1 ratio of real to synthetic data in the areas with larger amounts of real-world data.

In training, we add a single term to our loss function to constrain these generated signals. Term D in Equation 15.4 shows the loss term added. In this experiment, notation remains the same as in previous experiments, but we add the terms $\hat{\mathbf{c}}_{j\,gen}$ and $\mathbf{y}_{j\,gen}$

to be the jth condition prediction on generated data and the generated range condition for the synthesized data, respectively. Additionally, we write our decoder's output signal prediction as $D(\hat{\mathbf{z}}_i, \hat{\mathbf{c}}_i) = \hat{\mathbf{x}}_i$.

$$
\begin{aligned}
&Loss = \frac{1}{n}\sum_{i=1}^{n} \\
&A: \lambda_1 \left\| \mathbf{y}_i - \hat{\mathbf{c}}_{i,1} \right\|_F^2 + \\
&B: \lambda_2 \left\| \mathbf{x}_i - \hat{\mathbf{x}}_i \right\|_F^2 + \\
&C: \lambda_3 \left\| \mathbf{y}_i - \hat{\mathbf{c}}_{i,2} \right\|_F^2 \\
&\frac{1}{m}\sum_{j=1}^{m} \\
&D: \lambda_4 \left\| \mathbf{y}_{j\,gen} - \hat{\mathbf{c}}_{j\,gen} \right\|_F^2 \\
&where\ n + m = batch\ size
\end{aligned}
\tag{15.4}
$$

Term D in Equation 15.4 enforces an internal consistency upon the hidden layer of the neural networks by ensuring that any signal generated from the decoder will correspond to a range prediction from the encoder consistent with the generated range. Further, we select points and generated ranges from our hidden layers to be consistent with the hidden layer representation of range. Enforcing the consistency of the hidden layer during training is the reason we require a well-defined range condition in our work.

15.5.5 Experiment 5: Iterative Model with Oracle Conditioning

Our fifth experiment is composed of a five-iterative model conditioned upon source localization range. With the goal of producing a model which predicts well upon completion of its fifth iteration, we utilize a weighted loss function that penalizes errors at later iterations more harshly than errors at earlier iterations. We also utilize continuation within the loss function to ensure convergence.

The loss function when training an iterative model becomes necessarily more complicated, as error is calculated at each iterative step. To avoid needless length in our equations, we write the loss function for a single iteration in terms of its signal reconstruction and range prediction. However, it is important to note that each iterative prediction is based upon the predictions made at previous steps. The function for the loss of an iterative model is shown in Equation 15.5, where j refers to the iteration step, m refers to the maximum number of iterations, $\hat{\mathbf{c}}_{i,j}$ refers to the source localization prediction made at the jth iteration, and $\hat{\mathbf{x}}_{i,j}$ refers to the signal reconstruction made at the jth iteration.

$$Loss = \sum_{j=1}^{m} \frac{1}{n} \sum_{i=1}^{n} \tag{15.5}$$
$$A: 10^j \lambda_1 \left\| \mathbf{y}_i - \hat{\mathbf{c}}_{i,j} \right\|_F^2 +$$
$$B: 10^j \lambda_2 \left\| \mathbf{x}_i - \hat{\mathbf{x}}_{i,j} \right\|_F^2$$

When training this model, we do not use oracle conditioning. While it is possible to utilize oracle conditioning with an iterative model, there are many parameters to consider, including but not limited to the following: which iterations will utilize oracle conditioning and whether oracle conditioning interacts with continuation. At the time of writing, we do not have conclusive evidence as to the best method of addressing these concerns, and as such have chosen to omit oracle conditioning from our iterative experiment herein.

15.5.6 Results

Before discussing our results, it is important to note that our testing procedure for Experiments 2, 3, and 4 does not utilize cycles. In testing these three experiments, we simply pass real-world signals through the encoder and evaluate performance based upon the range predictions produced. Figure 15.1 illustrates this testing structure. As we do not utilize cycles in our testing for Experiments 2, 3, and 4, we do not make use of the multitask nature of our encoders. On the other hand, we utilize iterations for Experiment 5. The results for Experiment 5 detailed herein are the prediction results of the fifth and final iteration. All neural networks are tested exclusively on the same set of real-world data.

When testing the algorithms, we see that performance between the unconditioned regressor and the cyclic regressor with standard conditioning is comparable within the gap. Figures 15.7 and 15.8 show training and testing range prediction performance for the five experiments, while Table 15.1 shows the test metrics for the five experiments both inside and outside the low-data gap. Outside the gap, the iterative model has the lowest Root Mean Squared Error (RMSE) and standard deviation of all five experiments while the unconditioned regressor has the lowest bias. This performance demonstrates the utility of our iterative model in situations where enough training data is available. The neural network employed in Experiment 3, with cyclic oracle conditioning, shows improved test performance in the low-data gap when compared to Experiments 1 and 2, having the lowest inside-gap RMSE by nearly a factor of 10. The inside-gap bias for Experiment 3 is similarly small. As the difference between Experiments 2 and 3 is solely the addition of oracle conditioning, the inside-gap test performance improvement can be attributed to oracle conditioning.

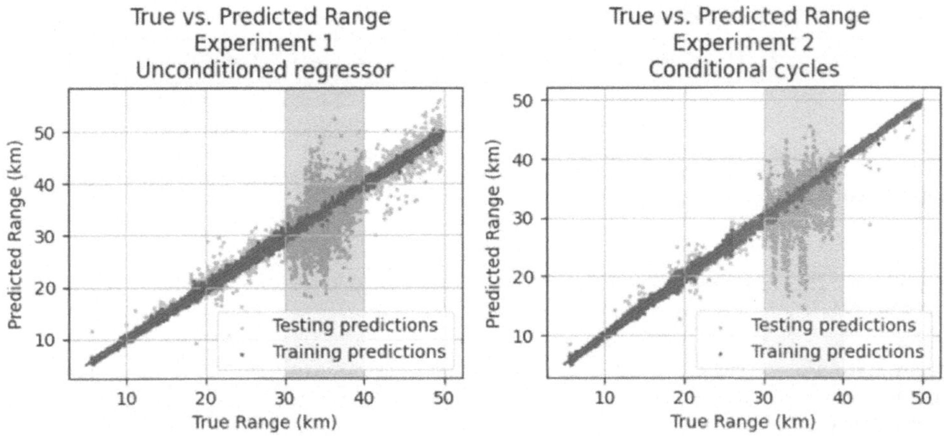

FIGURE 15.7 True range versus predicted range for the first two experiments. Note that the neural network range predictions are poor in the low-data gap.

The neural network trained while generating data in Experiment 4 does not perform the best by any metric, inside or outside the low-data gap. However, it has the second lowest inside-gap bias and RMSE of all five experiments, outperformed only by the neural network in Experiment 3 with identical architecture but no data generation steps in training. Outside of the gap, the data generation neural network demonstrates the worst bias but an RMSE comparable to, but slightly better than, the RMSE of the unconditioned regressor.

Our fifth experiment, concerning the iterative neural network without oracle conditioning, shows useful results. This model has the lowest standard deviation and RMSE in the outside-gap region. While the model overfits in the low-data region, it still performs better than the models in Experiments 1 and 2, despite our training process for

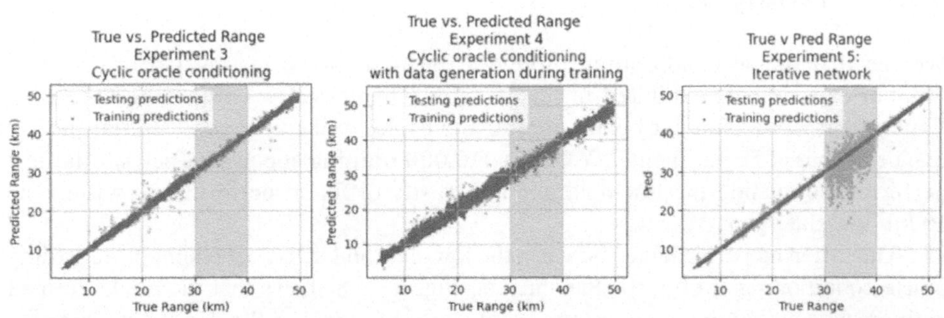

FIGURE 15.8 True range versus predicted range for the final three experiments. Note that Experiment 3, with oracle conditioning, performs much better in the low-data gap than the first two experiments. Further, the data generation experiment also performs reasonably well within the low-data gap in comparison to its training performance. The iterative neural network performs best in the high-data region.

TABLE 15.1 Testing performance metrics

	OUTSIDE GAP			*INSIDE GAP*		
	BIAS (M)	*STANDARD DEVIATION (M)*	*RMSE (M)*	*BIAS (M)*	*STANDARD DEVIATION (M)*	*RMSE (M)*
Experiment 1: Unconditioned regressor	18.435	1,573.656	1,576.389	−1,100.815	3,822.484	3,987.014
Experiment 2: Conditional cycles	−52.338	835.376	858.533	−1,223.285	3,448.404	3,660.241
Experiment 3: Cyclic oracle conditioning	60.878	885.420	906.799	−5.519	392.178	398.850
Experiment 4: Cyclic oracle conditioning with data generation	125.886	1,198.943	1,213.559	−110.713	1,006.269	1,018.035
Experiment 5: Iterative model	67.800	547.915	552.094	−1,099.633	2,476.604	2,709.753

Note that our iterative model performs best outside of the low-data gap, while cyclic oracle conditioning improves performance within the gap. Furthermore, the data generation results are comparable to the results of the unconditioned regressor adapted from our previous work [11].

the iterative model, not including oracle conditioning. It is likely that, with tuning, the incorporation of oracle conditioning into the iterative model's training process would yield performance improvement in the low-data region.

15.5.7 Analysis

We see that oracle conditioning produces the best in-gap testing performance of the four experiments exhibited in this chapter. Theoretically, with enough training, Experiment 2 with standard conditioning and cycles will converge to similar performance as that of Experiment 3. Yet after 110,000 training epochs, we see similar test performance outside the low-data gap and vastly different performance within the 10 km low-data gap.

The differing performance between the low-data and dense data domains regarding oracle conditioning is of particular interest. Figure 15.8 shows that the model trained with oracle conditioning demonstrates some level of overfit in the dense data domains but exhibits no signs of overfit within the low-data domain. Rather, the oracle conditioned neural network's performance in the low-data gap is notably better than performance outside of the gap, suggesting that oracle conditioning not only serves to allow for good test predictions in low-data domains but also serves as a regularizor and mitigates overfit under conditions where overfit is likely to occur.

The regression prediction results when trained with generated data, shown in Figure 15.8, are also notable. When we train our neural networks with the loss function in Equation 15.4, we see poor prediction performance on training data when compared to the other three experiments. However, the testing performance when trained with generated data is comparable to the performance of the other experiments shown in this chapter. Future work will include refinements to our data generation methods and longer training times to produce test prediction results comparable to cyclic oracle conditioning without data generation.

Perhaps surprisingly, our iterative model performed the best of all five models in the outside-gap region. Our model in Experiment 5 has the same number of parameters as the models in Experiments 2, 3, and 4, yet the alternate training scheme utilizing iterations causes it to perform much better in testing. To further support this result, we explore the performance of a three-iterative neural network in the following section on a different physical problem. Further research is needed to determine how much predictive power iterations can provide a neural network and how to combine such iterative neural networks with other machine learning techniques.

15.6 LUNAR LANDER EXPERIMENTAL DESIGN

Our work evaluating lunar lander trajectories is not as expansive as our work on EM signals. However, we include this section to demonstrate the generalizability of our model to alternate domains within the physical sciences.

In our lunar lander experiment, we attempt to distinguish trajectories from "Company A" [16] and "Company B" [17] based only upon the x and y trajectory coordinates. To ensure our models can iterate, we concatenate our trajectory data with a two-dimensional vector containing a uniform prior distribution over the two classes ($[0.5, 0.5]$) and require our neural network to return a reconstruction of the trajectory and a more accurate probability that the trajectory belongs to each class. This concatenation of trajectory and probability distribution as input allows us to transform our loss function in unusual ways. However, we adhere to the loss functions put forward in this chapter to be consistent. Future work will investigate different inputs, outputs, and loss functions of iterative neural networks and how to optimize these features.

Our loss function for these experiments consists of the weighted sum of reconstruction error and class prediction error: a simple modification of Equation 15.5, where $\hat{\mathbf{c}}_{i,j}$ refers to a binary class prediction rather than continuous range prediction error. As our lunar lander data is synthetic and less complicated than the real-world EM data, our neural network is much smaller, and we use only a three-iterative neural network. Our model structure can be seen below.

```
iterative_model = nn.Sequential(nn.Linear(202 + number_classes, 2),
                                nn.ELU(),
                                nn.Linear(2, 202 + number_classes))
```

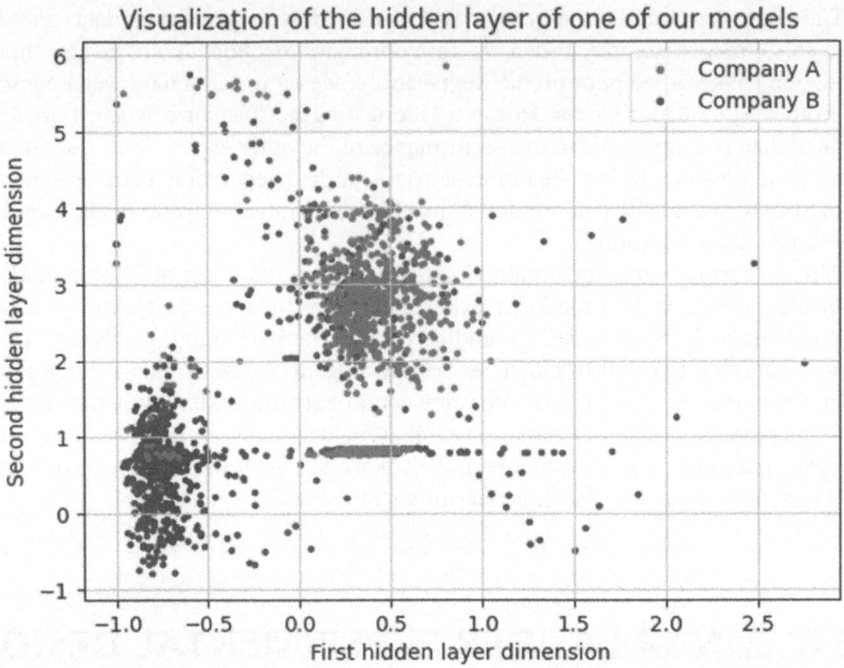

FIGURE 15.9 Hidden layer for one of our lunar lander iterative models. Note that the two classes of trajectories, while not completely separated, are separated well enough to yield clearly defined spaces in which one class or the other is viable for data generation.

We train our model for 1,000 epochs and utilize continuation in our loss function.

Our iterative model has the benefit of retaining important aspects of previous models: easy interpretability of the output as well as hidden layers that can be leveraged to generate more data if necessary. In our hidden layers, we wish to separate the two classes into different clusters rather than produce a continuous gradient, as in our EM application. While our dataset contains some ambiguity, we see in Figure 15.9 a hidden layer characteristic of our lunar lander neural networks. While this hidden layer does not separate the two classes entirely, it still shows areas associated with only one class or the other, which are suitable for data generation.

Naturally, we do not need to use these models to generate more data, as we are already utilizing open-source models to create trajectories. However, once we have acquired real-world trajectory data, we will be limited in the amount collected. In this circumstance, the ability to generate synthetic data from our models trained on the real-world data will be necessary.

15.6.1 Results

In the brief time working with this problem, we have achieved improvements in accuracy when using iterations. Our three-iterative model begins with a test accuracy

TABLE 15.2 Test performance metrics for our iterative lunar lander model

	ACCURACY	*COMPANY A PRECISION*	*COMPANY B PRECISION*	*COMPANY A RECALL*	*COMPANY B RECALL*
Iteration 1	0.84	0.95	0.77	0.72	0.96
Iteration 2	0.86	0.94	0.81	0.77	0.95
Iteration 3	0.88	0.94	0.83	0.81	0.95

Company B precision and Company A recall demonstrate notable improvement with iterations. Note that iteration improves accuracy, Company B precision, and Company A recall. Further, the two other metrics (Company A precision and Company B recall) are not notably hindered by iteration. This demonstrates that iterations can be useful in improving performance while not detracting from performance in areas it does not help.

of 0.84 but improves to 0.88 accuracy after the third iteration. Similarly, iteration improves precision for Company B and recall for Company A. Perhaps just as important, we have found no evidence that iteration *harms* testing performance. A noniterative model may produce performance metrics like those of our iterative model in its first iteration. Our iterative models, with the same number of parameters and little additional computational time, can improve test performance notably and, in cases where they do not notably improve performance, will not cause harm. When iterating did not help a metric, test performance fell a fractional amount but not enough to cause concern: Company A precision fell from 0.95 to 0.94 while Company B recall fell from 0.96 to 0.95. Table 15.2 shows the full test metrics for our model at each iteration.

15.7 CONCLUSION

The results reported in this chapter were motivated by the goal of producing models which can reliably predict well on real-world data in low-data domains. We demonstrated methods of training neural networks which lead to better prediction performance than other state-of-the-art models can achieve.

Oracle conditioning allows us to produce neural networks capable of reliable prediction in low-data environments. Further, oracle conditioning combined with data generation produces neural networks with only slightly diminished prediction capabilities while producing synthetic data suitable for training neural networks. Utilizing iterative neural networks leads to even better performance in our EM application, demonstrating the strength of our model and techniques in real-world domains.

To show the versatility of our methods, we also applied our techniques to lunar lander trajectory analysis. In the lunar lander case, we showed again that iteration is a powerful tool which can lend itself to test prediction performance while only requiring the same number of trainable parameters as some noniterative models. Furthermore, testing iterative models only adds fractions of a second to the testing process for each iteration.

We intend to continue our work with iterations and conditioning, applying these concepts to other real-world domains while also exploring the theoretical aspects more thoroughly. Future work will investigate the predictive power of adding iterations compared to the predictive power of adding trainable parameters to neural networks.

ACKNOWLEDGMENTS

We would like to thank Pat Bidigare and Ilana Heintz at Synoptic Engineering for funding this research through the Messina project. We would also like to thank Kevin Vanslette and Alex Lay at Raytheon BBN as well as Rick Brown at WPI for their invaluable help.

This research was sponsored in part by the DEVCOM Analysis Center and was accomplished under Cooperative Agreement Number W911NF-22-2-0001. The views and conclusions contained in this document are those of the authors and should not be interpreted as representing the official policies, either expressed or implied, of the Army Research Office or the U.S. Government. The U.S. Government is authorized to reproduce and distribute reprints for Government purposes notwithstanding any copyright notation herein.

This material is based upon work supported by the NRT CEDAR at Worcester Polytechnic Institute (WPI), which is supported by the National Science Foundation under Grant NRT-HDR-2021871.

This research was performed using computational resources supported by the Academic & Research Computing group at Worcester Polytechnic Institute.

REFERENCES

1. M. Barger, R. Paffenroth and H. Pathak, “Conditioned Cycles in Sparse Data Domains: Applications to Electromagnetics,” in *2023 International Conference on Machine Learning and Applications (ICMLA)*, 2023.
2. S. Greenland, J. A. Schwartzbaum and W. D. Finkle, “Problems due to Small Samples and Sparse Data in Conditional Logistic Regression Analysis,” *American Journal of Epidemiology*, vol. 151, pp. 531–539, March 2000.
3. D. D. Jackson, “Interpretation of Inaccurate, Insufficient and Inconsistent Data,” *Geophysical Journal International*, vol. 28, pp. 97–109, June 1972.
4. J.-Y. Zhu, T. Park, P. Isola and A. A. Efros, “Unpaired Image-to-Image Translation Using Cycle-Consistent Adversarial Networks,” in *Proceedings of the IEEE International Conference on Computer Vision*, 2017.
5. A. M. Moore, R. C. Paffenroth, K. T. Ngo and J. R. Uzarski, “Cycles Improve Conditional Generators: Synthesis and Augmentation for Data Mining,” in *Advanced Data Mining and Applications*, Cham: Springer Nature Switzerland, 2022.

6. E. Perez, F. Strub, H. de Vries, V. Dumoulin and A. Courville, "FiLM: Visual Reasoning with a General Conditioning Layer," in *Proceedings of the AAAI Conference on Artificial Intelligence*, vol. 32, April 2018.
7. N. Agarwal, P. Awasthi and S. Kale, "A Deep Conditioning Treatment of Neural Networks," in *Proceedings of the 32nd International Conference on Algorithmic Learning Theory*, 2021.
8. Y.-H. Kwon and M.-G. Park, "Predicting Future Frames Using Retrospective Cycle GAN," in *Proceedings of the IEEE/CVF Conference on Computer Vision and Pattern Recognition (CVPR)*, 2019.
9. W. Li and J. Wang, "Residual Learning of Cycle-GAN for Seismic Data Denoising," *IEEE Access*, vol. 9, pp. 11585–11597, 2021.
10. G. Shafer, "Conditional Probability," *International Statistical Review/Revue Internationale de Statistique*, vol. 53, p. 261–275, 1985.
11. E. Witz, M. Barger and R. Paffenroth, "Deep Learning for Range Localization via Over-Water Electromagnetic Signals," in *2021 20th IEEE International Conference on Machine Learning and Applications (ICMLA)*, 2021.
12. X. Zhao, "Evaporation Duct Height Estimation and Source Localization From Field Measurements at an Array of Radio Receivers," *IEEE Transactions on Antennas and Propagation*, vol. 60, pp. 1020–1025, 2012.
13. K. S. Zaidi, V. Jeoti, M. Drieberg, A. Awang and A. Iqbal, "Fading Characteristics in Evaporation Duct: Fade Margin for a Wireless Link in the South China Sea," *IEEE Access*, vol. 6, pp. 11038–11045, 2018.
14. P. Bidigare and I. Heintz, Radio Frequency Multipath Characterization & Transmitter Localization of Overwater Emitters, private communications, September 2022.
15. G. Brockman, V. Cheung, L. Pettersson, J. Schneider, J. Schulman, J. Tang and W. Zaremba, "Openai gym (2016)," arXiv preprint arXiv:1606.01540, vol. 476, 2016.
16. araffin, *ppo-LunarLander-v2,* GitHub. araffin, *ppo-LunarLander-v2*, GitHub, https://huggingface.co/araffin/ppo-LunarLander-v2.
17. sb3, *a2c-LunarLander-v2,* GitHub. sb3, *a2c-LunarLander-v2*, GitHub, https://huggingface.co/sb3/a2c-LunarLander-v2.
18. L. Gondara, "Medical Image Denoising Using Convolutional Denoising Autoencoders," in *2016 IEEE 16th International Conference on Data Mining Workshops (ICDMW)*, 2016.
19. Q. Hershey, R. Paffenroth, H. Pathak and S. Tavener, Rethinking the Relationship between Recurrent and Non-Recurrent Neural Networks: A Study in Sparsity, arXiv preprint arXiv:2404.00880, 2024.
20. S. L. Richter and R. A. Decarlo, "Continuation Methods: Theory and Applications," *IEEE Transactions on Systems, Man, and Cybernetics*, vol. SMC-13, pp. 459–464, 1983.
21. Q. Hershey, "Exploring Neural Network Structure through Iterative Neural Networks: Connections to Dynamical Systems". MS thesis, Worcester Polytechnic Institute, 2023.
22. Q. Hershey, R. Paffenroth and H. Pathak, "Exploring Neural Network Structure through Sparse Recurrent Neural Networks: A Recasting and Distillation of Neural Network Hyperparameters," in *2023 International Conference on Machine Learning and Applications (ICMLA)*, 2023.
23. H. N. Pathak, R. Paffenroth and Q. Hershey, "Sequential2D: Organizing Center of Skip Connections for Transformers," in *2023 International Conference on Machine Learning and Applications (ICMLA)*, 2023.
24. E. L. Allgower and K. Georg, *Numerical Continuation Methods: An Introduction*, vol. 13, Springer Science & Business Media, 2012.
25. H. N. Pathak and R. Paffenroth, "Parameter Continuation Methods for the Optimization of Deep Neural Networks," in *2019 18th IEEE International Conference On Machine Learning And Applications (ICMLA)*, 2019.

26. L. Fang, T. Zeng, C. Liu, L. Bo, W. Dong and C. Chen, Transformer-based Conditional Variational Autoencoder for Controllable Story Generation, arXiv preprint arXiv:2101.00828, 2021.
27. M. Simonovsky and N. Komodakis, "Dynamic Edge-Conditioned Filters in Convolutional Neural Networks on Graphs," in *Proceedings of the IEEE Conference on Computer Vision and Pattern Recognition (CVPR)*, 2017.
28. S. Zheng et. al., "Conditional Random Fields as Recurrent Neural Networks," in *Proceedings of the IEEE International Conference on Computer Vision (ICCV)*, 2015.
29. H. White, "A Reality Check for Data Snooping," *Econometrica*, vol. 68, pp. 1097–1126, 2000.
30. J. P. Romano, A. M. Shaikh and M. Wolf, "Formalized Data Snooping Based on Generalized Error Rates," *Econometric Theory*, vol. 24, pp. 404–447, 2008.

16 Enhancing Aerial Combat Tactics through Hierarchical Multiagent Reinforcement Learning

Ardian Selmonaj, Oleg Szehr, Giacomo Del Rio, Alessandro Antonucci, Adrian Schneider, and Michael Rüegsegger

16.1 INTRODUCTION

This work investigates deep *reinforcement learning* (RL) for exploring preset air defense scenarios in a cost-effective and safe-to-fail environment. RL agents have shown strong performance in finding *courses of action* (CoAs) in various environments, including combinatorial scenarios like Chess and Go, real-time continuous control tasks in video games, and combined control and strategic decision making in modern wargames.

Applying RL to air combat faces challenges like complex dynamics, large state and action spaces, large planning horizons, and stochasticity. The vast game tree (possible CoAs) in defense scenarios complicates search methods. In real-world operations,

DOI: 10.1201/9781003570882-20

the complexity of defense scenarios is managed by a hierarchical organization of the decision-making process. Low-level decisions (firing, evading) are made by individual units, while high-level planning (attacking, retreating) is done by commanders. This hierarchy allows task-specific training and delegation of authority for autonomous control and strategic planning. Effective information flow and filtration are essential, with troops using local information and commanders gathering broader observations.

The proposed hierarchical *multiagent reinforcement learning* (MARL) framework analyzes simulated air defense scenarios with heterogeneous agents. Hierarchical MARL systems allow decentralized control, enabling targeted group behavior and guidance. The decision-making process is divided into two levels: low-level policies control individual units, trained on preset scenarios like attack or evade, and high-level policies set these flags based on mission targets. Training for low-level policies follows a curriculum with increasing complexity and league-based self-play, while high-level policies train to align subordinate policies with mission objectives. This structure eases training by exploiting policy symmetries and targeting information flow to relevant decision points, separating control and command exercises for dedicated training procedures.

This work presents a hierarchical MARL framework for air combat simulations and extends a previous contribution from a conference paper [1]. Training incorporates a fictitious self-play mechanism with curriculum learning to improve combat performance. Our platform supports rapid simulation of core agent dynamics and interactions, focusing on MARL system design and aircraft dynamics. We develop neural network architectures for each hierarchy level. Deep learning systems typically have a black-box nature, posing evaluation risks. Our structured approach enhances research by dividing the system and providing component analysis for explainability. This hierarchical MARL framework lends itself to the hierarchical decision-making process in defense organizations, supporting MARL training and integration for simultaneous combat maneuvering and tactical decisions.

16.2 RELATED WORK

16.2.1 General Single-Agent Techniques

Aerial combat tactics have been extensively studied, particularly in 1-vs-1 combat scenarios. Research on small engagements focuses on unit control, examining how to maneuver a combat unit to gain an advantageous position [2]. Methods include expert systems [3–6], control laws for pursuit/evasion [7–10], game theoretic approaches [11–13], machine learning (ML) [14–17], and hybrid systems [18–24]. RL techniques have gained interest, especially with the success of deep RL in games like Atari [25], Go and chess [26], Starcraft II [27], and Dota II [28]. RL systems offer a flexible approach to developing strong CoAs without relying on extensive human expert data. For example, the success of *deep q-networks* (DQNs) in Atari games [25] suggests potential for broader control tasks, including air combat, where a DQN approach with a situation evaluation function has been proposed in [29]. Other approaches involve deep deterministic policy gradient (DDPG) [30, 31] and A3C [32] for learning unmanned aerial vehicles (UAVs) combat maneuvers. Cascade learning schemes to increase combat complexity are introduced in [33, 34]. League play prevents agents from overfitting to specific opponent strategies, because in self-play, this overfitting can lead to training cycles without overall performance improvement [27, 35, 36].

16.2.2 Multiagent Techniques

In multiagent scenarios, the state and action spaces grow exponentially with the number of agents. MARL systems address this "curse of dimensionality" by exploiting symmetries within individual agents. For example, interchangeable agents can be equipped with identical policies, simplifying the training and execution processes [37]. Methods like *centralized training with decentralized execution* (CTDE) [38] and the *multiagent deep deterministic policy gradient* (MADDPG) algorithm have shown competitive performance in coordinating agents [39, 40]. In defense modeling, multiagent engagements often focus on weapon-target assignment [41], pilot-like decision making [42], and high-level tactical decisions [43], emphasizing planning of CoAs over individual unit control. A maneuvering strategy for UAV swarms using MADDPG is discussed in [44], limited to one-to-one or multi-to-one combat. Attention-based neural networks are applied in [45] and [46] for air combat, the former uses a two-stage attention mechanism for coordination, while the latter calculates the relative importance of surrounding aircraft with opponents controlled by scripts. To better reflect real-world behavior, CTDE systems can incorporate structural constraints like agent type and properties. One constraint is agent attrition, where agents are removed before termination, requiring dynamic adaptation by the remaining agents. This challenge leads to the study of hierarchical MARL.

16.2.3 Hierarchical Techniques

There appears to be relatively little research in defense modeling that combines *hierarchical reinforcement learning* (HRL) and MARL. Combining HRL within MARL could facilitate the adoption of advanced strategies by structuring decision making. A hierarchical MARL approach is proposed in [47] to train agents in situations where agent attrition is important. The authors employ an attention mechanism and self-play with a DQN high-level policy trained with QMIX [48]. An approach similar to ours, incorporating heterogeneous agents, was explored in [49]. The high-level target allocation agents are trained using DQN, and the low-level cooperative attacking agents are based on independent asynchronous proximal policy optimization. However, they follow the goal of *suppression of enemy air defense* (SEAD). Unlike the concept of SEAD, which focuses on neutralizing enemy air defense systems, dogfighting aims at defeating enemy aircraft in direct air combat.

This article examines air combat scenarios, using hierarchical MARL to develop coordinated dogfighting strategies with heterogeneous agents. It employs a cascaded league-play training scheme to add realistic complexity and enhance strategic depth. The approach aims to improve tactical effectiveness and understanding of cooperative strategies in dynamic environments.

16.3 FOUNDATIONAL CONCEPTS

16.3.1 Aircraft Dynamics

We base our modeling on the dynamics of the Dassault Rafale fighter aircraft.[1] We assume a constant altitude in a 2D environment. The 2D model simplifies analysis by omitting the dynamics associated with the third dimension yet retains the essential characteristics of air combat scenarios. This simplification aids in modeling and analyzing key interactions and strategies, such as positioning, maneuvering, and timing, which are critical for studying and simulating air combat tactics without the computational overhead of a 3D model.

Air combat engagements are classified into *beyond visual range* (BVR) and *within visual range* (WVR) scenarios [50]. BVR involves long-range engagements beyond line of sight, while WVR involves close-range combat where opponents are visible to each other, necessitating higher frequency maneuvering. In this work, we focus on the latter.

Our approach considers two specific types of aircraft with different capabilities and dynamics, introducing a level of heterogeneity that mirrors real-world scenarios. The first aircraft (AC1) is agile and equipped with rockets, while the second aircraft (AC2) has no rockets but a longer cannon range. The dynamics of AC1 and AC2 are detailed in Table 16.1, and their attacking mechanisms are illustrated in Figure 16.1. Cannon shots are modeled with a conical *weapon engagement zone* (WEZ), where units within this area are destroyed based on a hit probability parameter. Rockets deterministically destroy a unit upon reaching an aircraft's position, with an error margin of approximately 10 m.

TABLE 16.1 Aircraft control parameters

PARAMETER	*SYMBOL*	*UNIT*	*AC1*	*AC2*
Angular Velocity	ω_{AC}	[°/s]	[0, 5]	[0, 3.6]
Speed	v_{AC}	[kn]	[100, 900]	[100, 600]
WEZ	$\omega_{WEZ,AC}$	[°]	[0, 10]	[0, 7]
Range	$d_{a,AC}$	[km]	[0, 2]	[0, 4.5]
Hit Probability	p_{AC}	[%]	0.70	0.85

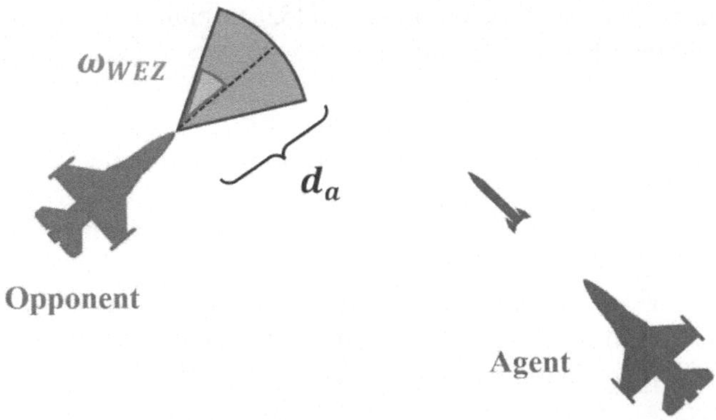

FIGURE 16.1 Dark aircraft are RL agents; bright aircraft denote opponents. Shooting with cannon is represented by WEZ, and a rocket with a corresponding symbol.

16.3.2 Multiagent Reinforcement Learning

RL is a computational approach in the field of ML and is used to solve sequential decision-making problems. Through trial-and-error interactions within an environment, an agent learns a behavior that is evaluated through a reward function to ensure it aligns with the environment's specified goals. The decision-making function, called policy, maps states to a distribution over actions. In MARL, there are multiple agents interacting in a cooperative or competing fashion, or both. Figure 16.2 illustrates the interaction cycle. With $i \in \{1,2,\ldots,N\}$ representing the index of each agent, the mathematical model for the interactions of N agents is formulated as a *partially observable Markov game* (POMG) [51], which is defined by the tuple $(\mathcal{S},\mathcal{O},\mathcal{A}_1,\ldots,\mathcal{A}_N,P,R_1,\ldots,R_N,\gamma)$.

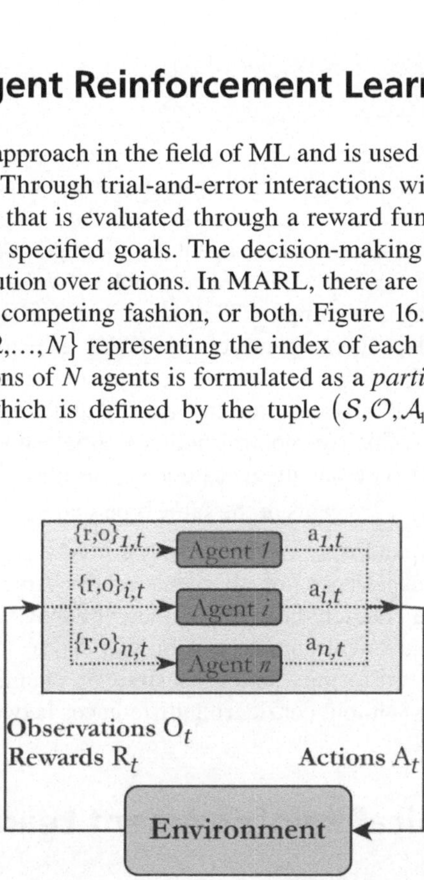

FIGURE 16.2 MARL interaction cycle. $r_{i,t}$ denotes rewards, $o_{i,t}$ denotes observations, and $a_{i,t}$ denotes actions.

POMGs generalize Markov-decision processes [52] to multiple agents that simultaneously interact within a shared environment.

- $\mathcal{S}$ is the state-space of the environment.
- $\mathcal{O} \subset \mathcal{S}$ is the set of (partial) observations of the environment state.
- $\mathcal{A}_i$ is the action-space for player i.
- $P(s' \mid s, a_1, \ldots, a_N)$ represents the probability of transitioning to state s' when players take actions $a_1, \ldots, a_N$ in state s.
- $R_i(s, a_1, \ldots, a_N, s')$ defines the immediate reward for player i when the system transitions from state s to state s' with players taking actions $a_1, \ldots, a_N$.
- γ is the discount factor, used to discount future rewards.

The objective in MARL is to find a set of policies $\pi^* = (\pi_1^*, \ldots, \pi_N^*)$ for all N agents such that each policy π_i^* maximizes the expected return for agent i while considering the policies of other agents:

$$\pi_i^* = \underset{\pi_i}{\operatorname{argmax}} \mathbb{E}\left[\sum_{t=0}^{\infty} \gamma^t R_i(s_t, a_t, s_{t+1}) \middle| \pi_i, \pi_{-i}\right], \tag{16.1}$$

where π_i is the policy of agent i and $\boldsymbol{\pi}_{-i}$ represents the policies of all agents except i. Finding the optimal joint policy $\boldsymbol{\pi}^*$ is challenging because each agent's reward and optimal policy depends on the actions of other agents. This interdependence makes the environment nonstationary for any single agent as other agents' policies evolve.

To address the nonstationary environment problem, we use CTDE, a state-of-the-art framework for multiagent settings. CTDE handles nonstationarity, enhances coordination, and allows independent agent actions during execution. Agents are trained centrally with global information to develop coordinated strategies but rely on local observations during execution for scalability and adaptability. Our modeled POMG with the CTDE scheme trains two control policies: a fight policy (π_f) and an escape policy (π_e). Each aircraft type has distinct policies, resulting in four control policies: $[\pi_{f,AC1}, \pi_{f,AC2}, \pi_{e,AC1}, \pi_{e,AC2}]$. Agents of the same type share the same policies, meaning all AC1 agents use $\pi_{f,AC1}$ and $\pi_{e,AC1}$, and similarly for AC2 agents. This shared policy approach leverages the experiences of all agents of the same type, leading to faster training convergence and coherent behavior. Coherent behavior is defined as agents of the same type having equivalent combat capabilities due to identical knowledge and maneuvering skills. This uniformity simplifies strategic planning for the commander, who can issue commands without considering differences between agents.

16.3.3 Hierarchical Reinforcement Learning

HRL enhances learning efficiency by using temporal abstraction, breaking tasks into a hierarchy of subtasks. Higher-level commands control policies over limited time spans, and exploits low-level symmetries for similar subtasks, such as controlling similar aircraft. This reduces state and action space dimensions, improving scalability and generalization by combining subtasks into new skills [53]. Like military strategy, HRL systematically manages complex tasks by dividing them into manageable subtasks controlled by different model layers. Our hierarchical system is modeled as a *partially observable semi-Markov decision process* (POSMDP). Our POSMDP includes options, which expand the standard concept of actions to include temporally extended CoA. These options can be thought of as macros (commands) that consist of multiple primitive (control) actions. Options last for varying lengths of time, which is characteristic of semi-Markov processes, where transitions between states do not necessarily occur at regular time intervals. The POSMDP is defined by the tuple $(\mathcal{S},\mathcal{O},\mathcal{A},\mathcal{T},P,R,T_l,\gamma)$, where:

- $\mathcal{S}$ is the state-space of the environment.
- $\mathcal{O} \subset \mathcal{S}$ is the set of observations.
- $\mathcal{A}$ is the action-space.
- $\mathcal{T}$ is the set of options, where each option $\tau \in \mathcal{T}$ is defined by a triplet $(I_\tau,\pi_\tau,\beta_\tau)$:
  - $I_\tau \subseteq \mathcal{S}$ initiation set specifies the states from which the option can be initiated.
  - $\pi_\tau(a \mid s)$ the policy associated with the option, responsible for action selection $a \in \mathcal{A}$. In our setting, π_τ is one of the low-level control policies.
  - $\beta_\tau(s)$ the termination condition specifies the probability of the option terminating at each state s.
- $P(s,\tau,s')$ defines the probability of landing in state s' from state s after the execution of option τ.
- $R(s,\tau,s')$ is the reward function, providing the immediate reward received after transitioning from state s to s'.
- $T_l(s,a)$ defines the execution time function, in which an option τ is active.
- $\gamma \in [0,1)$ is the discount factor, used to discount future rewards.

We again use a shared CTDE approach to train a single high-level commander policy π_c for all agents and aircraft types, without differentiating between AC1 and AC2, reflecting policy symmetries where the same strategy applies to any agent. While this makes the commander's behavior homogeneous, individual agents maintain their heterogeneity in operations and capabilities. Figure 16.3 shows the relationship between high-level commander policy and low-level control policies. Based on the option τ chosen by the commander, one of the low-level policies π_f or π_e is activated per agent for T_l time-steps or until termination condition β_τ is met. When an option terminates, the commander reassesses and determines new tactics.

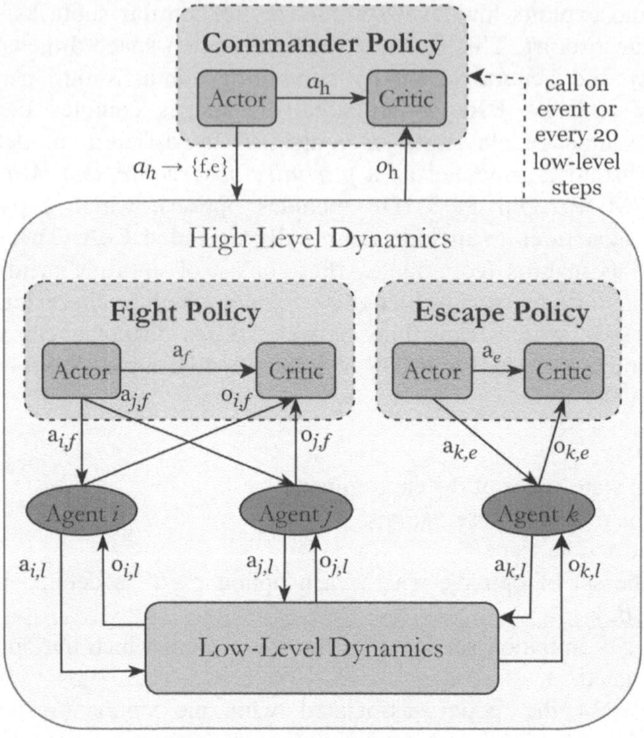

FIGURE 16.3 The commander manages high-level strategic planning with broader situational awareness, while fight and escape policies control low-level aircraft dynamics.

16.4 METHOD

16.4.1 Structural Overview

The training loop of our hierarchical MARL algorithm has two main stages (Figure 16.4a). First, low-level policies π_f and π_e are trained using observations O_l and rewards R_l, with either the fight or escape policy activated based on the training level. Scripted behavior is applied only to opponents. In the next stage, these low-level policies are fixed and used by the high-level commander π_c, which is trained using observations O_h and a combined reward signal $R_l + R_h$. The commander controls policy activation switches for agents, while the frequency of π_f and π_e selection for opponents is predetermined manually. The training of low-level policy π_f follows a five-level curriculum learning strategy, with increasing complexity by facing more competitive opponents (Table 16.2). After completing training at one level, the policy transfers to the next. The escape policy π_e is trained directly on L3 and then against $\pi_{f,L5}$.

TABLE 16.2 Curriculum learning levels

LEVEL 1 (L1)	*LEVEL 2 (L2)*	*LEVEL 3 (L3)*	*LEVEL 4 (L4)*	*LEVEL 5 (L5)*
Static hovering	Random maneuvers	Rule-based strategies	Previous L3 policy	one of L1–L4 policies

Our neural network is based on actor-critic [54] (Figure 16.4b). Low-level AC1 and AC2 agents have distinct neural networks with different input and output dimensions, but each aircraft type uses an identical instance of the neural network. Both instances incorporate a shared layer used by the actor and critic components within each network, enhancing coordination and collaborative decision-making among agents [55].

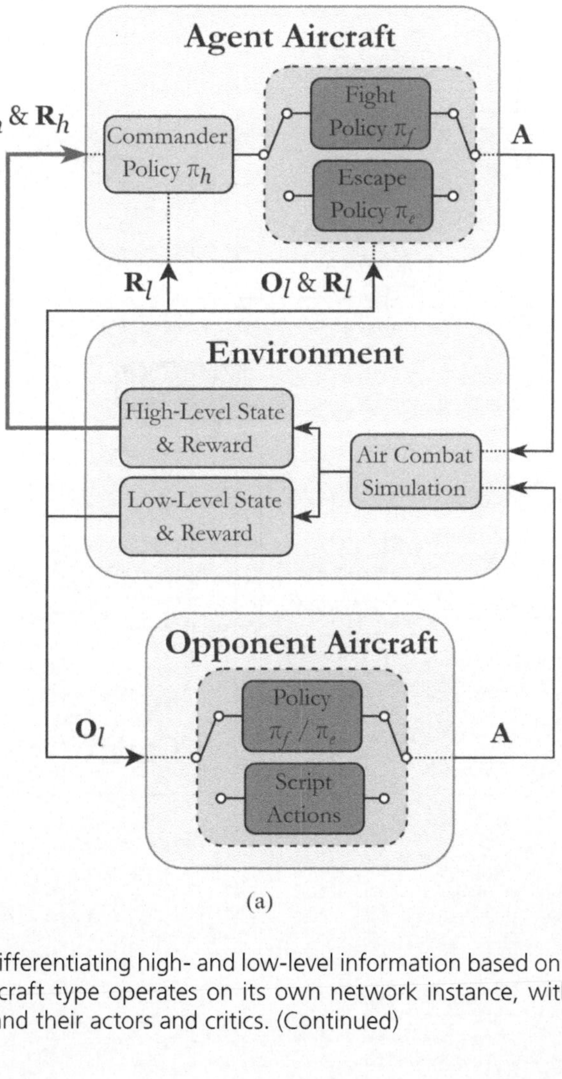

(a)

FIGURE 16.4 (a) Differentiating high- and low-level information based on the policy being trained. (b) Each aircraft type operates on its own network instance, with a shared layer between instances and their actors and critics. (Continued)

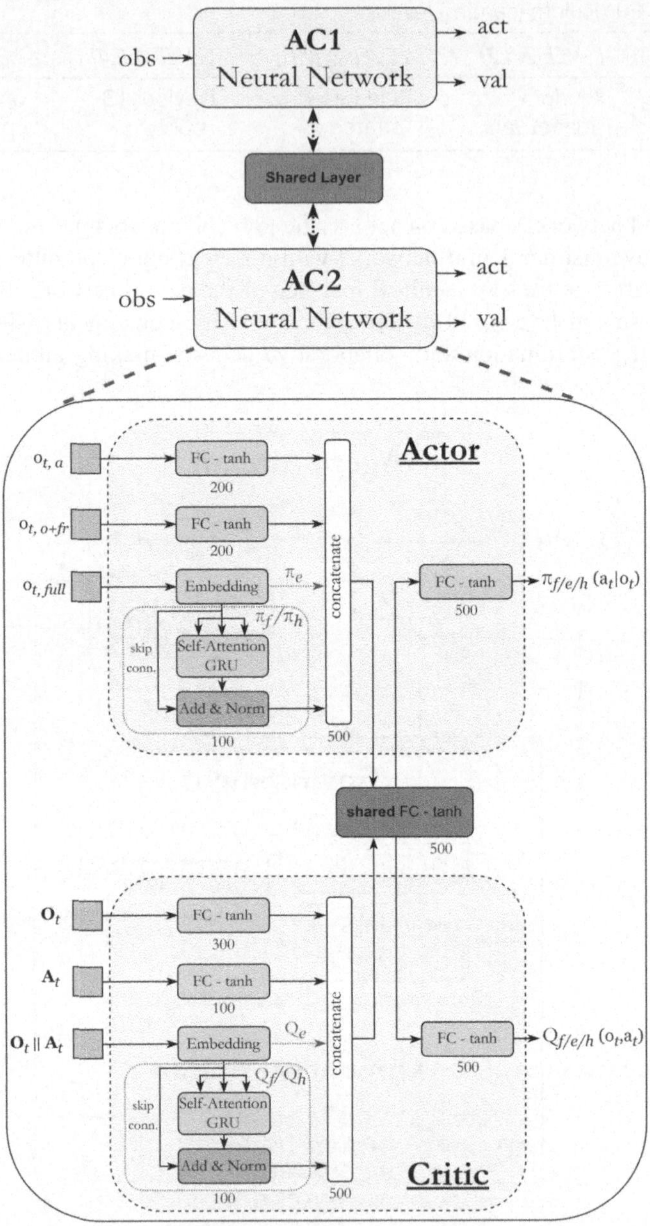

(b)

FIGURE 16.4 (Continued)

The architecture modifications are marked for the three policy types: π_f incorporates a *self-attention* (SA) module [56], π_c employs a *gated-recurrent-unit* (GRU) module [57], and π_e does not utilize either module. The embedding layer is a linear layer comprising 100 neurons and *tanh* activation function. The high-level commander policy π_c has only one network instance, irrespective of aircraft types.

Using the CTDE approach, the critic receives inputs consisting of the observations and actions of all interacting agents. Parameter sharing and a fully observable critic enhance coordination among heterogeneous agents [37]. Network parameters are updated using *proximal policy optimization* (PPO) [58] (Algorithm 1). PPO maximizes the expected return over trajectories τ by optimizing a clipped surrogate objective function for stable policy updates and has shown robust performance in multiagent games [59].

Algorithm 1 PPO Training Procedure for π_f, π_e and π_c

Set number of episodes N, time horizon T, batch size b, and levels L
Initialize buffer $D \leftarrow \{\}$, policy parameters θ, value function parameters φ
for level $l = 1$ to L do
    for episode $n = 0$ to N do
        Initialize state s_0
        for $t = 0$ to T do
            Get agent actions $A_{t,ag}$ by current policy π_θ
            Get opp action $A_{t,o}$: script if $l \leq 3$ else $\pi_{\theta_{l-1}}$
            Execute $\left(A_{t,ag}, A_{t,o}\right)$, obtain R_t and S_{t+1}
            Store $D \leftarrow \left(S_t, A_{t,ag}, A_{t,o}, R_t, S_{t+1}\right)$
        end for
        if $|D| \geq b$ then
            compute advantage estimation A^{π_θ}
            for update iteration k = 1 to K do
                update policy parameters

$$\theta_{k+1} = \underset{\theta}{\operatorname{argmax}}\, \mathbb{E}_{\tau \sim D}\left[\sum_{t=0}^{T}\left[\min\left(\frac{\pi_\theta}{\pi_{\theta k}}, clip\left(\frac{\pi_\theta}{\pi_{\theta k}}, 1-\epsilon, 1+\epsilon\right)\right)A^{\pi_{\theta_k}}\right]\right]$$

                update value function parameters

$$\varphi_{k+1} = \underset{\varphi}{\operatorname{argmin}}\, \mathbb{E}_{\tau \sim D}\left[\sum_{t=0}^{T}\left(V_\varphi\left(s_t\right) - R_t\right)^2\right]$$

            end for
            empty buffer $D \leftarrow \{\}$
            update parameters $\theta \leftarrow \theta_{k+1}, \varphi \leftarrow \varphi_{k+1}$
        end if
    end for
end for

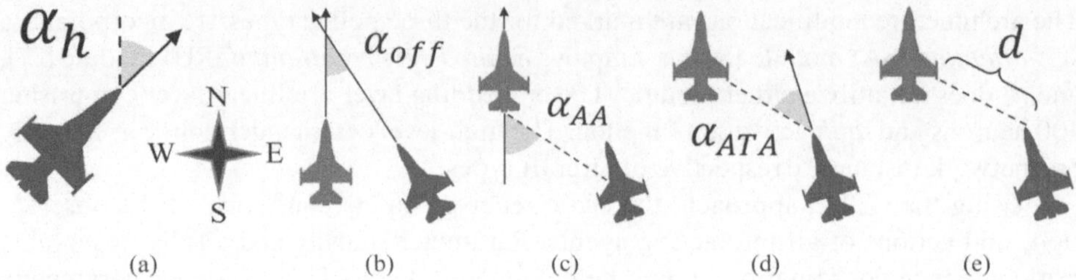

FIGURE 16.5 Aircraft metrics: (a) heading, (b) heading off, (c) aspect angle, (d) antenna train angle, (e) distance.

16.4.2 Air Combat Metrics

The observation values are illustrated in Figure 16.5. The actual heading angle, α_h, is defined with respect to the north direction. The difference in heading angles between an agent and opponent is captured by α_{off}. The aspect angle, α_{AA}, is defined as the angle from the opponent's tail to the position of the agent aircraft, whereas the antenna train angle, $\alpha_{ATA,}$ is the difference of the agent's current heading direction to the opponent's position. The parameter d measures the actual distance between two aircraft. Further observations include map position (x, y), current speed (s), remaining cannon ammunition (c_1) and remaining rockets (c_2). Indicator (w) defines if the next rocket is ready to be fired and (s_r) indicates if the aircraft is currently shooting. Subscript a indicates agent, o opponent, and fr friendly aircraft. A subscript in a value (e.g., $\alpha_{off,o}$) defines the angle-off with respect to the opponent. All observation values are normalized to fall within the range $[0,1]$ and actions of all policies are discrete.

16.4.3 Fight Policy

The objective of the fight policy π_f is to control the aircraft for attacking maneuvers. π_f observes its closest opponent and closest friendly aircraft. The sign || denotes vector concatenation:

$$o_{t,a} := \left[x, y, s, \alpha_h, \alpha_{off,o}, \alpha_{AA,o}, \alpha_{ATA,o}, d_o, c_1, \overbrace{c_2, w}^{\text{AC1}}, s_r \right]$$

$$o_{t,o} := \left[x, y, s, \alpha_h, \alpha_{off,a}, \alpha_{AA,a}, \alpha_{ATA,a}, d_a, s_r \right]$$

$$o_{t,fr} := \left[x, y, s, \alpha_{ATA,a}, \alpha_{ATA,fr}, d_a \right]$$

$$o_{t,full} := o_{t,a} \| o_{t,o} \| o_{t,fr}$$

The control maneuvers (actions) are defined in Table 16.3.

In air combat, facing the opponent's tail is a favorable situation for shooting. We therefore define the reward function based on $\alpha_{ATA,a}$ of the opponent to the agent. We

TABLE 16.3 Control actions of π_f

PARAMETER	*VALUE*
Relative heading maneuvers in range $[-90°, 90°]$	$h \in \{-6,\ldots,6\}, \alpha_h = 15 \cdot h + \alpha_h$
Velocity mapping of v to range of AC1 or AC2	$v \in \{0,\ldots,8\}$
Shooting with cannon	$c \in \{0,1\}$
Shooting with rocket	$r \in \{0,1\}$

further encourage the combat efficiency by incorporating the remaining ammunition $(c_{rem} = c_1 + c_2)$, yielding the following killing reward:

$$r_k = \alpha_{ATA,a} + \frac{c_{max} - c_{rem}}{c_{max}} \in [1,2]. \tag{16.2}$$

Punishing rewards are given when getting destroyed by an opponent $r_d = -2$, destroying a friendly aircraft $r_{fr} = -2$, and flying out of environment boundaries $r_b = -5$. We explicitly design the reward values such that r_d and/or r_{fr} have a greater negative impact compared to r_k. Conversely, it would be counterproductive if the combination of penalties for failure and the rewards for successful kills would still result in a positive total. The total reward for π_f is:

$$r_{fight} = r_k + r_d + r_{fr} + r_b\,. \tag{16.3}$$

16.4.4 Escape Policy

The goal of π_e is to remain alive by carefully steering the aircraft and avoiding dangerous situations. π_e senses the two closest opponents and its closest friendly aircraft.

$$\begin{aligned}
o_{t,a} &:= \left[x, y, s, \alpha_h, c_1, \overbrace{c_2}^{\text{AC1}}\right] \\
o_{t,o} &:= \left[x, y, s, \alpha_h, \alpha_{off,a}, \alpha_{AA,a}, \alpha_{ATA,a}, d_a\right] \\
o_{t,fr} &:= \left[x, y, s, \alpha_{ATA,a}, \alpha_{ATA,fr}, d_a\right] \\
o_{t,full} &:= o_{t,a} \| o_{t,o_1} \| o_{t,o_2} \| o_{t,fr}
\end{aligned}$$

The actions remain the same as for π_f (shooting and maneuvering) but there is no reward for kills. Instead, we design the reward function to purely survive the battle. The definitions of r_d, r_{fr}, and r_b remain the same as for training, π_f. The total nonpositive reward for π_e is:

$$r_{esc} = r_d + r_{fr} + r_b. \tag{16.4}$$

16.4.5 Commander Policy

The task for the high-level commander policy, π_c, is to provide strategic commands to the low-level policies. The individual observations are based on the two closest opponents and the two closest friendly aircraft of the calling low-level agent.

$$
\begin{aligned}
o_{t,a} &:= \left[x, y, s, \alpha_h\right] \\
o_{t,o} &:= \left[x, y, s, \alpha_h, \alpha_{off,a}, \alpha_{AA,a}, \alpha_{AA,o}, \alpha_{ATA,a}, \alpha_{ATA,o}, d_a\right] \\
o_{t,fr} &:= \left[x, y, s, \alpha_{ATA,a}, \alpha_{ATA,fr}, d_a\right] \\
o_{t,full} &:= o_{t,a} \| o_{t,o_1} \| o_{t,o_2} \| o_{t,fr_1} \| o_{t,fr_2}
\end{aligned}
$$

The tactics the commander learns dictate which low-level policy each agent must activate, as π_c gets invoked by every agent separately. The action set (options) is $a_c \in \{0,1,2\}$, where 0 activates π_e and π_f otherwise. If π_f is activated, the chosen option (1 or 2) determines which of the two observable opponents the agent should attack. The agent then gets the corresponding observation for its low-level policy. Our model allows altering this sensing strategy (i.e., the commander can be adjusted to sense three instead of two close opponents).

For training, the action assessment reward function, r_{act}, encourages the commander to exploit favorable situations. A favorable situation for an agent is defined as being in an advantageous position to attack the opponent with a high chance of destroying it. r_{act} is defined as:

$$
r_{act} = \begin{cases} +0.1 & d_o < 5\text{km} \wedge \alpha_{ATA,a-o} < 15° \wedge a_c = i_o \in \mathcal{A}_o \\ +0.1 & d_o < 5\text{km} \wedge \alpha_{ATA,o-a} < 15° \wedge \alpha_{ATA,a-o} > 30° \wedge a_c = 0 \\ -0.1 & a_c = i_o \notin \mathcal{A}_o \\ 0 & \text{else.} \end{cases} \tag{16.5}
$$

In Equation (16.5), the conditions are evaluated for all agents with respects to each opponent. d_o defines the distance and $\alpha_{ATA,a-o}$ the antenna train angle, both from the perspective of an agent to an opponent, whereas $\mathcal{A}_o$ denotes the set of active opponents. r_{act} defines a favorable situation in the first case, that is characterized as having a short distance for firing and approximately facing the opponent. If these conditions are met and the commander has chosen the corresponding opponent index i_o, the commander gets an incentive reward. The same reasoning holds for the second case, if the commander sets the agent into escaping mode while the opponent is in a favorable situation and the agent faces away from the opponent. Contrary, selecting an opponent index, $i_o \notin \mathcal{A}_o$, will punish the commander. We also include the killing reward, $r_k = 1$, if an agent killed an opponent and $r_d = -1$ if the agent got destroyed. The out-of-boundary reward is $r_b = -2$ and the friendly kill punishment is omitted. The total reward for π_c is:

$$
r_c = r_k + r_d + r_b + r_{act}. \tag{16.6}
$$

Based on the visualization in Figure 16.3, an algorithmic description of the commander training procedure is given in Algorithm 2. The commander gets invoked dynamically on termination of the low-level horizon T_l or of events, which are defined as:

- any aircraft got destroyed.
- an agent approaches the map boundary ($d < 5$ km).
- an agent *or* an opponent is in a favorable situation (first case of Equation (16.5)).

Algorithm 2 Simulation Procedure for Training Commander Policy π_c

Set high-level time horizon T_h, low-level time horizon T_l, max number of episodes N, opponent fight policy assignment probability p_o
for episode $n = 0$ to N do
  Sample a combat scenario
  for $t_h = 0$ to T_h do
    Get Commander actions for all n agents: $A_{h,t_h} = \{a_{h,1}, \dots, a_{h,n}\}$
    Agents: activate low-level policies $\pi_{a,f/e} \leftarrow a_{h,i}$
    Opponents: policy assignment $\pi_{o,f/e} \leftarrow p_o$
    for $t = 0$ to T do
      Execute π_f or π_e for each agent or opponent
      if $t \geq 10$ or event then
        break
      end if
    end for
    Get $R_{t,h}$ and $S_{t,h}$
  end for
  update π_c according to Algorithm 1
end for

16.4.6 Rule-Based Opponents

The antenna train angle, $\alpha_{ATA} \in [0,180]°$, defines the difference of the current heading angle to the target. Applying $\alpha_h + \alpha_{ATA}$ orients the adversaries to the right, as α_h is measured with respect to the Northern Hemisphere, as depicted in Figure 16.5a. However, this is not always optimal, as it might provoke superfluous rotations. To mitigate this, we refer to Figure 16.6 and construct a unit vector v_o from the opponent's location (bright aircraft) and determine if the target is located on the left or right side of v_o. The sign s of the determinant of the 2D vectors ($\overrightarrow{AB}$, $\overrightarrow{AC}$) indicates the orientation of the transformed vector space relative to the original and determines whether to increment or decrement α_h in $\alpha_h = \alpha_h + s \cdot \alpha_{ATA}$, or whether to turn left or right:

$$s = \text{sign}\big((B_x - A_x)(C_y - A_y) - (B_y - A_y)(C_x - A_x)\big). \tag{16.7}$$

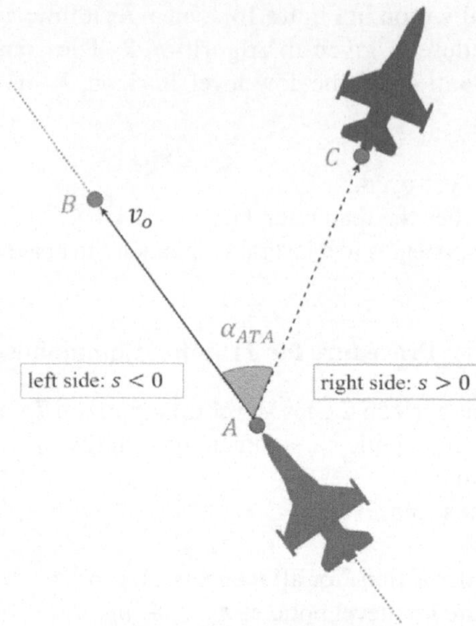

FIGURE 16.6 Calculation of turning direction of opponent (bright) to face agent (dark). The sign *s* determines if an agent is located to the right or to the left of an opponent.

To introduce variability, the heading calculation is infused with some randomness $r \sim \text{Uniform}(0,1)$, yielding the heading formula as:

$$\alpha_h = \alpha_h + s \cdot r \cdot \alpha_{ATA}. \tag{16.8}$$

Additional considerations for adversary strategy include velocity modulation based on proximity to the target and the likelihood of firing with a cannon or rocket increases inversely with α_{ATA}.

16.5 EXPERIMENTS

In this section, we describe our simulation setup and present experimental results. Our code for running simulations is publicly available.[2]

16.5.1 Simulation Settings

16.5.1.1 Environment and libraries

A notable aspect of our work is the development of a dedicated 2D (Python) simulation platform to have full control and low inertia. During a fixed time frame (typically 0.1

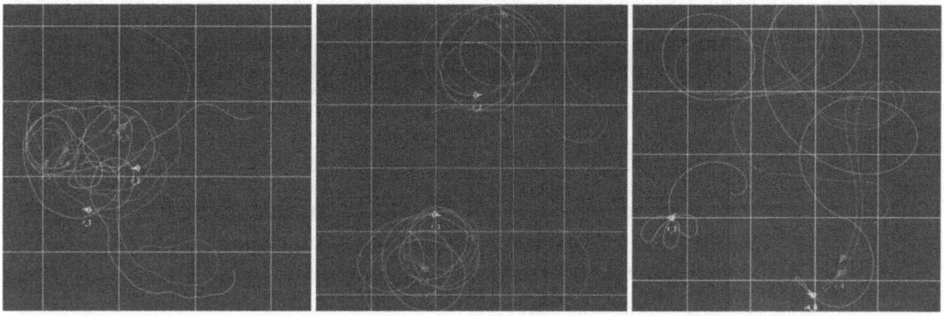

(a) (b) (c)

FIGURE 16.7 Different policy trajectories in the simulation platform. (a) Commander π_c. (b) Fight π_f. (c) Escape π_e.

seconds), the next aircraft positions and distinct events (e.g., shots or kills) are computed. Movements of aircraft are computed with respect of geodesics (WGS84), so even on larger maps the movements are plausible. Our platform is lightweight, fast, and accurately simulates the dynamics of our aircraft. Figure 16.7 shows some rendered scenes of our environment under different combat settings. Trajectories of each aircraft can be visualized, and a landmark is set at the position where an aircraft got destroyed. Map size and number of aircraft can be specified, highlighting the diversity properties of our model. We refer to time-step t as one simulation round.

16.5.1.2 Training and evaluation configuration

Our shared CTDE configuration allows simulations with a variable number of agents. For each episode, teams are randomly assigned a side of the map, with aircraft given random initial positions and headings. To ensure heterogeneity, aircraft types are randomly selected, with at least one of each type per group. Map sizes per axis are 30 km for low-level and 50 km for high-level training. Unless otherwise stated, the learning curves showing mean rewards include the training performance of all agents. Low-level agents, sensing one opponent and one friendly aircraft per time step, are trained in a 2-vs-2 setting. The commander policy π_c is trained in a 3-vs-3 setting, regardless of sensing strategy. Evaluations are done for 1,000 episodes and in the same configuration as in low-level or high-level training. Low-level training and evaluations are compared using L3, where opponents exhibit the most deterministic behavior for consistent comparisons. For inspecting the commander performance, the pretrained policies, $\pi_{f,L5}$ and π_e, are employed. *Win* is when all opponents are destroyed, *Loss* if all agents are destroyed, and *Draw* if at least one agent per team remains alive after the episode ends. An episode ends when either the time horizon is reached or a loss occurs. An aircraft is destroyed when getting hit by cannon, rocket, or map boundary.

The PPO parameters are kept constant for all training procedures: learning rate for actor and critic is $lr = 0.0001$, discount factor is $\gamma = 0.95$, clip parameter is $\epsilon = 0.2$,

Adam as optimizer, batch size of 2,000 for low-level policies, and 1,000 for high-level policy. We employ the popular libraries Ray RLlib and Pytorch for training our model.

16.5.2 Fight Policy π_f

The aircraft is configured with an ammunition of 200 cannon shots and 5 rockets (only AC1). We make the opponents stronger by giving them ammunition of 400 cannons and 8 rockets. As levels increase in curriculum learning, we extend the episode time horizon by $T = 50$, starting from $T = 150$ at L1.

16.5.2.1 Reward inspection

Within the framework of CTDE, the distribution of a global reward among agents is a common practice to foster cooperation. Nevertheless, this approach may precipitate the credit assignment problem, wherein it becomes challenging for individual agents to discern their specific contribution to the team's success [37]. To address this, our study explores three distinct reward structures. The first reward function is the fight reward, r_{fight}, defined in Equation (16.3), denoted as baseline. The other two functions are:

$$r_{FriPun} = r_{fight} + r_{fp},\ r_{fp} = -2, \tag{16.9}$$

$$r_{ShFrac} = r_{i,fight} + \rho \sum_{k=1/i}^{n} r_{k,fight}. \tag{16.10}$$

The reward mechanism in Equation (16.9) is termed *friendly punishment* (FriPun). It accounts for scenarios where an agent inadvertently causes damage to a friendly unit and extends r_{fight} to evaluating the impact of such penalties on *both*, the offending (r_{fr}) and the victim (r_{fp}) agent. This should encourage the agents to exercise greater caution and avoid such incidents. The other reward function in Equation (16.10) is called *shared fraction* (ShFrac). The base fight reward of agent i, $r_{i,fight}$, is augmented by a fraction of the total rewards of the remaining friendly agents $\{1,2,\ldots,n\} \setminus \{i\}$ controlled by the parameter ρ, which we set to $\rho = 0.5$ for this ablation study. The results of training and evaluation across these rewards are depicted in Figure 16.8. Training data does not decisively indicate which reward yields superior performance. However, during evaluation, the base reward, r_{fight}, demonstrates a higher efficacy compared to the alternatives, where the shared reward performed worst.

To understand the reduced performance of shared rewards (ShFrac) compared to the base reward, we analyze the average rewards of agent types AC1 and AC2. Figure 16.9a shows AC1 performing better with shared rewards, while Figure 16.9b shows AC2 performing worse. Both AC1 and AC2 accrue similar average rewards (approximately 1) under the base reward mechanism, yet a notable disparity emerges with shared rewards, indicating the presence of the credit assignment problem due to AC1's superior and AC2's inferior performance. The observed reward mismatch between AC1 and AC2

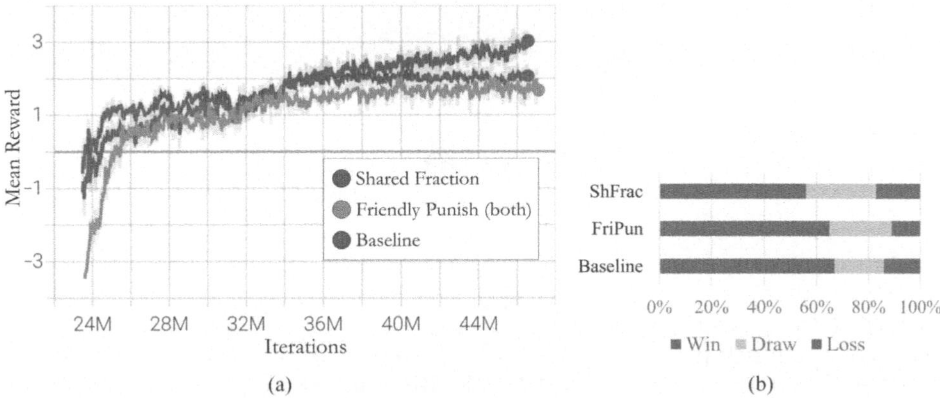

FIGURE 16.8 Comparison on performance at L3 under different fight rewards. (a) Training π_f. (b) Evaluation π_f.

under ShFrac is not desirable, suggesting the potential for lazy agents [37]. We deduce that assigning rewards based solely on an agent's direct combat achievements engenders optimal performance outcomes.

We refer to Figure 16.10 for investigating the performance impact of the reward r_{FriPun}. These statistics from the 2-vs-2 evaluations show the different contributions of agents and opponents under both the baseline and FriPun reward functions. The combat skills of AC1 surpass those of AC2, which is most likely due to rockets as further equipment and having more agile dynamics. Upon closer examination, we observe a notable reduction of friendly fire events and an increase of draws under the FriPun reward. This indicates a more cautious behavior and suggests that accepting the risk of friendly fire can still enhance combat performance, thereby supporting the appropriateness of the reward function r_{fight}.

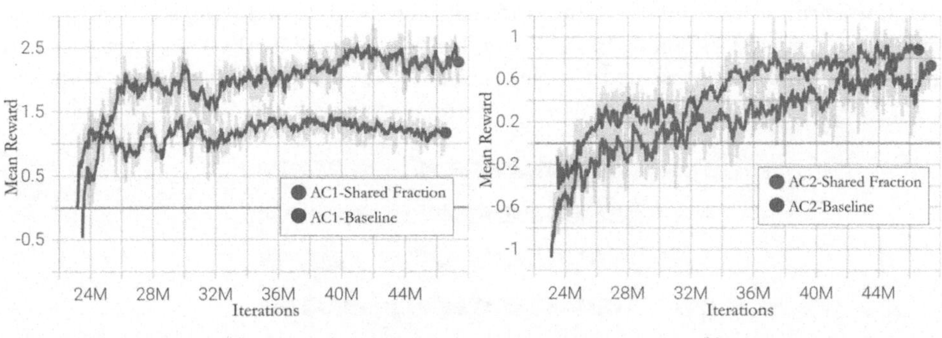

FIGURE 16.9 Training of AC1 and AC2 at L3 under different fight rewards: individual rewards (Baseline) and a shared reward (Shared Fraction). (a) Training of AC1. (b) Training of AC2.

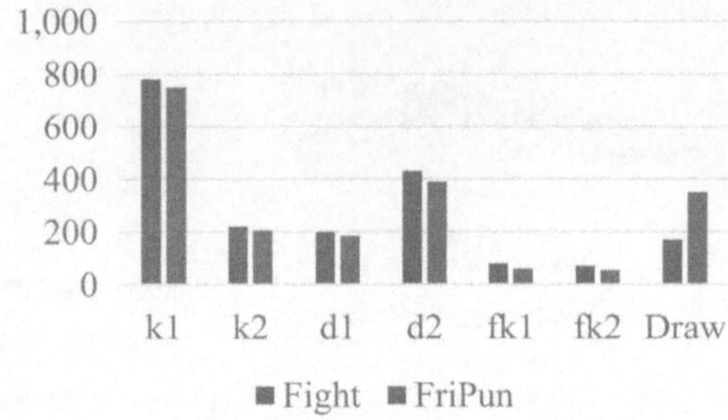

FIGURE 16.10 Evaluation statistics of π_f under r_{fight} and r_{FriPun}. The *y*-axis displays the number of counted events per category. Destroying an opponent is abbreviated with *k*; getting destroyed with *d* and *fk* indicates "friendly" kills. The attached numbers indicate the aircraft types (e.g., *k*1 is killed by AC1).

After validating r_{fight}, we assess low-level training performance in Figure 16.11. Agents use $\pi_{f,L5}$ and opponents use $\pi_{f,L4}$. We infer that agents can improve combat performance during L5 training, as $\pi_{f,L5}$ outperforms $\pi_{f,L4}$. Additionally, our agents can engage in scenarios up to 5-vs-5, even though training was conducted in a 2-vs-2 scheme. Interestingly, the proportion of combat outcomes remains constant and there is a slight increase in the number of wins as the number of aircraft grows. The improved performance with more agents likely results from a higher probability of eliminating opponents due to more L5 agents. The saturated learning curve in Figure 16.20a during the L5 stage together with the solid evaluation performance indicates that agents have achieved peak combat capabilities. An illustrative example of a combat scenario is presented in Figure 16.7b, showing the circular trajectories to reach the tail of opponents.

16.5.2.2 Architecture inspection

To underscore the effectiveness of our network architecture and the influence of curriculum learning, we train π_f using different architectures and compare their performances as depicted in Figure 16.12. We take a *fully connected* (FC) neural network to compare against our presented SA-Net. This FC-Net comprises two layers with 500

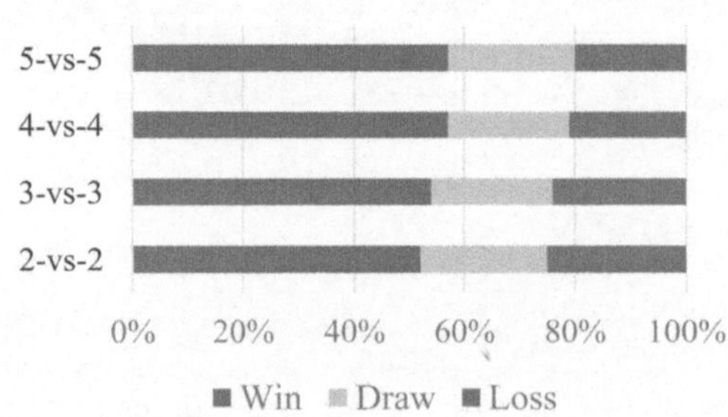

FIGURE 16.11 Evaluation of $\pi_{f,L5}$-vs-$\pi_{f,L4}$ under r_{fight} after completing L5 training.

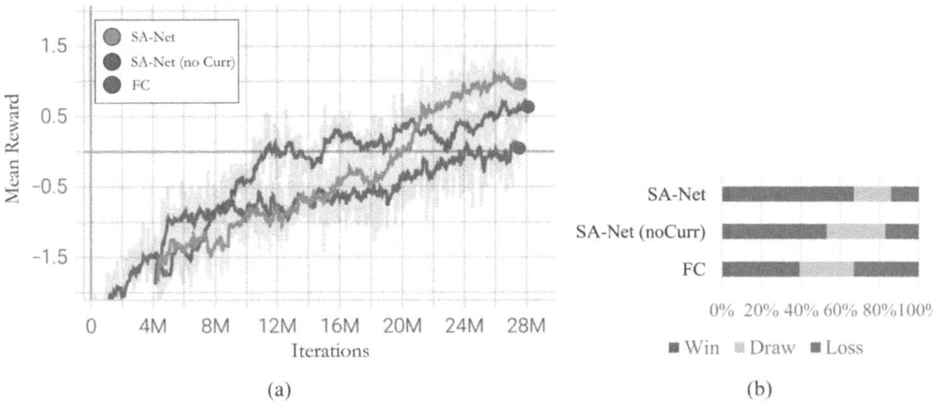

FIGURE 16.12 Training performance of π_f at L3 under different architectures. (a) Training π_f. (b) Evaluation π_f.

neurons each and uses *tanh* activation. Additionally, we trained the SA-Net without curriculum stages, opting instead for direct training on L3. Notably, the results indicate a reasonable enhancement in both training and evaluation performance with our proposed methodology.

16.5.2.3 MARL framework inspection

As the superiority of CTDE is not universally applicable [60], we examine two further training paradigms: *centralized training with centralized execution* (CTCE) and *decentralized training with decentralized execution* (DTDE) [37]. In the DTDE approach, each agent operates its distinct network without information sharing and the CTCE strategy integrates a single network to process the aggregated observations and actions of all agents. Training and evaluation outcomes are depicted in Figure 16.13. We infer that CTCE trails in both training and evaluative measures. DTDE showcases a solid training performance, like CTDE, but falls short during evaluation. The losses with CTDE training predominantly occur in situations where the agent and its opponent are directly facing each other and simultaneously firing, thereby creating a scenario where both have an equal likelihood of eliminating the other. Nonetheless, our analysis confirms that CTDE emerges as the most proficient framework.

16.5.3 Escape Policy π_e

The training process of π_e is configured similarly to that of π_f, utilizing the same ammunition and time horizons.

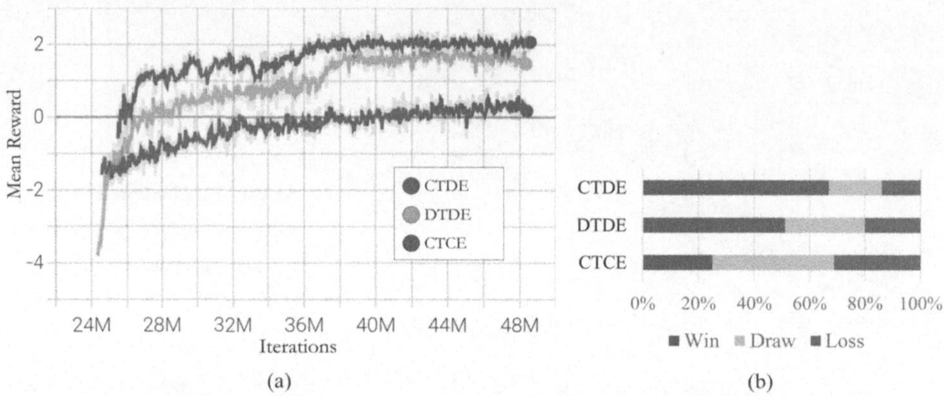

FIGURE 16.13 Performance comparison of different MARL frameworks at L3. (a) Training of π_f. (b) Evaluation of π_f.

16.5.3.1 Reward inspection

In escape mode, the agents' main objective is to evade opponent aircraft. We compare the training reward defined in Equation (16.4) with two distinct reward functions:

$$r_{dist,t} = r_{esc} + r_{p,t}, \tag{16.11}$$

$$r_{dist-speed,t} = r_{esc} + r_{pv,t}. \tag{16.12}$$

The baseline function, r_{esc}, is designed strictly as nonpositive, with the primary goal to avoid penalties. The reward functions in Equation (16.11) and (16.12) further include two per-time-step rewards, where d_o is the distance to opponents and v_{AC} is the velocity defined in Table 16.1.

$$r_{p,t} = \begin{cases} -0.1 & d_o < 6\text{km} \\ +0.1 & d_o > 13\text{km} \\ 0 & \text{else,} \end{cases} \tag{16.13}$$

$$r_{pv,t} = \begin{cases} -0.1 & d_o < 6\text{km} \wedge v_{AC} < 300 \text{ kn} \\ +0.1 & d_o > 13\text{km} \wedge v_{AC} > 600 \text{ kn} \\ 0 & \text{else.} \end{cases} \tag{16.14}$$

The functions in Equations (16.13) and (16.14) are designed to reward agents for avoiding proximity to opponents, with $r_{pv,t}$ further considering the agent's velocity. Rather than focusing on reward values, we analyze mean episode lengths during training as an indicator of agent survival duration. With $T = 300$ for L3 training, the data illustrates that agents trained with r_{esc} exhibit the longest survival times (Figure 16.14a).

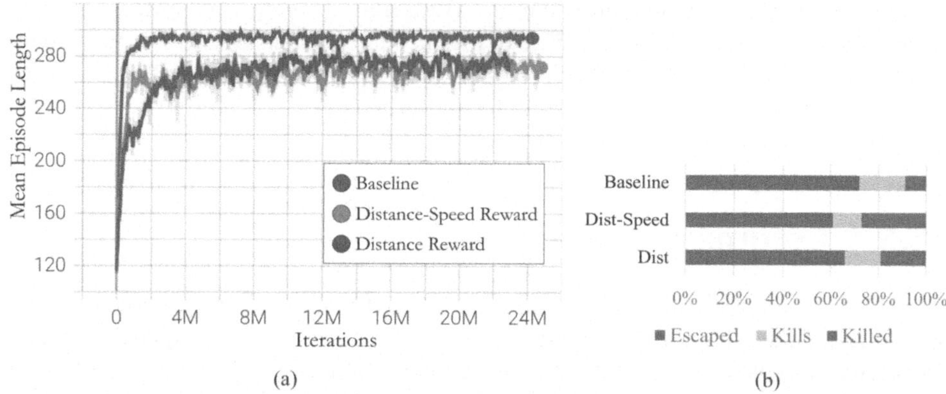

FIGURE 16.14 Comparison of performance at L3 under different escape rewards. *Escaped* means no agent got destroyed, *killed* is when at least one agent got killed, *kills* when at least one opponent got destroyed. (a) Training of π_e. (b) Evaluation of π_e.

Both per-time-step rewards yield similar, albeit slightly inferior, performance to the baseline function. Evaluation stages reveal an interesting increase in number of defeats, especially under $r_{dist-speed,t}$ (Figure 16.14b). Consequently, training with per-time-step rewards appears less effective, as agents can accumulate sufficient positive rewards such that penalties for eliminations are less impactful in the learning process. We conclude that r_{esc} reveals the most effective escaping performance.

16.5.3.2 Fleeing performance

We proceeded with training π_e against $\pi_{f,\,L5}$. We maintained the time horizon $T = 300$, as during L3 training. The fleeing agents, shown in Figure 16.15, did not come as close

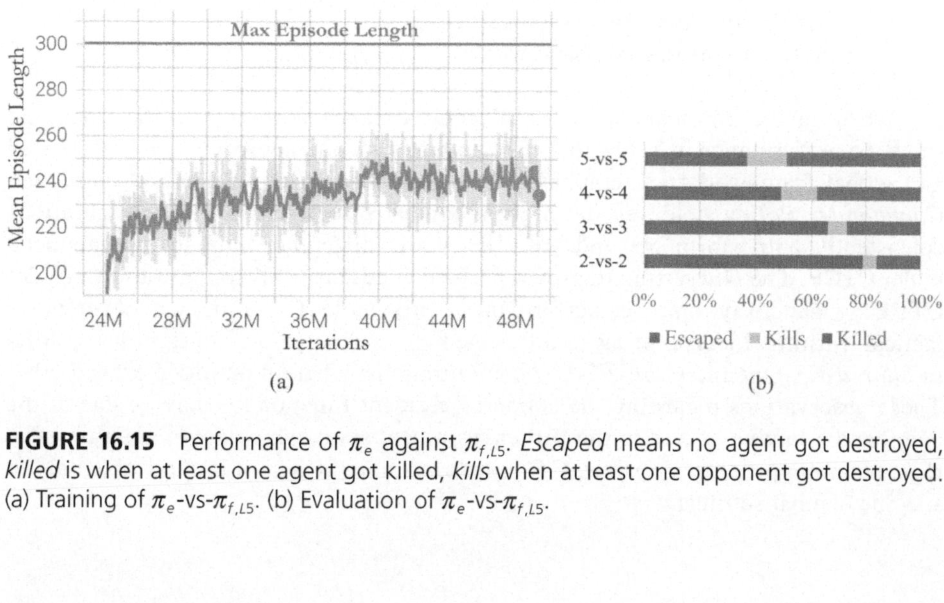

FIGURE 16.15 Performance of π_e against $\pi_{f,L5}$. *Escaped* means no agent got destroyed, *killed* is when at least one agent got killed, *kills* when at least one opponent got destroyed. (a) Training of π_e-vs-$\pi_{f,L5}$. (b) Evaluation of π_e-vs-$\pi_{f,L5}$.

to the maximum episode length as during L3 training. In scenarios up to 3-vs-3, we observe successful evasions in over 50% of the encounters but also a reasonable amount in more complex scenarios even though the likelihood of an agent being killed rises with the number of involved aircraft. Overall, the performance of π_e remains appealing. An example of fleeing trajectories is plotted in Figure 16.7c.

16.5.4 Commander Policy π_c

The objective of the commander policy π_c is to activate either $\pi_{f,L5}$ or π_e for each agent. Opponents predominantly engage in combat, with a minor propensity to flee, possessing a fight-to-escape ratio of approximately 3:1 ($p_o = 75\%$ in Algorithm 2). Compared to the training of π_f and π_e, we equip agents and opponents with an equal arsenal of 300 cannons and 8 rockets. This setup facilitates a focused comparison of tactical approaches.

16.5.4.1 Commander modifications

We modify some fundamental components in the hierarchical structures, as listed below:

- *Shared-vs-Global.* Comparison of a shared CTDE network for each agent versus a global CTCE network for all agents.
- *N2-vs-N3.* In the N2 configuration, the commander detects the two nearest opponents per agent, while in the N3 setting, it senses three opponents.
- *Opt-vs-noOpt.* The commander can either select which opponent to attack (*Opt*) or decide only to attack or flee (*noOpt*). In *noOpt*, the low-level policy targets the nearest opponent when attacking.
- *Assess-vs-noAssess.* We compare *Assess* that incorporates the intrinsic reward r_{act} to the raw combat rewards $r_k + r_d + r_b$, termed *noAssess*.

We train the commander using all combinations of these modifications and show the performance in Table 16.4. The best-performing approaches in both shared and global frameworks are emphasized in bold. Our method, detailed in section *Commander Policy*, achieves the highest combat effectiveness. N2 configurations consistently outperform N3 under CTDE, while differences are less pronounced under CTCE. The *Opt* setting improves performance across all configurations. Under CTDE, r_{act} generally improves performance, while under CTCE, it does not enhance tactical learning. Incorporating r_{act} alters the fight-escape ratio, with significant shifts in *Shared-N3-Opt* and *Glob-N3-Opt* configurations, where fight ratios exceed 90%. These observations regarding the action assessment function r_{act} may be due to the commander having access to the full environment state, which could simplify strategic planning process. However, CTCE restricts flexibility as it is applicable only to specific combat settings.

TABLE 16.4 Hierarchical configurations and results

CONFIGURATION	*WIN*	*LOSS*	*DRAW*	*FIGHT/ESC*
Shared-N2-Opt-Assess	**63.7**	**16.2**	**20.1**	**87.1/12.9**
Shared-N2-Opt-noAssess	55.9	20.2	23.9	82.2/17.8
Shared-N2-noOpt-Assess	53.5	18.4	28.1	71.1/28.9
Shared-N2-noOpt-noAssess	48.9	21.1	30.0	69.3/30.7
Shared-N3-Opt-Assess	48.2	26.3	25.5	84.9/15.1
Shared-N3-Opt-noAssess	47.4	26.7	25.9	91.2/8.8
Shared-N3-noOpt-Assess	43.3	28.6	28.1	76.5/23.5
Shared-N3-noOpt-noAssess	42.7	28.2	29.1	75.9/24.1
Glob-N2-Opt-Assess	59.1	16.5	24.4	88.9/11.1
Glob-N2-Opt-noAssess	**60.6**	**16.6**	**22.8**	**82.1/17.9**
Glob-N2-noOpt-Assess	53.4	18.2	28.4	78.7/21.3
Glob-N2-noOpt-noAssess	54.3	17.9	27.8	77.3/22.7
Glob-N3-Opt-Assess	59.9	16.9	23.2	93.7/6.3
Glob-N3-Opt-noAssess	55.5	18.8	25.7	84.8/15.2
Glob-N3-noOpt-Assess	48.8	17.4	33.8	72.3/27.7
Glob-N3-noOpt-noAssess	42.7	19.1	38.2	68.4/31.6

All numerical values are given in [%].

16.5.4.2 Reward inspection

We select the optimal shared and global commander policies, as highlighted in Table 16.4. We examine the N2 and N3 configurations, as these had the most significant performance impact, with a particular focus on the *Opt* setting, which showed the best performance overall. The training outcomes are depicted in Figure 16.16, where r_{act} is

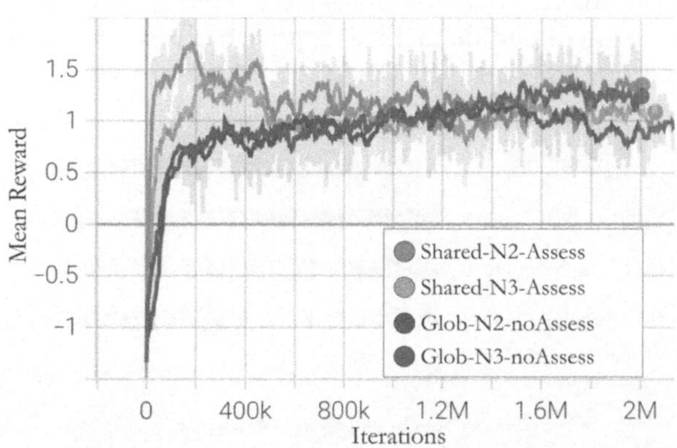

FIGURE 16.16 Comparison of the training rewards of the best-performing hierarchy configurations.

subtracted in the mean reward for the CTDE approach to have a consistent comparison. All four methods converge around a mean reward value of 1, with the N2 configurations slightly outperforming the N3 setting. This observation aligns with the results presented in Table 16.4. The superior performance of the N2 configurations may be attributed to the reduced complexity of the commander's task.

16.5.4.3 Opponent selection

We focus on the N2-vs-N3 configurations to evaluate the opponent selection strategies, depicted in Figure 16.17, with Opp1 being the nearest and Opp3 the furthest opponent. Predominantly, Opp1 is selected with the highest frequency, especially by *Glob-N2-NoAssess,* that indicates the most deterministic CoA. Nonetheless, a significant selection frequency is also observed for Opp2, especially for *Shared-N3-Assess. Glob-N3-NoAssess* seems to favor Opp3 more frequently than other strategies. These outcomes are contingent upon the specific combat scenarios, as the interplay between strategic decisions made by the commander and the low-level executions collectively influence tactical behavior. These findings suggest that letting the commander decide which opponent to attack enhances performance.

16.5.4.4 Architecture inspection

In the base training setup for the commander, we employ the GRU-Net. We compare this with the SA-Net, used to train π_f, and with a FC-Net (as for π_f). The performance of all three networks is depicted in Figure 16.18. Although all three networks achieve comparable mean rewards during training, the evaluation clearly demonstrates that GRU-Net delivers superior performance. The GRU module appears more proficient in mastering strategic planning by incorporating the last states in the observations.

16.5.4.5 Different combat scenarios

As validated in Table 16.4, *Shared-N2-Opt-Assess* demonstrated the best performance. We evaluate the strength of our hierarchical approach across various combat scenarios, where π_c is trained in 3-vs-3 settings. We extended the analysis to 2-vs-2 to 15-vs-15

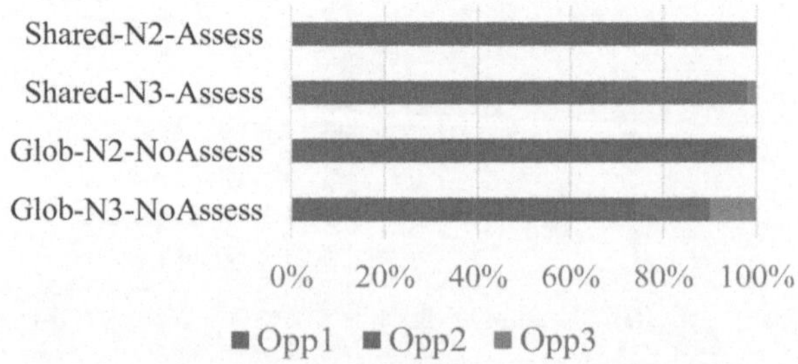

FIGURE 16.17 Opponent selection frequency, where Opp1 is closest and Opp3 furthest aircraft.

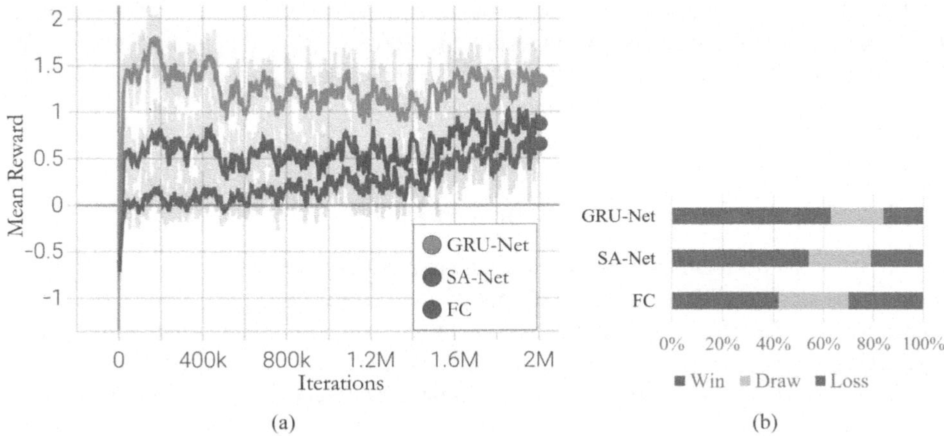

FIGURE 16.18 Performance of π_c at L3 with different neural network architectures. GRU-Net and SA-Net are defined in Figure 16.4b. (a) Training π_c. (b) Evaluation π_c.

scenarios and varied opponent tactics, shown in Figure 16.19. In smaller team configurations (2-vs-2 to 5-vs-5), our commander policy consistently achieves over 50% victory rates, with more draws as participants increase. In uneven setups (2-vs-4 and 3-vs-5), the distribution of wins, losses, and draws is balanced, showing the robustness of our method even when outnumbered. In pure-fight (PF) opponent configuration, the commander policy achieved nearly a 50% win rate. In pure-escape (PE) mode, the loss rate was minimal, and despite more draws, the win rate remained predominant. For larger

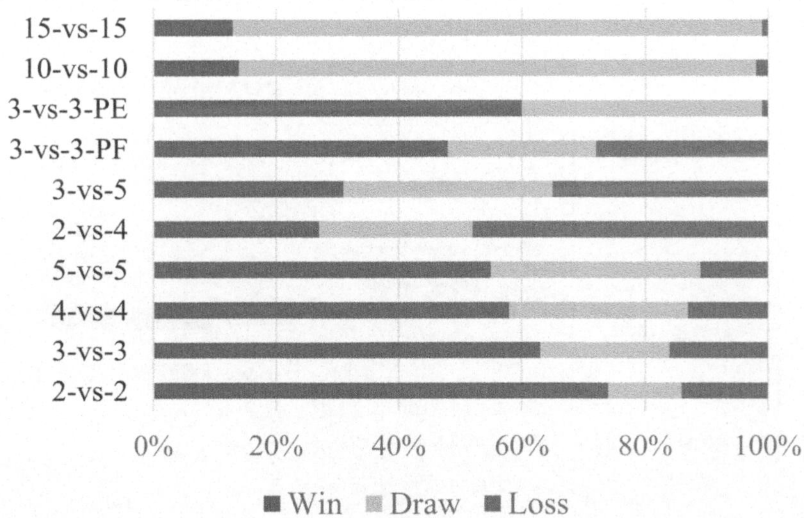

FIGURE 16.19 Examining different combat scenarios and opponent behaviors. For opponents, PF denotes *Pure-Fight* ($p_o = 100\%$), PE denotes *Pure-Escape* ($p_o = 0\%$).

configurations involving 10-vs-10 and 15-vs-15 entities, we extended the time horizon $T = 1{,}000$. Most of these engagements resulted in draws, attributable to the higher likelihood of an agent or opponent surviving until the end of an episode. The favorable win-loss ratio underscores that our approach can effectively succeed in more challenging air combat scenarios.

16.5.4.6 Commander-vs-standard RL

We compare the performance of our hierarchical model against a conventional RL technique. This standard RL approach with policy π_{std} resembles the CTCE configuration, where we introduce two principal modifications. First, omission of curriculum learning, opting instead for direct training at L3 to facilitate optimal comparison; second, integration of fight-and-escape rewards to simulate both behaviors. The total reward for training π_{std} is:

$$r_{std} = r_k + r_d + r_{fr} + r_b + r_{p,t}. \tag{16.15}$$

$r_{p,t}$ from Equation (16.13) is adjusted to avoid penalizing the agent for proximity to opponents, specifically, $r_{p,t} = +0.1$ if $d > 13$ km. The policy π_{std} is required to autonomously discern when evasive actions supersede attacking maneuvers. For consistency, π_{std} senses its closest opponent and friendly aircraft during training. In Figure 16.20a, we evaluate the 2-vs-2 training performance of π_{std} relative to our entire curriculum training sequence for π_f. Notably, π_{std} fails to achieve the performance that π_f reaches at only half the total training iterations. In evaluation, shown in Figure 16.20b, π_{std} does not match the superior performance of π_c, that incorporates $\pi_{f,L5}$ and π_e for distinct behaviors. This comparison underscores the strategic and operational advantages attributable to our design choices. An example of a 2-vs-4 combat scenario with the commander involved is shown in Figure 16.7a.

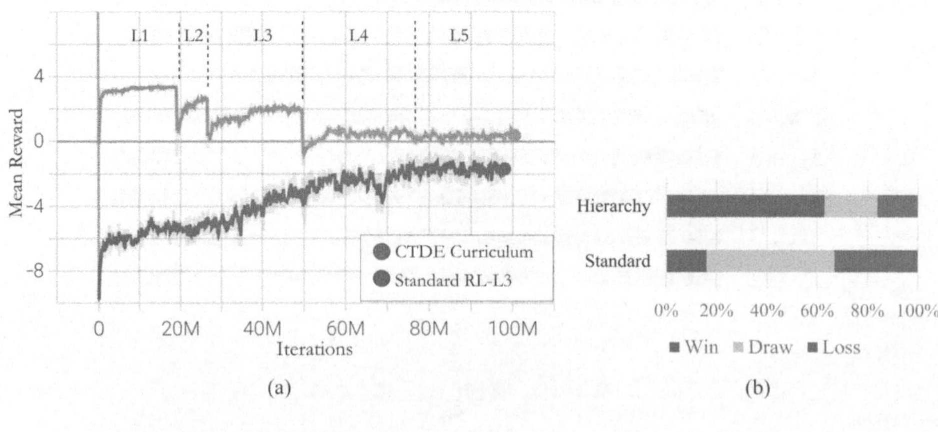

FIGURE 16.20 Performance comparison of standard RL-vs-hierarchical MARL. (a) Training π_{std}-vs-π_f. (b) Evaluation π_{std}-vs-π_c.

16.6 CONCLUSION

We introduced a heterogeneous hierarchical MARL approach for preset air combat simulations, integrating recent developments like learning curricula, fictitious self-play, and hierarchy-specific neural networks. Our agents showed effective air combat capabilities across various scenarios, even in large teams. Low-level agents made autonomous decisions, while a high-level commander improved performance through look-ahead planning. Curriculum-based training and tactical commands were effective for low-level agents. Individual combat rewards yielded better performance and intrinsic rewards positively affected the commander. Reduced sensing (N2) led to better tactics, and the CTDE framework effectively coordinated agent training, even with heterogeneous agents.

In some scenarios, agents fled under the fight policy π_f, especially when allies outnumbered opponents, indicating insufficient training in asymmetric situations. Future research will refine low-level policies for better behavioral distinctions and detailed commands beyond attack/evade. We aim to enhance commander training using algorithms like the Monte-Carlo Tree Search or AlphaZero [61] and include BVR scenarios with varied weapons and sensing. We're also developing a 3D simulation environment based on JSBSim[3] for realistic air combat simulations.

NOTES

1. https://dassault-aviation.com/en/defense/rafale.
2. Offcial implementation: https://github.com/IDSIA/hhmarl_2D.
3. https://jsbsim.net

REFERENCES

1. Selmonaj A, Szehr O, Rio GD, et al. Hierarchical Multi-Agent Reinforcement Learning for Air Combat Maneuvering. International Conference on Machine Learning and Applications (ICMLA), 1031–1038, IEEE, 2023.
2. Shaw RL. Fighter Combat: Tactics and Maneuvering. Naval Institute Press, Annapolis, Maryland, 1987.
3. Burgin G, Fogel L, Phelps JP. An adaptive Maneuvering logic Computer Program for the Simulation of one-on-one Air-to-Air Combat. NASA, Contractor Report CR-2582, 1975.
4. Burgin G, Sidor L. Rule-Based Air Combat Simulation. NASA Contractor Report 4160, 1988.
5. Jones R, Laird J, Nielsen P, et al. Automated intelligent Pilots for Combat Flight Simulation. AI Magazine, 20, 27–42, 1999.

6. Austin F, Carbone G, Falco M, et al. Automated Maneuvering Decisions for Air-to-Air Combat. American Institute of Aeronautics and Astronautics (AIAA) Guidance, Navigation, and Control Conference, Monterey, 1987.
7. Eklund J, Sprinkle J, Sastry S. Implementing and Testing a nonlinear Model Predictive Tracking Controller for aerial Pursuit/Evasion Games on a Fixed Wing Aircraft. Proceedings of the 2005 American Control Conference, IEEE, 1509–1514, 2005.
8. Virtanen K, Raivio T, Hamalainen RP. Modeling Pilot's Sequential Maneuvering Decisions by a Multistage Influence Diagram. Journal of Guidance, Control, and Dynamics, 27, 665–677, 2004.
9. Virtanen K, Karelahti J, Raivio T. Modeling Air Combat by a moving Horizon Influence Diagram Game. Journal of Guidance, Control, and Dynamics, 29, 1080–1091, 2006.
10. You DI, Shim DH. Design of an Aerial Combat Guidance Law using virtual Pursuit Point Concept. Proceedings of the Institution of Mechanical Engineers, Part G: Journal of Aerospace Engineering, 229, 792–813, 2014.
11. Jarmark B, Hillberg C. Pursuit-Evasion between two realistic Aircraft. Journal of Guidance, Control, and Dynamics, 7, 690–694, 1984.
12. Merz AW. To Pursue or to Evade - that is the Question. Journal of Guidance, Control, and Dynamics, 8, 161–166, 1985.
13. Greenwood N. A Differential Game in Three Dimensions: The Aerial Dogfight Scenario. Dynamics and Control, 2, 161–200, 1992.
14. McGrew JS, How JP, Williams B, et al. Air-Combat Strategy using approximate Dynamic Programming. Journal of Guidance, Control, and Dynamics, 33, 1641–1654, 2010.
15. Ma X, Xia L, Zhao Q. Air-Combat Strategy using Deep Q-Learning. Proceedings of the Chinese Automation Congress (CAC), 3952–3957, IEEE, 2018.
16. Day M, Strickland L, Squires E, et al. Responding to Unmanned Aerial Swarm Saturation Attacks with Autonomous Counter-Swarms. In Pham T. Kolodny MA, Wiegmann DM editors, Ground/Air Multisensor Interoperability, Integration, and Networking for Persistent ISR IX, SPIE, 2018.
17. Vlahov B, Squires E, Strickland L, et al. On Developing a UAV Pursuit-Evasion Policy using Reinforcement Learning. International Conference on Machine Learning and Applications (ICMLA), 859–864, IEEE, 2018.
18. Smith R, Dike B, Mehra R, et al. Classifier Systems in Combat: two-sided Learning of Maneuvers for Advanced Fighter Aircraft. Computer Methods in Applied Mechanics and Engineering, 186, 421–437, 2000.
19. Smith RE, Dike BA, Ravichandran B, et al. The Fighter Aircraft LCS: A Case of Different LCS Goals and Techniques. Lecture Notes in Computer Science, 1813, 271–282, Springer Berlin Heidelberg, 2000.
20. Smith RE, Dike BA, Ravichandran B, et al. Two-Sided, Genetics-Based Learning to Discover novel Fighter Combat Maneuvers. Lecture notes in Computer Science, 294–313, Springer Berlin Heidelberg, 2001.
21. Smith R, Dike B, Ravichandran B, et al. Discovering Novel Fighter Combat Maneuvers. In Bentley PJ and Corne DW editors, Creative Evolutionary Systems, 467–490, Morgan Kaufmann Publishers, San Francisco, 2002.
22. Smith R, El-Fallah A, Ravichandran B, et al. The fighter aircraft LCS: A real-world, Machine Innovation Application. In Bull L editors, Applications of Learning Classifier Systems, 113–142, Springer Berlin Heidelberg, 2004.
23. Toubman A, Roessingh JJ, Spronck P, et al. Transfer Learning of Air Combat Behavior. International Conference on Machine Learning and Applications (ICMLA), 226–231, IEEE, 2015.
24. Toubman A, Roessingh JJ, van Oijen J, et al. Modeling Behavior of Computer generated Forces with Machine Learning Techniques, the NATO Task Group Approach. International Conference on Systems, Man, and Cybernetics (SMC), 1906–1911, IEEE, 2016.

25. Mnih V, Kavukcuoglu K, Silver D, et al. Playing Atari with Deep Reinforcement Learning. arXiv preprint arXiv:1312.5602, 2013.
26. Schrittwieser J, Antonoglou I, Hubert T, et al. Mastering Atari, Go, Chess and Shogi by Planning with a Learned Model. Nature, 588, 604–609, 2019.
27. Vinyals O, Babuschkin I, Czarnecki WM, et al. Grandmaster Level in StarCraft II using Multi-Agent Reinforcement Learning. Nature, 575, 350–354, 2019.
28. OpenAI BC, Brockman G, et al. Dota 2 with large scale Deep Reinforcement Learning. arXiv preprint arXiv:1912.06680, 2019.
29. Zhang JD, Yu YF, Zheng LH, et al. Situational Continuity-Based Air Combat Autonomous Maneuvering Decision-Making. Defense Technology, 29, 66–79, 2022.
30. Guo J, Wang Z, Lan J, et al. Maneuver Decision of UAV in Air Combat based on Deterministic Policy Gradient. International Conference on Control & Automation, 767–772, IEEE, 2022.
31. Yuan W, Xiwen Z, Rong Z, et al. Research on UCAV Maneuvering Decision Method based on Heuristic Reinforcement Learning. Computational Intelligence and Neuroscience, 1477078, 2022.
32. Fan Z, Xu Y, Kang Y, et al. Air Combat Maneuver Decision Method based on A3C Deep Reinforcement Learning. Machines, 10, 1033, 2022.
33. Yang Q, Zhang J, Shi G, et al. Maneuver Decision of UAV in short-range Air Combat based on Deep Reinforcement Learning. IEEE Access, 2020.
34. Bae JH, Jung H, Kim S, et al. Deep Reinforcement Learning based Air-to-Air Combat Maneuver Generation in a Realistic Environment. IEEE Access, 8, 360–373, 2023.
35. Balduzzi D, Garnelo M, Bachrach Y, et al. Open-Ended Learning in Symmetric Zero-Sum Games. International Conference on Machine Learning (ICML), 434–443, 2019.
36. Wang Z, Li H, Wu H, et al. Improving Maneuver Strategy in Air Combat by alternate freeze Games with a Deep Reinforcement Learning Algorithm. Mathematical Problems in Engineering, 7180639, 2020.
37. Gronauer S, Diepold K. Multi-Agent Deep Reinforcement Learning: A Survey. Artificial Intelligence Review, 55, 895–943, 2022.
38. Lowe R, Wu Y, Tamar A, et al. Multi-Agent Actor-Critic for mixed cooperative-competitive Environments. Conference on Neural Information Processing Systems (NeurIPS), 6379–6390, 2017.
39. Chu X, Ye H. Parameter Sharing Deep Deterministic Policy Gradient for cooperative Multiagent Reinforcement Learning. arXiv preprint arXiv:1710.00336, 2017.
40. Foerster J, Farquhar G, Afouras T, et al. Counterfactual multi-agent policy gradients. Proceedings of the AAAI Conference on Artificial Intelligence, 2974–2982, 2017.
41. den Broeder G, Ellison R, Emerling L. On Optimum Target Assignments. Operations Research, 7, 322–326, 1959.
42. Tidhar G, Heinze C, Selvestrel M. Flying Together: Modelling Air Mission Teams. Applied Intelligence, 8, 195–218, 1998.
43. Day MA. Multi-Agent Task negotiation among UAVs to defend against Swarm Attacks. Thesis, Naval Postgraduate School, Monterey, 2012.
44. Wang L, Hu J, Zhao C. Autonomous Maneuver Strategy of Swarm Air Combat based on DDPG. Autonomous Intelligent Systems, 1, 15, 2021.
45. Zhang T, Qiu T, Liu Z, et al. Multi-UAV Cooperative Short-Range Combat via Attention-Based Reinforcement Learning using Individual Reward Shaping. International Conference on Intelligent Robots and Systems (IROS), 13737–13744, IEEE, 2022.
46. Jiang F, Xu M, Li Y, et al. Short-range Air Combat Maneuver Decision of UAV Swarm based on Multi-Agent Transformer introducing Virtual Objects. Engineering Applications of Artificial Intelligence, 123, 106358, 2023.

47. Kong Wr, Zhou Dy, Du Yj, et al. Hierarchical Multi-Agent Reinforcement Learning for Multi-Aircraft Close-Range Air Combat. IET Control Theory and Applications, 16, 179–188, 2022.
48. Rashid T, Samvelyan M, Schroeder de Witt C, et al. QMIX: Monotonic Value Function Factorisation for Deep Multi-Agent Reinforcement Learning. International Conference on Machine Learning (ICML), 4295–4304, 2018.
49. Yue L, Yang R, Zou J, et al. Unmanned Aerial Vehicle Swarm Cooperative Decision-Making for Sead Mission: A Hierarchical Multiagent Reinforcement Learning Approach. IEEE Access, 10, 10635–10646, 2022.
50. Hu J, Wang L, Hu T, et al. Autonomous Maneuver Decision-Making of Dual-UAV cooperative Air Combat based on Deep Reinforcement Learning. Electronics, 11, 467, 2022.
51. Littmann ML. Markov Games as a framework for Multi-Agent Reinforcement Learning. International Conference on Machine Learning (ICML), 157–163, Morgan Kaufmann Publishers Inc., 1994.
52. Bellman R. A Markovian Decision Process. Journal of Mathematics and Mechanics, 6, 679–684, 1957.
53. Flet-Berliac Y. The Promise of Hierarchical Reinforcement Learning. The Gradient, 2019.
54. Konda V, Tsitsiklis J. Actor-Critic Algorithms. Conference on Neural Information Processing Systems (NeurIPS),1008–1014, 1999.
55. Gupta J, Egorov M, Kochenderfer M. Cooperative Multi-Agent Control using Deep Reinforcement Learning. Adaptive Learning Agents Workshop at AAMAS, 16, 2017.
56. Vaswani A, Shazeer N, Parmar N, et al. Attention Is All You Need. Conference on Neural Information Processing Systems (NeurIPS), 5998–6008, 2017.
57. Cho K, van Merrienboer B, Gulcehre C, et al. Learning Phrase Representations using RNN Encoder–Decoder for Statistical Machine Translation. arXiv preprint arXiv:14061078, 2014.
58. Schulman J, Wolski F, Dhariwal P, et al. Proximal Policy Optimization Algorithms. arXiv:170706347v2, 2017.
59. Yu C, Akash V, Vinitsky E, et al. The Surprising Effectiveness of PPO in Cooperative Multiagent Games. 36th Conference on Neural Information Processing Systems (NeurIPS),24611–24624, 2022.
60. Yihe Z, Shunyu L, Qing Y, et al. Is Centralized Training with Decentralized Execution Framework Centralized enough for MARL? arXiv preprint arXiv:2305.17352, 2023.
61. Silver D, Hubert T, Schrittweiser J, et al. Mastering Chess and Shogi by Self-Play with a General Reinforcement Learning Algorithm. arXiv preprint arXiv:1712.01815, 2017.

Index

Note: – *Italicized* page references refer to figures and **bold** references refer to tables.

N

O

P

Q

R

S

Y

Z